Supercritical Fluids

ACS SYMPOSIUM SERIES **329**

Supercritical Fluids

Chemical and Engineering Principles and Applications

Thomas G. Squires, EDITOR
Iowa State University

Michael E. Paulaitis, EDITOR
University of Delaware

Developed from a symposium sponsored by
the Division of Fuel Chemistry
at the 190th Meeting
of the American Chemical Society,
Chicago, Illinois,
September 8-13, 1985

American Chemical Society, Washington, DC 1987

Library of Congress Cataloging-in-Publication Data

Supercritical fluids.
  (ACS symposium series, ISSN 0097-6156; 329)

  "Developed from a symposium sponsored by the Division of Fuel Chemistry at the 190th Meeting of the American Chemical Society, Chicago, Illinois, September 8-13, 1985."

  Bibliography: p.
  Includes index.

  1. Separation (Technology)—Congresses. 2. High pressures (Technology)—Congresses.
  I. Squires, Thomas G., 1938-   . II. Paulaitis, Michael E. III. American Chemical Society. Division of Fuel Chemistry. IV. American Chemical Society. Meeting (190th: 1985: Chicago, Ill.) V. Series.

TP156.S45S872   1987   660.2'842   86-26480
ISBN 0-8412-1010-1

Copyright © 1987

American Chemical Society

All Rights Reserved. The appearance of the code at the bottom of the first page of each chapter in this volume indicates the copyright owner's consent that reprographic copies of the chapter may be made for personal or internal use or for the personal or internal use of specific clients. This consent is given on the condition, however, that the copier pay the stated per copy fee through the Copyright Clearance Center, Inc., 27 Congress Street, Salem, MA 01970, for copying beyond that permitted by Sections 107 or 108 of the U.S. Copyright Law. This consent does not extend to copying or transmission by any means—graphic or electronic—for any other purpose, such as for general distribution, for advertising or promotional purposes, for creating a new collective work, for resale, or for information storage and retrieval systems. The copying fee for each chapter is indicated in the code at the bottom of the first page of the chapter.

The citation of trade names and/or names of manufacturers in this publication is not to be construed as an endorsement or as approval by ACS of the commercial products or services referenced herein; nor should the mere reference herein to any drawing, specification, chemical process, or other data be regarded as a license or as a conveyance of any right or permission, to the holder, reader, or any other person or corporation, to manufacture, reproduce, use, or sell any patented invention or copyrighted work that may in any way be related thereto. Registered names, trademarks, etc., used in this publication, even without specific indication thereof, are not to be considered unprotected by law.

PRINTED IN THE UNITED STATES OF AMERICA

ACS Symposium Series

M. Joan Comstock, *Series Editor*

Advisory Board

Harvey W. Blanch
University of California—Berkeley

Alan Elzerman
Clemson University

John W. Finley
Nabisco Brands, Inc.

Marye Anne Fox
The University of Texas—Austin

Martin L. Gorbaty
Exxon Research and Engineering Co.

Roland F. Hirsch
U.S. Department of Energy

Rudolph J. Marcus
Consultant, Computers &
  Chemistry Research

Vincent D. McGinniss
Battelle Columbus Laboratories

Donald E. Moreland
USDA, Agricultural Research Service

W. H. Norton
J. T. Baker Chemical Company

James C. Randall
Exxon Chemical Company

W. D. Shults
Oak Ridge National Laboratory

Geoffrey K. Smith
Rohm & Haas Co.

Charles S. Tuesday
General Motors Research Laboratory

Douglas B. Walters
National Institute of
  Environmental Health

C. Grant Willson
IBM Research Department

Foreword

The ACS SYMPOSIUM SERIES was founded in 1974 to provide a medium for publishing symposia quickly in book form. The format of the Series parallels that of the continuing ADVANCES IN CHEMISTRY SERIES except that, in order to save time, the papers are not typeset but are reproduced as they are submitted by the authors in camera-ready form. Papers are reviewed under the supervision of the Editors with the assistance of the Series Advisory Board and are selected to maintain the integrity of the symposia; however, verbatim reproductions of previously published papers are not accepted. Both reviews and reports of research are acceptable, because symposia may embrace both types of presentation.

Contents

Preface ix

PHYSICOCHEMICAL PROPERTIES

1. Effect of Critical Phenomena on Transport Properties in the Supercritical Region 2
   Es. Gulari, H. Saad, and Y. C. Bae

2. Transport and Intermolecular Interactions in Compressed Supercritical Fluids 15
   J. Jonas and D. M. Lamb

3. Application of Solvatochromic Probes to Supercritical and Mixed Fluid Solvents 29
   S. L. Frye, C. R. Yonker, D. R. Kalkwarf, and R. D. Smith

4. Effects of Supercritical Solvents on the Rates of Homogeneous Chemical Reactions 42
   Sunwook Kim and K. P. Johnston

CHEMICAL REACTIONS

5. Organic Chemistry in Supercritical Fluid Solvents: Photoisomerization of *trans*-Stilbene 58
   Tetsuo Aida and Thomas G. Squires

6. Solvent Effects During the Reaction of Coal Model Compounds 67
   Martin A. Abraham and Michael T. Klein

7. Heterolysis and Homolysis in Supercritical Water 77
   Michael Jerry Antal, Jr., Andrew Brittain, Carlos DeAlmeida, Sundaresh Ramayya, and Jiben C. Roy

PHASE EQUILIBRIA

8. A Statistical Mechanics Based Lattice Model Equation of State: Applications to Mixtures with Supercritical Fluids 88
   Sanat K. Kumar, R. C. Reid, and U. W. Suter

9. Van der Waals Mixing Rules for Cubic Equations of State 101
   E. H. Benmekki, T. Y. Kwak, and G. A. Mansoori

10. High-Pressure Phase Equilibria in Ternary Fluid Mixtures with a Supercritical Component 115
    A. Z. Panagiotopoulos and R. C. Reid

11. Solubilities of Five Solid *n*-Alkanes in Supercritical Ethane 130
    Iraj Moradinia and Amyn S. Teja

12. Solubility of *meso*-Tetraphenylporphyrin in Two Supercritical Fluid Solvents 138
    T. R. Bergstresser and Michael E. Paulaitis

CHROMATOGRAPHY

13. **Supercritical Fluid Adsorption at the Gas–Solid Interface** 150
    Jerry W. King

14. **Mechanism of Solute Retention in Supercritical Fluid Chromatography** ... 172
    C. R. Yonker, R. W. Wright, S. L. Frye, and R. D. Smith

15. **Supercritical Fluid Extraction and Chromatography of Nonpolar Nonvolatile Coal-Derived Products** 189
    J. W. Jordan, R. J. Skelton, and L. T. Taylor

FRACTIONATION AND SEPARATION

16. **Supercritical Carbon Dioxide Extraction of Lemon Oil** 202
    Steven J. Coppella and Paul Barton

17. **Near-Critical Separation of Butadiene–Butene Mixtures with Mixtures of Ammonia and Ethylene** 213
    D. S. Hacker

18. **Fractional Destraction of Coal-Derived Residuum** 229
    Robert P. Warzinski

FUEL APPLICATIONS

19. **Isotope Effects in Supercritical Water: Kinetic Studies of Coal Liquefaction** 242
    David S. Ross, Georgina P. Hum, Tiee-Chyau Miin,
    Thomas K. Green, and Riccardo Mansani

20. **Effect of Solvent Density on Coal Liquefaction Kinetics** 251
    G. V. Deshpande, G. D. Holder, and Y. T. Shah

21. **Extraction of Australian Coals with Supercritical Aqueous Solvents** 266
    John R. Kershaw and Laurence J. Bagnell

22. **Hydrotreating in Supercritical Media** 281
    J. Y. Low

Author Index 295

Subject Index 295

Preface

SCIENTISTS HAVE BEEN AWARE of the novel properties of supercritical fluids for more than a century. Early investigators were fascinated by the "schizophrenic" behavior of this gaslike, liquidlike state, and it is disconcerting but coincidental that research activities also split into two disparate areas. Fundamental interactions in simple systems were meticulously investigated to correlate and predict phase behavior. At the other extreme, scientists applied supercritical fluids with Edisonian zeal, seeking miraculous solutions to complex problems in extraction and fractionation.

The explosion of interest in supercritical fluids during the past decade has seen a bridging of these extremes, a movement toward balance. Investigators have sought to understand and develop applications on the basis of underlying physicochemical principles, and recent reports of semiempirical treatment of supercritical solvent properties have shifted these fluids squarely into the mainstream of chemical research.

We have sought to maintain this balance by probing the principles underlying supercritical fluid behavior in the first three sections and examining applications from the perspective of these principles in the last three sections. Within each section, there is also a flow from fundamentals to applications; the initial chapter provides the basis and the focus for ensuing chapters.

The first section features new approaches to investigating physicochemical properties. Its final two chapters facilitate the transition to the second section, on chemical reactions, a new topic of fundamental importance. Phase equilibria are described in the final section of principles. Here initial chapters are devoted to modeling, and the final chapters report solubility studies. The final three sections are devoted to important applications of supercritical fluids: chromatography, fractionation and separation, and fuel applications. The chapters in each of these sections are also arranged so that there is a transition to more applied topics in the later chapters.

The contributions to this volume are representative of the exciting work under way in supercritical fluid research and attest that there is still much to be done. It is clear that there are abundant opportunities for enhancing our understanding of supercritical fluid behavior and extending the useful application of their properties.

THOMAS G. SQUIRES[1]
Ames Laboratory
Iowa State University
Ames, IA 50011

MICHAEL E. PAULAITIS
Department of Chemical Engineering
University of Delaware
Newark, DE 19716

August 28, 1986

[1]Current address: Associated Western Universities, 142 East 200 South, Suite 200, Salt Lake City, UT 84111

PHYSICOCHEMICAL PROPERTIES

Chapter 1

Effect of Critical Phenomena on Transport Properties in the Supercritical Region

Es. Gulari, H. Saad, and Y. C. Bae

Department of Chemical and Metallurgical Engineering, Wayne State University, Detroit, MI 48202

The transport properties of supercritical fluid mixtures are strongly affected by critical fluctuations. As a first step in testing the validity and the applicability range of the existing theories of critical phenomena, we have used photon correlation spectroscopy to measure the decay rate of density fluctuations in dilute supercritical solutions of heptane, benzene, and decane in CO_2. The results along critical isochores were analyzed in terms of the mode-coupling and the dynamic renormalization-group theories. The values of ν, the temperature exponent of the size of the fluctuations, were in very good agreement with its ideal value. The relative magnitudes of the background contributions in different systems indicated the effects of molecular diffusion and solute-solvent interactions.

Introduction and Background
In supercritical extraction, the dissolution step is diffusion-controlled and the transport properties of the supercritical phase govern the rate of extraction. Due to difficulties in measuring time-dependent phenomena at high pressures, there are very few data and therefore there is a need for experimental investigations of transport properties of dense supercritical fluids which can be correlated for use in practical applications (1, 2).
The supercritical phase is a solution of highly asymmetric components, e.g., CO_2 and a hydrocarbon, at temperatures and pressures very close to the critical point of the solution. For example, for pure CO_2, $T_c = 31°C$ and $P_c = 7.3$ MPa, but for a mixture of CO_2 with 2 mol % benzene, $T_c = 39.3°C$ and $P_c = 8.1$ MPa, and for a mixture of CO_2 with 2 mol % decane, $T_c = 46.8°C$ and $P_c = 8.9$ MPa. Therefore, frequently a process operating at $40°C$ and at a certain pressure in order to attain a given solubility is close to a critical line or surface.
In the vicinity of a critical point, even at equilibrium, domains of different density or concentration exist in a fluid.

These fluctuations, which are referred to as order-parameter fluctuations in studies of critical phenomena (3), comprise the driving forces for transport in the system. For liquid mixtures near a critical mixing point, the order parameter is concentration, and for pure gases near the vapor-liquid critical point, the order parameter is density. For gas mixtures such as supercritical solutions near the critical line, the order parameter is again density, which is a function of composition and temperature compared to a pure gas where density is a function of only temperature at constant pressure.

These fluctuations manifest themselves as optical inhomogeneities in the fluid which can be most easily detected by light scattering techniques (4). Relatively far away from a critical point, $T - T_c \approx 10°C$, the characteristic sizes of domains are ≈ 1 nm and they grow to ≈ 100 nm in the immediate vicinity of a critical point, $T - T_c \leq 1°C$. If the domains are comparable to or less than the visible wavelengths and the refractive index increment is appreciable, the light scattering techniques become ideal nonintrusive probes. In applying these techniques to supercritical solutions two types of experiments can be performed, namely the time-averaged intensity measurements to determine the magnitude of the order-parameter fluctuations and the Photon Correlation Spectroscopy (PCS) to measure their decay rate.

The order-parameter fluctuations are temperature- and system-dependent and their decay rate is related to the transport coefficients (5). Usually the magnitude of the fluctuations are characterized by a correlation length ξ. Along a critical isochore or isopleth, the correlation length diverges as

$$\xi = \xi_0 \varepsilon^{-\nu} \tag{1}$$

where $\varepsilon = (T - T_c)/T_c$, ξ_0 is a system-dependent amplitude and $\nu = 0.63$ is a universal exponent.

The decay rate of the order-parameter fluctuations is proportional to the thermal diffusivity in case of pure gases near the vapor-liquid critical point and is proportional to the binary diffusion coefficient in case of liquid mixtures near the critical mixing point (6). Recently, we reported (7) single-exponential decay rate of the order-parameter fluctuations in dilute supercritical solutions of liquid hydrocarbons in CO_2 for $T - T_c \leq 10°C$. This implied that the time scales associated with thermal diffusion and mass diffusion are similar in these systems.

The mode-coupling theory (8) and the dynamic renormalization-group theory (9) are the two theoretical approaches for the interpretation of the decay rate of the order-parameter fluctuations. The mode-coupling theory yields a relation among the decay rate or the transport coefficient, the viscosity, the correlation length, and the temperature of the system. The dynamic renormalization-group theory predicts how transport coefficients will diverge or converge on various paths of approach to the critical point.

The transport properties of a near-critical system contain an enhancement or a reduction due to critical fluctuations in addition to the contributions of molecular transport processes which are strictly a function of the thermodynamic state. Therefore, the transport coefficients in the critical region are usually partitioned into a critical part and a background which corresponds to the value of the transport coefficient extrapolated from a region far away from the critical point (10). The binary diffusion coefficient can be expressed as $D = L/\chi$, where L is mass conductivity and $\chi = (\delta C/\delta \mu)_T$, the chemical potential derivative of concentration. Then L can be written as

$$L = \tilde{L} + \Delta L \quad \text{or} \quad D = \tilde{L}/\chi + \Delta D \tag{2}$$

where $\tilde{D}$, $\tilde{L}$, and ΔL, ΔD denote the background and the critical contributions respectively. The prediction of the mode-coupling theory for ΔD is

$$\Delta D = k_B T/(6\pi \eta \xi) \Omega(q\xi) \tag{3}$$

where k_B is Boltzmann's constant, η is viscosity, $q = 2q_o \sin(\theta/2)$, with q_o and θ being the wave number of the incident light and the scattering angle. In the hydrodynamic limit, $q\xi \ll 1$, the universal dynamic scaling function $\Omega(q\xi) = 1.03$.

The effect of the background contributions are quite significant for gases and must be determined (3). Unfortunately, diffusivity data at higher temperatures and supercritical pressures are not available for mixtures of hydrocarbons with CO_2 and therefore there is no easy way of estimating the background contribution independently. Using the scaling prediction for ε-dependence of $(\delta C/\delta \mu)_T$, we introduced the temperature dependence of the background term in the following form:

$$\tilde{D} = \tilde{L}/\chi = A_o \varepsilon^{2\nu} \tag{4}$$

where A_o is a system-dependent amplitude which can be treated as a constant over the temperature range of our measurements.

After substituting the background and the critical terms from equations 3 and 4 into equation 2, the diffusion coefficient in the supercritical region is given by

$$D = A_o \varepsilon^{2\nu} + k_B T/(6\pi\eta\xi) \tag{5}$$

The above equation provides a basis for correlating the temperature dependence of a transport coefficient such as mass diffusivity in the supercritical region. The effects of composition, solute, and solvent characteristics can also be introduced into the correlations via ξ_o and A_o which are system-dependent amplitudes. However, a rigorous test of the applicability of equation 5 requires independent measurements of the decay rate of the order-parameter fluctuations, the correlation length, and the viscosity.

In this study, we employed PCS to measure the decay rate of the order-parameter fluctuations in dilute supercritical solutions of heptane, benzene, and decane in CO_2. The refractive index increment with concentration is much larger than the refractive index increment with temperature in these systems. Therefore the order-parameter fluctuations detected by light scattering are mainly concentration fluctuations and their decay rate Γ is proportional to the binary diffusion coefficient, $D = \Gamma/q^2$. The results along critical isochores were analyzed in terms of equation 5. The values of the exponent ν and the significance of the background term are discussed.

Experimental

The light-scattering spectrometer was similar to the one described in a previous paper (11). The design of the high-pressure cell allowed the range of the scattering angles to be 6-11 degrees with homodyne detection. The sample cell was machined out of a solid brass block, fitted with flanged 1.25 cm thick quartz windows, and tested up to 20 MPa. The optical path length of the cell was 7 cm. The inside surfaces of the cell were blackened and a red glass tube was inserted into the cell to absorb the reflected light. The cell was fitted into a brass jacket. The temperature of the circulating water in the jacket was controlled within $0.002°C$ in two steps by using a Neslab RTE-8 circulating bath and Tronac PTC40 temperature controller in series. The temperature was measured by a thermistor embedded in the cell block. The pressure was monitored by a Dynisco Model PT520-5M pressure transducer, the sensitivity and the reproducibility of which were ± 0.005 MPa and ± 0.03 MPa, respectively.

The amounts of the hydrocarbon and CO_2 charged into the cell were determined gravimetrically. For a fixed amount of the hydrocarbon, the CO_2 charge was adjusted carefully so that the relative amounts of the liquid and gas phases were equal as the system crossed from the two-phase region to the dense gas region. Then the process was reversed to check whether the meniscus appeared at the center position of the cell. Based on the validity of the law of rectilinear diameter, this procedure ensured such a loading to be very close to the critical point of the mixture.

Results and Discussion

Correlation function measurements were made along four critical isochores for each of the three systems: CO_2-n-heptane, CO_2-benzene, and CO_2-n-decane. The critical densities and the corresponding compositions are plotted in Figure 1. The three hydrocarbons in order of higher to lower solubility in CO_2 were heptane, benzene, and decane. The measured binary diffusion coefficients or the decay rates of the order-parameter fluctuations at various temperatures and pressures are listed in Tables I, II, and III for CO_2-heptane, CO_2-benzene, and CO_2-decane systems respectively. In Figure 2, the critical lines of the three binary systems in the dilute hydrocarbon range are shown in the pressure-temperature space. dP/dT along the critical lines of CO_2-heptane and CO_2-benzene systems are similar and lower than dP/dT along the critical line of CO_2-decane system, which indicates that CO_2 and decane form more asymmetric mixtures relative to CO_2 with heptane or benzene.

The measured decay rates were analyzed in terms of equation 5, which becomes

$$\Gamma/q^2 = A_o \varepsilon^{2\nu'} + (k_B T/6\pi\eta\xi_o^-)\varepsilon^{\nu'} \tag{6}$$

when the temperature dependence of ξ is substituted in. η was taken to be the viscosity of pure CO_2 evaluated at the system

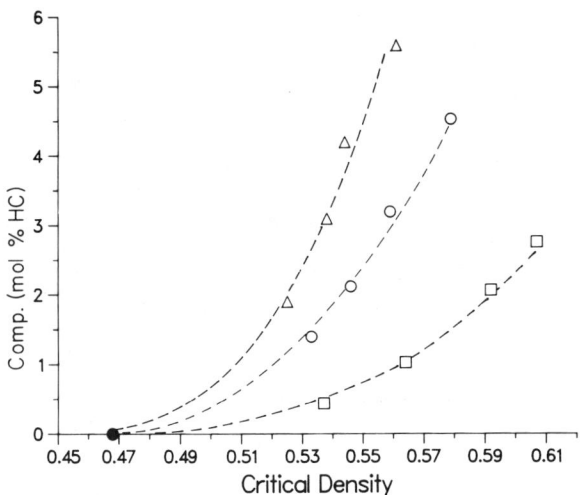

Figure 1: Plot of the hydrocarbon content of various mixtures versus their critical densities. Triangles, circles, and squares denote heptane-, benzene-, and decane-CO_2 solutions respectively. The solid circle marks the critical density of pure CO_2.

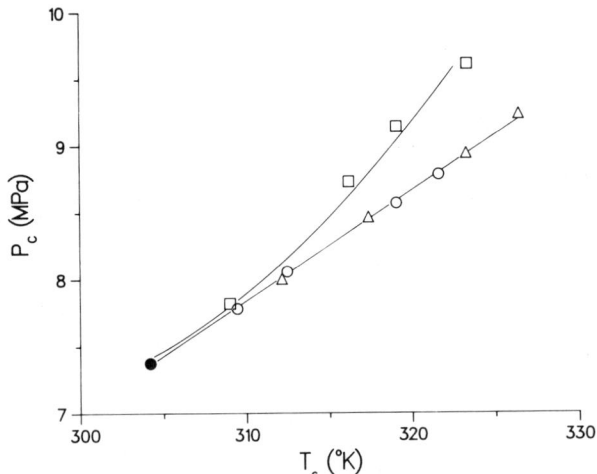

Figure 2: The critical lines of heptane-CO_2 (triangles), benzene-CO_2 (circles), and decane-CO_2 (squares) systems. The solid circle marks the critical temperature and pressure of pure CO_2.

Table I: Measured Γ/q^2 at Various Temperatures and Pressures: CO_2-Heptane System

T(°C)	P(MPa)	$\Gamma/q^2 \cdot 10^5$ (cm²/s)
\multicolumn{3}{l}{94.4 mol % CO_2, ρ_c = 0.561 g/cm³}		
\multicolumn{3}{l}{T_c = (53.31 $\pm$ 0.02) °C, P_c = 9.24 MPa}		
53.74	9.29	1.03 $\pm$ 0.03
54.12	9.35	1.30 $\pm$ 0.02
54.76	9.45	1.52 $\pm$ 0.02
55.58	9.57	1.92 $\pm$ 0.02
56.69	9.75	2.49 $\pm$ 0.03
58.15	9.97	3.39 $\pm$ 0.04
60.76	10.41	4.71 $\pm$ 0.04
63.39	10.84	5.96 $\pm$ 0.05
\multicolumn{3}{l}{95.8 mol % CO_2, ρ_c = 0.544 g/cm³}		
\multicolumn{3}{l}{T_c = (50.14 $\pm$ 0.02) °C, P_c = 8.94 MPa}		
50.61	9.02	1.04 $\pm$ 0.03
50.90	9.07	1.28 $\pm$ 0.02
51.57	9.18	1.63 $\pm$ 0.02
52.18	9.27	2.13 $\pm$ 0.02
52.72	9.37	2.40 $\pm$ 0.02
53.65	9.52	3.02 $\pm$ 0.02
54.38	9.64	3.52 $\pm$ 0.03
55.45	9.83	4.16 $\pm$ 0.03
56.52	10.01	4.83 $\pm$ 0.05
57.98	10.26	5.62 $\pm$ 0.05
59.98	10.61	6.74 $\pm$ 0.07
\multicolumn{3}{l}{96.9 mol % CO_2, ρ_c = 0.538 g/cm³}		
\multicolumn{3}{l}{T_c = (44.24 $\pm$ 0.02) °C, P_c = 8.46 MPa}		
44.49	8.50	0.74 $\pm$ 0.01
44.94	8.57	1.16 $\pm$ 0.01
45.36	8.64	1.48 $\pm$ 0.01
46.37	8.79	2.25 $\pm$ 0.01
47.70	8.96	2.97 $\pm$ 0.02
48.47	9.08	3.61 $\pm$ 0.02
49.52	9.25	4.41 $\pm$ 0.02
50.95	9.49	5.33 $\pm$ 0.03
52.48	9.74	6.36 $\pm$ 0.06
54.31	10.04	7.51 $\pm$ 0.05
\multicolumn{3}{l}{98.1 mol % CO_2, ρ_c = 0.525 g/cm³}		
\multicolumn{3}{l}{T_c = (39.13 $\pm$ 0.01) °C, P_c = 8.00 MPa}		
39.43	8.05	1.13 $\pm$ 0.02
40.15	8.17	1.83 $\pm$ 0.02
41.11	8.32	2.71 $\pm$ 0.02
42.26	8.51	3.78 $\pm$ 0.02
43.51	8.74	4.81 $\pm$ 0.03
45.14	9.00	6.17 $\pm$ 0.05
47.64	9.33	7.71 $\pm$ 0.05
49.12	9.57	8.82 $\pm$ 0.05

Table II: Measured Γ/q^2 at Various Temperatures and Pressures: CO_2-Benzene System

T (°C)	P(MPa)	$\Gamma/q^2 \cdot 10^5$ (cm^2/s)
	95.5 mol % CO_2, $T_c = (48.48 \pm 0.02)$ °C, $P_c = 8.78$ MPa	$\rho_c = 0.579$ g/cm^3
48.81	8.86	0.88 ± 0.04
49.37	8.95	1.32 ± 0.03
49.98	9.05	1.87 ± 0.03
50.64	9.16	2.34 ± 0.03
51.15	9.22	2.84 ± 0.03
52.76	9.51	4.03 ± 0.02
54.38	9.82	5.02 ± 0.04
56.57	10.19	6.69 ± 0.03
58.19	10.48	7.58 ± 0.05
	96.8 mol % CO_2, $T_c = (45.98 \pm 0.05)$ °C, $P_c = 8.61$ MPa	$\rho_c = 0.560$ g/cm^3
46.00	8.62	1.12 ± 0.03
46.72	8.69	1.24 ± 0.03
48.56	8.84	2.39 ± 0.03
49.29	8.95	2.87 ± 0.03
50.03	9.06	3.38 ± 0.02
51.25	9.22	4.30 ± 0.02
52.29	9.35	4.98 ± 0.02
53.36	9.50	5.84 ± 0.04
54.41	9.64	6.61 ± 0.05
55.97	9.85	7.78 ± 0.05
	97.9 mol % CO_2, $T_c = (39.30 \pm 0.02)$ °C, $P_c = 8.06$ MPa	$\rho_c = 0.546$ g/cm^3
39.40	8.08	0.99 ± 0.03
39.51	8.10	1.19 ± 0.02
39.79	8.15	1.34 ± 0.02
40.08	8.19	1.65 ± 0.03
40.48	8.23	1.95 ± 0.03
41.25	8.31	2.51 ± 0.02
44.00	8.70	5.40 ± 0.02
46.96	9.08	8.18 ± 0.02
49.34	9.32	10.0 ± 0.10
51.74	9.56	12.3 ± 0.10
	98.6 mol % CO_2, $T_c = (36.31 \pm 0.02)$ °C, $P_c = 7.78$ MPa	$\rho_c = 0.533$ g/cm^3
36.56	7.82	1.05 ± 0.02
37.12	7.91	1.72 ± 0.02
37.63	8.00	2.28 ± 0.02
38.48	8.14	3.18 ± 0.03
39.48	8.28	4.20 ± 0.02
41.14	8.55	5.82 ± 0.04
43.03	8.84	7.62 ± 0.04
46.32	9.35	10.45 ± 0.05

Table III: Measured Γ/q^2 at Various Temperatures and Pressures: CO_2-Decane System

T (°C)	P(MPa)	$\Gamma/q^2 \cdot 10^5$ (cm²/s)
	97.2 mol % CO_2, $T_c = (50.22 \pm 0.03)$ °C, $P_c = 9.61$ MPa	$\rho_c = 0.607$ g/cm³
50.31	9.63	0.80 ± 0.06
50.77	9.70	0.99 ± 0.04
51.02	9.75	1.15 ± 0.02
51.42	9.84	1.37 ± 0.02
52.42	10.06	1.76 ± 0.02
53.56	10.26	2.14 ± 0.03
54.34	10.39	2.50 ± 0.02
55.15	10.56	2.80 ± 0.02
56.60	10.86	3.32 ± 0.03
57.88	11.12	3.83 ± 0.04
59.29	11.42	4.30 ± 0.09
59.88	11.54	4.50 ± 0.06
	97.9 mol % CO_2, $T_c = (45.95 \pm 0.03)$ °C, $P_c = 8.90$ MPa	$\rho_c = 0.592$ g/cm³
46.20	9.91	0.86 ± 0.04
46.65	9.27	1.10 ± 0.01
47.59	9.37	1.40 ± 0.01
48.28	9.52	1.62 ± 0.02
49.94	9.81	2.52 ± 0.02
51.42	10.10	3.11 ± 0.06
52.84	10.37	3.78 ± 0.06
55.50	10.88	4.80 ± 0.10
	99.0 mol % CO_2, $T_c = (43.05 \pm 0.02)$ °C, $P_c = 8.73$ MPa	$\rho_c = 0.564$ g/cm³
43.38	8.77	1.15 ± 0.02
44.32	8.95	1.76 ± 0.02
45.30	9.09	2.30 ± 0.02
46.33	9.29	2.97 ± 0.02
48.35	9.55	3.78 ± 0.05
50.06	9.86	4.86 ± 0.05
51.44	10.12	5.63 ± 0.06
52.83	10.37	6.40 ± 0.10
	99.6 mol % CO_2, $T_c = (35.86 \pm 0.02)$ °C, $P_c = 7.82$ MPa	$\rho_c = 0.537$ g/cm³
35.99	7.84	1.01 ± 0.02
36.35	7.91	1.35 ± 0.02
36.85	8.00	1.80 ± 0.02
37.71	8.14	2.61 ± 0.04
38.75	8.32	3.64 ± 0.03
39.81	8.51	4.52 ± 0.02
40.99	8.71	5.66 ± 0.09
42.31	8.95	6.73 ± 0.12
45.60	9.28	9.00 ± 0.10

density using $\tilde{\eta} = 42.58\rho + 661.7\rho^2 + 82.89\rho^3 + 152 + 125\epsilon$ from reference (3). In the first part of the analysis, the background contribution was neglected and a two-parameter fit for ν' and ξ_o was performed. In the second part, ν' was fixed at 0.67 and the background term was included. Again a two-parameter fit for A_o and ξ_o was performed. The results are summarized in Table IV.

When the background term was neglected, we obtained $\nu' = 0.71 \pm 0.04$ for heptane in CO_2, $\nu' = 0.76 \pm 0.05$ for benzene in CO_2, and $\nu' = 0.66 \pm 0.05$ for decane in CO_2. We have denoted the temperature exponent of the correlation length ν' in equation 6 instead of ν, and ν' values should be compared to a universal value of 0.67 instead of 0.63 because in our analysis ν' includes the ξ-dependence of the viscosity. When we used the background viscosity $\tilde{\eta}$ for pure CO_2, we did not take into account the weak divergence of viscosity which is $\eta \alpha \xi^z$, where z is a universal exponent. In fact, there is a postulate that the viscosity ratio $\eta/\tilde{\eta}$ exhibits a power-law behavior as

$$\eta = \tilde{\eta} (Q\xi)^z \tag{7}$$

and Q, which is a system-dependent amplitude, is related to the background contribution of the conductivity $\tilde{L}$ (12). If the divergence of the viscosity is taken into account, ΔD should go to zero as $\xi^{-(1+z)}$ or $\epsilon^{\nu(1+z)}$. The theoretical prediction for z is 0.05 and the experimentally determined value for CO_2 is 0.056 ± 0.005 (13). Therefore, $\nu' = \nu(1.06) = 0.67$.

Still slightly higher ν' values can be attributed to the effects of impurities on the critical exponents which were analyzed on general grounds by Fisher (14). When the perturbed critical point of a pure species is studied at a constant impurity concentration x, the renormalization of the critical exponents is expected. For example, ν governing the temperature variation of ξ when $x = 0$ becomes normalized to $\nu_x = \nu/(1 - \alpha)$. α is the temperature exponent of the specific heat and the theoretical and experimental predictions for α give ≈ 0.1. If the dilute hydrocarbon content in our systems were treated as an impurity perturbing the critical point of pure CO_2, ν' would have to be normalized to $\nu'_x = \nu'/(1 - \alpha)$ or $\nu'_x = 0.67/0.9 = 0.74$. The new value of ν' should become evident only for $T - T_c < \Delta T_x$ where the crossover temperature ΔT_x becomes smaller with x and vanishes as x tends to zero. Although there isn't an easy way for calculating ΔT_x, it is expected to vary as a high power, such as $1/\alpha \approx 10$, of the concentration x. Primarily for CO_2-benzene systems and also for CO_2-heptane systems we observed ν' values which were larger than the ideal ν' value. This was in qualitative agreement with the renormalization of the critical exponents at constant x, particularly if the hydrocarbon content in various systems was taken into account. The maximum benzene and heptane content dissolved in CO_2 was 5 mol %, whereas the maximum decane content in CO_2 was 2.8 mol %. However, the accuracy of our measurements is not sufficient and the temperature and composition ranges are not wide enough to determine the changeover from the ideal power-law behavior to the renormalized power-law behavior quantitatively.

The relative magnitudes of the background contributions for the three systems were evaluated by fixing ν' at 0.67 and

Table IV: Values of Parameters Obtained
from Theory of Critical Fluctuations

	Without Background		ν' fixed at 0.67	
	$\xi'_o \times 10^8$ cm	ν'	$\xi'_o \times 10^8$ cm	$A_o \times 10^3$ (cm^2/s)

System: CO_2-n-Heptane

98.1 mol% CO_2, $\rho_c = 0.525$ g/cm^3 — 0.68 ± 0.06 — 0.69 ± 0.02 — 0.76 ± 0.04 — 0.51 ± 0.15

96.9 mol% CO_2, $\rho_c = 0.538$ g/cm^3 — 0.60 ± 0.07 — 0.76 ± 0.03 — 1.09 ± 0.08 — 1.80 ± 0.25

95.8 mol% CO_2, $\rho_c = 0.544$ g/cm^3 — 0.79 ± 0.07 — 0.71 ± 0.02 — 1.06 ± 0.06 — 1.09 ± 0.20

94.4 mol% CO_2, $\rho_c = 0.561$ g/cm^3 — 0.98 ± 0.17 — 0.68 ± 0.04 — 1.10 ± 0.13 — 0.48 ± 0.20

System: CO_2-Benzene

98.6 mol% CO_2, $\rho_c = 0.533$ g/cm^3 — 0.53 ± 0.04 — 0.72 ± 0.02 — 0.74 ± 0.02 — 2.07 ± 0.14

97.9 mol% CO_2, $\rho_c = 0.546$ g/cm^3 — 0.44 ± 0.07 — 0.75 ± 0.04 — 0.76 ± 0.09 — 2.23 ± 0.54

96.8 mol% CO_2, $\rho_c = 0.559$ g/cm^3 — 0.42 ± 0.11 — 0.85 ± 0.07 — 1.21 ± 0.03 — 2.95 ± 0.07

95.5 mol% CO_2, $\rho_c = 0.579$ g/cm^3 — 0.58 ± 0.05 — 0.74 ± 0.02 — 0.92 ± 0.01 — 1.17 ± 0.05

System: CO_2-n-Decane

99.6 mol% CO_2, $\rho_c = 0.537$ g/cm^3 — 0.62 ± 0.08 — 0.69 ± 0.03 — 0.73 ± 0.02 — 0.82 ± 0.11

99.0 mol% CO_2, $\rho_c = 0.564$ g/cm^3 — 0.91 ± 0.19 — 0.67 ± 0.04 — 0.95 ± 0.12 — 0.23 ± 0.55

97.9 mol% CO_2, $\rho_c = 0.592$ g/cm^3 — 1.16 ± 0.28 — 0.67 ± 0.05 — 1.23 ± 0.18 — 0.50 ± 0.52

97.2 mol% CO_2, $\rho_c = 0.607$ g/cm^3 — 1.60 ± 0.17 — 0.59 ± 0.02 — 1.02 ± 0.02 — -0.82 ± 0.05

determining the corresponding values of A_o and ξ_o which are also listed in Table IV. The background contribution was more significant for benzene in CO_2 than heptane and decane in CO_2. The relative contributions to the background and the critical part of the decay rate are shown in Figures 3 and 4 along two isochores, a benzene-CO_2 isochore with $\rho_c = 0.546$ g/cm^3 and a decane-CO_2 isochore with $\rho_c = 0.592$ g/cm^3 respectively, which were at the same overall composition. Along any critical isochore the background contribution is largest for data points farthest away from the critical point. For the benzene-CO_2 mixture shown in Figure 3, the background amounts to about 25% of the total decay rate at about 10°C away from the critical point and diminishes as the critical point is approached. However, for the decane-CO_2 mixture shown in Figure 4, the background contribution is smaller at the same temperature distance to the critical point and amounts to about 10% of the total decay rate. The different background contributions for the three solutes can be explained as follows: Since benzene is the most compact solute relative to heptane and decane, one would expect the diffusion of benzene in the same solvent to be faster, resulting in the higher value for the background term. On the same basis, the diffusion of heptane should be faster than decane and hence yield a higher A_o for heptane than decane.

In our analysis ξ'_o is a fitted parameter. It is related to ξ_o in equation 1 by $\xi'_o = \xi_o Q^z$. Values of ξ'_o in the range 1.0×10^{-8} to 1.2×10^{-8} cm were obtained from the fits and are in fair agreement with $\xi_o = 1.5 \times 10^{-8}$ cm reported for pure CO_2 in the literature (3) and indicates that Q is an amplitude of order one. In the limited composition range of this study, ξ'_o appeared to increase with the hydrocarbon content in the mixture. ξ_o can be independently determined from turbidity measurements (15) and these measurements are in progress and will enable us to determine the composition dependence of ξ_o and the relation between Q and A_o quantitatively.

Independent measurements of the viscosity, the decay rate, and the correlation length for the same fluid are necessary in order to test the theory of critical fluctuations and determine its range of validity in supercritical fluid mixtures. As a first step, we have measured the decay rate of dilute mixtures of three hydrocarbons with CO_2. The results showed that the critical fluctuations dominated the overall transport processes within a 10°C distance to the critical point of a mixture. The size of the critical fluctuations described by the correlation length ξ decayed with increased temperature distance from the critical point. The values of the temperature exponent ν' were in very good agreement with the ideal value of 0.67. If the dilute amounts of hydrocarbon were considered as impurities, slightly higher values of ν' could be explained by the renormalization of ν' due to impurities. The results also indicated that the relative magnitudes of the background contributions could be used to correlate the effects of molecular diffusion and solute-solvent interactions.

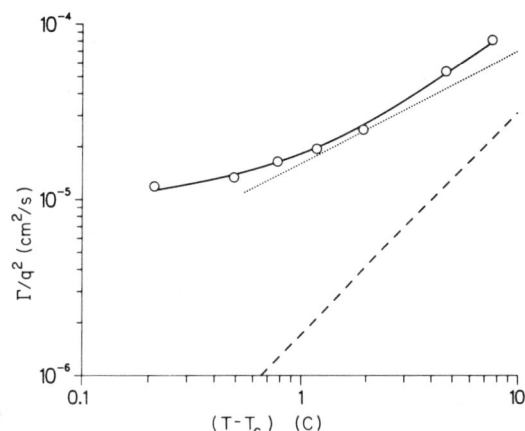

Figure 3: Plot of the measured decay rates and the calculated critical and background contributions along a critical isochore of the benzene-CO_2 system at 97.9 mol % CO_2 and $\rho_c = 0.546$ g/cm^3.

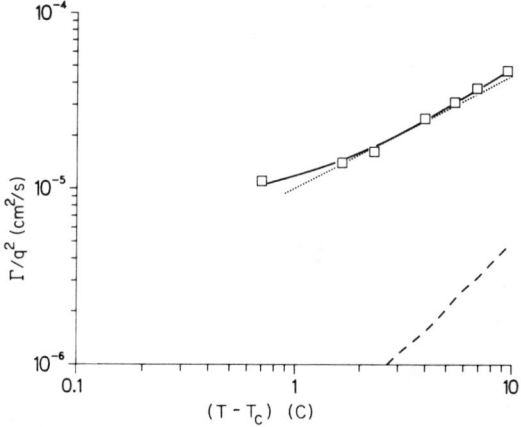

Figure 4: Plot of the measured decay rates and the calculated critical and background contributions along a critical isochore of the decane-CO_2 system at 97.9 mol % CO_2 and $\rho_c = 0.592$ g/cm^3.

Acknowledgments

This work was supported by National Science Foundation Grant No. CBT-8419755.

Literature Cited
1. Franck, E. U.; Schneider, G. M. Ber. Bunsenges. Phys. Chem. 1984, 88, 784.
2. Paulaitis, M. E.; Krukonis, V. J.; Kurnik, R. T.; Reid, R. C. Rev. in Chem. Engineering 1983, 1, 179.
3. Swinney, H. L.; Henry, D. L. Phys. Rev. A 1973, 8, 2586.
4. Goldburg, W. I. In "Light Scattering Near Phase Transitions"; Cummins, H. Z.; Levanyuk, A. P., Eds.; North Holland: Amsterdam, 1983; p. 531.
5. Sengers, J. V. Int. J. of Thermophysics 1985, 6, 203.
6. Mountain, R. D.; Deutch, J. M. J. Chem. Phys. 1969, 50, 1103.
7. Saad, H.; Gulari, Es. Ber. Bunsenges. Phys. Chem. 1984, 88, 834.
8. Kawasaki, K. In "Phase Transitions and Critical Phenomena"; Domb, C.; Green, M. S., Eds.; Academic Press: New York, 1976; Vol. 5A, p. 165.
9. Hohenberg, P. C.; Halperin, B. I. Rev. Mod. Phys. 1977, 49, 435.
10. Sengers. J. V. Ber. Bunsenges. Phys. Chem. 1972, 76, 234.
11. Saad, H.; Gulari, Es. J. Phys. Chem. 1984, 88, 136.
12. Ohta, T. J. Phys. C 1977, 10, 791.
13. Bruschi, L.; Torzo, G. Phys. Lett. 1983, 98A, 257.
14. Fisher, M. E. Phys. Rev. 1968, 176, 257.
15. Kopelman, R. B.; Gammon, R. W.; Moldover, M. R. Phys. Rev. A 1984, 29, 2084.

RECEIVED July 17, 1986

Chapter 2

Transport and Intermolecular Interactions in Compressed Supercritical Fluids

J. Jonas and D. M. Lamb

Department of Chemistry, School of Chemical Sciences, University of Illinois, Urbana, IL 61801

>General overview of several studies of transport and intermolecular interactions in compressed supercritical fluids is presented. The unique aspects of the instrumentation used in these studies are emphasized. First, the results of NMR studies of self-diffusion in supercritical ethylene and toluene are discussed. These experiments used the fixed field gradient NMR spin-echo technique. Second, the novel NMR technique for the determination of solubility of solids in supercritical fluids is described.

Research on the properties of supercritical fluids and supercritical fluid mixtures has become very important in recent years due to the great promise of supercritical fluid extraction techniques. These techniques and their applications have been reviewed by several authors (1-4). There are many advantages of using supercritical fluid extraction over conventional extraction techniques. Many low volatility molecular solids show greatly enhanced solubilities in supercritical dense fluids. Solvent recovery is easily accomplished by manipulating the density, and therefore the solvating power, of the supercritical fluid to precipitate the solid. In addition, although the densities of the supercritical fluids are comparable to liquid densities, the viscosities are generally an order of magnitude smaller, and diffusivities an order of magnitude larger than liquids. A more efficient separation can therefore be achieved.
   Unfortunately, there is a lack of fundamental data on transport and relaxation in model fluids at supercritical conditions. Not surprisingly, there is a corresponding lack of theoretical models to explain the dynamics of supercritical fluids on a molecular level, particularly at the intermediate densities.

The main purpose of our work is the improvement of molecular level understanding of solute-solvent interactions under supercritical conditions. Unique nuclear magnetic resonance (5) techniques are employed to obtain new information about dynamics of molecules in supercritical fluids at high pressures.

The main results of several of our studies will be discussed. First, the results of NMR studies of self-diffusion in supercritical ethylene (6) and toluene (7) will be discussed. These experiments used the fixed field gradient NMR spin-echo technique. Second, the movel NMR technique (8) for the determination of solubility of solids in supercritical fluides will be described.

Experimental

The self-diffusion coefficients in supercritical ethylene were measured using the pulsed NMR spectrometer described elsewhere (9,10), automated for the measurement of diffusion coefficients by the Hahn spin echo method (11). The measurements were made at the proton resonance frequency of 60 MHz using a 14.2 kG electromagnet.

The pressure was generated using the gas compression system described previously (12). A Heise-Bourdon pressure gauge was installed between the compression system and the high pressure vessel to supplement the 30,000 psi pressure transducer. The oxygen scavenger system was bypassed as the amount of oxygen in the ethylene was below the minimum detection level (10 ppm) of the oxygen analyzer (Beckman Instruments, Inc.). In order to depress the extremely long T_1 values of pure ethylene (13) at the experimental conditions studied, small quantities (< 1000 ppm) of oxygen were mixed with the ethylene before measurement of the diffusion coefficient. The addition of oxygen brought the T_1 values down to 2-3 sec, but should not affect the value of the diffusion coefficient. The shorter T_1 values allowed a much shorter measurement time.

The self-diffusion coefficients in supercritical toluene-d_8 were measured at the deuterium resonance frequency of 9.21 MHz, using a 14.1 kG electromagnet with a wide gap (3.8") to accommodate the high pressure vessel. The pulsed NMR spectrometer and receiver system were described in detail elsewhere (10). The argon pressurized high pressure, high temperature NMR probe (14) was used previously for studies of relaxation (15) and diffusion (16) in compressed supercritical water. It consists of two high pressure vessels: the primary vessel, containing an internal furnace, two thermocouples and the RF coil and sample, and the secondary vessel, containing the stainless steel sample bellows. Quartz sample cells were used rather than ceramic cells, as corrosion is not a problem. The RF coil was constructed by winding 14 1/2 turns of 22 gauge nichrome (Chromel A) wire. The coil was silver soldered to nichrome conductor coaxial high pressure leads. The tuning circuit consisted of a six foot impedance transforming coaxial cable terminated with a tapped-parallel capacitor box with both fixed and variable capacitors totalling 70 pF in series and 10 pF in parallel. The observed signal peak to rms noise ratio in liquid toluene-d_8 (30°C) was 60:1 after one scan.

The solubilities for naphthalene in supercritical carbon dioxide (8) were measured at 60 MHz using the NMR spectrometer described elsewhere (10). The high pressure, high temperature NMR probe and gas compression system were the same as that used in the supercritical ethylene experiment (6). The solubility sample cell as shown in Figure 1 was of cylindrical design with 0.250 in. inner diameter and was machined from a high temperature polyimide plastic (Vespel, DuPont Co.). An excess of solid naphthalene was loaded into the cell before a solubility determination and the cell was closed with a close-fitting piston. Pressurized CO_2 entered the sample region through two small holes (0.016 in.) drilled through the sample cell walls. To assure that equilibrum solubilities were obtained, enough solid naphthalene was initially placed in the sample cell so that an excess would be present after dissolution. This made it necessary to separate the contribution to the NMR signal from the dissolved naphthalene and the remaining solid. This separation is easily accomplished due to the radically different spin-spin relaxation rates of dissolved and solid material (T_2, solid << T_2, dissolved). We used the $90°-\tau-180°$ spin-echo sequence with a pulse separation of $\tau = .007$ s; this ensured that no contribution to the NMR echo signal could result from the quickly relaxing protons of solid naphthalene. In this way we were able to monitor the NMR signal from the naphthalene dissolved in the supercritical solution exclusively. This experimental approach for separating the signal from mobile and immobile nuclei has been used previously in our laboratory (17).

The use of NMR spectroscopy as an analytical technique is well established (18). In order to quantitate our spin-echo height to the number of protons present, we performed an independent calibration using standard solutions of naphthalene in carbon tetrachloride. Concentrations for the standards were chosen to correspond to the anticipated supercritical CO_2 solubilities, and all calibration measurements were performed using a sample cell of the same dimensions as the solubility sample cell previously described. The response of our spectrometer to the standard solutions was linear over the concentration range. The reproducibility for independent measurements of the calibration curve was ± 3%. Throughout the experiment, all spectrometer conditions (pulse lengths, phases, receiver amplifier gain, etc.) were closely monitored, and frequent checks on the calibration of the spectrometer were performed. In this way we were able to obtain the molar solubility of solid naphthalene in supercritical carbon dioxide to an estimated experimental accuracy of ± 6%.

Our NMR technique for the determination of the solid-liquid-gas phase line that ends at the UCEP again makes use of the fact that the NMR signal from nuclei with different relaxation rates can be easily separated. In this case we are distinguishing between the signal from the naphthalene-rich liquid phase formed when the S-L-G line is crossed and solid naphthalene.

Since the spin-spin relaxation of the liquid phase naphthalene protons is much longer than that of the solid naphthalene protons, by once more using the spin-echo sequence with $\tau = .007$ s we are able to monitor the NMR signal from the liquid phase only.

The technique was implemented as follows. With solid naphthalene in our solubility cell we brought the system to a desired temperature and CO_2 pressure such that solid-supercritical gas equilibrium existed. The temperature was then slowly increased (heating rate approximately 1°C/hour) at constant pressure, while the NMR signal was monitored. When the S-L-G line was intersected, the solid naphthalene in the cell would melt with the formation of the naphthalene-rich liquid phase, and this resulted in a large and rapid increase in our NMR signal. The temperature at which we saw this discontinuous jump in our NMR signal gave the location of the phase line at that pressure.

The main source of error for this method is the use of a finite rate of heating. We estimate the error in the determination of the S-L-G line to be ± 0.4°C and ± 2 bar.

Results and Discussion

Self-Diffusion in Compressed Supercritical Ethylene. The main purpose of our work (6) was to provide transport data on dense supercritical ethylene and to analyze the data in terms of currently available theories. Ethylene was chosen for the study for a number of reasons. First, it is one of the most widely used solvents in industry due to its easily accessible critical temperature, its relatively low cost and wide availability. Highly accurate compressibility data are available (19-21) in the literature over a wide range of temperature and pressures. These data are necessary for a complete analysis of the transport data.

Some measurements have been made of self diffusion in pure ethylene and in ethylene-sulfur hexafluoride mixtures (22), but these measurements were made very close to the critical temperature and up to pressures of only about 100 bar. Proton spin-lattice relaxation times (T_1) of ethylene have been measured at temperatures from 0°C to 50°C and pressures up to about 2300 bar (13). The relaxation time values were 40-50 sec for much of the region studied. Several relaxation mechanisms contribute to this long relaxation time and make both the measurement and analysis of the relaxation times very difficult. For these reasons, we decided to limit our study to the measurement of the self-diffusion coefficient in supercritical ethylene (6).

The measurements were made as a function of density for pressures from 1 - 2000 bar and at 50°, 75°, 100° and 125°C. The temperatures chosen correspond to those for which density data (19-21) for ethylene and ethylene-CO_2 are available. The experimental self-diffusion coefficients of ethylene as a function of temperature and density are presented in Figure 2. The pressures corresponding to the chosen densities were determined from the compressibility data of Michels and Geldermans (20).

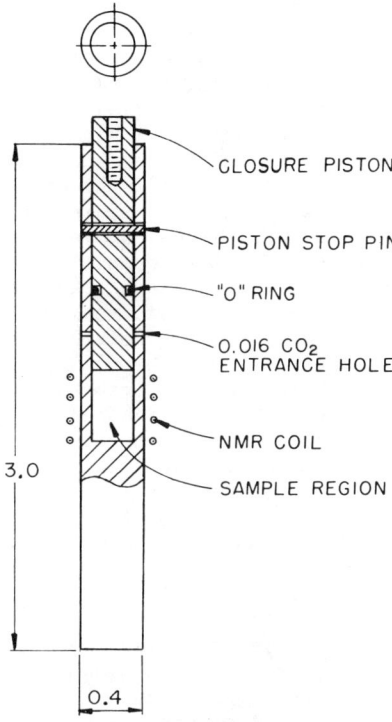

Figure 1. Schematic drawing of NMR sample cell for supercritical solubility measurements.

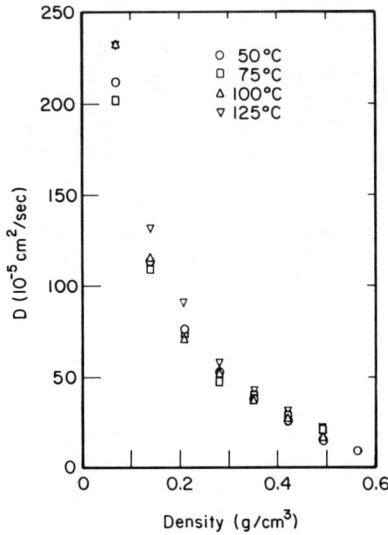

Figure 2. Density and temperature dependence of the experimental self-diffusion coefficients of compressed supercritical ethylene.

The variation in the self-diffusion coefficient is primarily determined by the change in density varied over a wide range. There is no significant temperature dependence within the error of the measurements. The data cover a range of densities $0.34 \leq \rho/\rho_c \leq 2.5$, and temperature $1.14 \leq T/T_c \leq 1.41$.

The density dependence of the ethylene self-diffusion coefficients was analyzed using the Enskog theory (23) of diffusion of hard spheres corrected for the effects of correlated motion (24). The corrected Enskog theory considers only repulsive forces between molecules, but it has proved to be in excellent agreement with experiment for the transport coefficients of supercritical dense gases such as argon, krypton, methane and carbon tetrafluoride (25). At densities less than the critical density (ρ_c), the attractive forces between the molecules become important and deviations from the theory are observed. We calculated the hard sphere diameter from the compressibility data based on the concept that at high density and high temperature, the van der Waal's model of transport in fluids is equivalent to the hard sphere model (26). Values of PV/NkT at a given density are plotted against reciprocal temperature. The high temperature intercept of the tangent to this curve at a given temperature is equated to the hard sphere value of PV/NkT. From this value and the known hard sphere equation of state, the hard sphere diameter is calculated.

The hard sphere diameters were then used to calculate the theoretical Enskog coefficients at each density and temperature. The results are shown in Figure 3 as plots of the ratio of the experimental to calculated coefficients vs. the packing fraction, along with the molecular dynamics results (24) for comparison. The agreement between the calculated ratios and the molecular dynamics results is excellent at the intermediate densities, especially for those ratios calculated with diameters determined from PVT data. Discrepancies at the intermediate densities can be easily accounted for by errors in measured diffusion coefficients and calculated diameters. The corrected Enskog theory of hard spheres gives an accurate description of the self-diffusion in dense supercritical ethylene.

The biggest differences are obtained for the lower ($\rho \leq \rho_c$) density points at each temperature, where the calculated ratios are up to approximately 20% smaller than the molecular dynamics results. This result is not surprising; at the lower densities, the effects of attractive forces become important and cause diffusion in a real gas to be slower. Dymond (25) has found that the theory predicted the experimental self-diffusion coefficients for densities down to about $0.7\ \rho_c$ for $T > T_c$.

The more obvious and consistent deviations from the hard sphere theory occur, at the low density values, due to the effects of attractive forces in the real system. We can attempt to correct for these effects using a method described previously (27-30) for the analysis of angular momentum correlation times in supercritical CF_4 and CF_4 mixtures with argon and neon. We replace the hard sphere radial distribution function at contact $g_{hs}(\sigma)$ with a function $g_p(\sigma)$ which uses the more realistic

Lennard-Jones potential. The new radial distribution function is estimated using the exponential approximation to the optimized cluster theory:(31)

$$g_p(\sigma) = g_{hs}(\sigma) \, e^{C_L(\sigma)} \qquad (1)$$

where $C_L(\sigma)$ is the renormalized intermolecular potential as described and given in Reference (32). This correction was successfully used to extend the rough hard sphere model (29) to the intermediate and low density regions for the interpretation of angular momentum correlation times.

If one corrects for the effects of attractive forces, the corrected values are much smaller than the observed values; the correction overestimates the importance of attractive forces on the value of the diffusion coefficient. Although attractive forces have an effect on the diffusion coefficient at low density, they are not nearly as important as they are in determining the value of the angular momentum correlation time. Diffusion is primarily determined by the repulsive forces between molecules, even at the lowest densities.

<u>Self-Diffusion in Compressed Supercritical Toluene-d_8.</u> Our investigation was motivated by the interest in supercritical toluene as a solvent in the extraction of thermally generated coal liquids (33,34). Typically, coal is heated to temperatures between 350° and 450°C in the presence of a supercritical fluid at a pressure of 100-200 atm. As the large molecular weight components depolymerize thermally, the resulting hydrogen rich material dissolves in the supercritical solvent and is removed. Toluene is a convenient solvent to use for the extraction, as its critical temperature is 319°C, and critical pressure is 41 atm. Recent experimental studies of supercritical fluid extraction in the process of coal liquefication investigate the basic steps involved by varying solvents, pressure and temperatures (35). The goal of our experiment was to provide fundamental data on transport in supercritical toluene-d_8. These data should help in the design and interpretation of extraction processes using supercritical toluene.

We have measured the self-diffusion coefficient in supercritical toluene-d_8 for temperatures from 300°C to 450°C (0.97 ≤ T_r ≤ 1.22) for pressures of 100, 500 and 1000 bar.

Self diffusion coefficients of deuterated toluene were measured, rather than protonated toluene in order to minimize the experimental difficulties associated with very long proton spin lattice relaxation times (T_1). Since the value of the T_1 determines the length of time between pulse sequences, a long relaxation time leads to prohibitively long measurement times. Previous measurements (36-38) of proton and deuterium relaxation times in liquid toluene have been made as a function of temperature and pressure. The relaxation is due to dipolar interactions in protonated toluene and quadrupolar interactions in toluene-d_8. Therefore, the relaxation times can be expected to increase with increasing temperature. However, the quadrupolar relaxed deuterium T_1 values are smaller than the proton T_1

values, resulting in a comparatively shorter measurement time. The self diffusion coefficients of deuterated toluene should not be significantly different from those of protonated toluene at the temperatures of the measurement and will provide an excellent estimate.

Our previous study (16) of self diffusion in compressed supercritical water compared the experimental results to the predictions of the dilute polar gas model of Monchick and Mason (39). The model, using a Stockmayer potential for the evaluation of the collision integrals and a temperature dependent hard sphere diameters, gave a good description of the temperature and pressure dependence of the diffusion. Unfortunately, a similar detailed analysis of the self diffusion of supercritical toluene is prevented by the lack of density data at supercritical conditions. Viscosities of toluene from 320°C to 470°C at constant volumes corresponding to densities from ρ/ρ_c = 0.5 to 1.8 have been reported (40). However, without PVT data, we cannot calculate the corresponding values of the pressure.

The diffusion data at 100 bar are compared in Figure 4 to values obtained using various estimation schemes for the self diffusion coefficient of toluene. No attempt was made to estimate coefficients at 500 and 1000 bar, as these correspond to reduced pressures (P_r = 11.9 and 24.7) well beyond the range of the approximative methods. The results obtained using the generalized charts of Slattery and Bird (41) and of Takahashi (42) do not give a very good agreement with the experimental results. The predicted values are consistently lower than the measured values, and the temperature dependence is not strong enough. The experimental temperature dependence is much more closely reproduced by the empirical correlations of Dawson et al. (43) and of Mathur and Thodos (44). The exact experimental values are not reproduced by any of the methods. However, considering the difference in molecular weight between toluene and toluene-d_8, the approximations involved, and the error in the experimental values (which gets higher as the density decreases), the correlation of Mathur and Thodos gives a very good estimation of the self diffusion coefficient in supercritical toluene.

NMR Measurements of Naphthalene Solubility in Supercritical Carbon Dioxide. The particular phase behavior of the naphthalene-CO_2 system makes it a viable choice for studies of solid solubilities in supercritical fluids. The system is known as a class III mixture (45); its pressure-temperature diagram is depicted in Figure 5. This type of phase behavior occurs for highly asymmetric binary mixtures where the melting point of the heavy component is greater than the critical temperature of the volatile component. Here the three phase solid-liquid-gas (S-L-G) freezing point depression curve intersects the mixture critical line at two locations: the lower critical end point (LCEP) and the upper critical end point (UCEP). The LCEP is typically very near the critical point of the pure light component; the location of the UCEP must be experimentally determined. The existence of the end points is of great importance: between these two critical end points and the associated

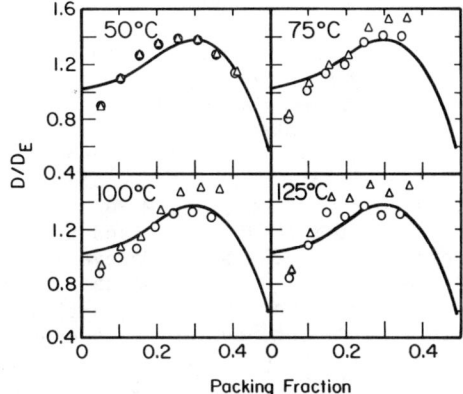

Figure 3. The ratio D/D_E as a function of packing fraction for supercritical ethylene. Δ indicates ratios calculated using hard sphere diameters determined from diffusion data. O indicates ratios calculated using hard sphere diameters determined from compressibility data. The solid lines are the molecular dynamics results, extrapolated to infinite systems, of Alder, Gass and Wainwright (Ref. 24).

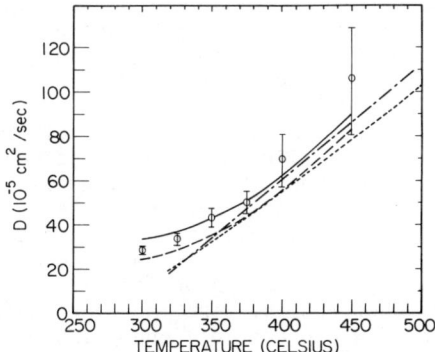

Figure 4. Comparison of experimental diffusion coefficients (o) for toluene-d_8 at 100 bar with predictions of toluene diffusion: (- - - -), method of Slattery and Bird; (- —— - ——), method of Takahashi; (——— ——— ———), method of Dawson et al.; (———————), method of Mathur and Thodos. The error bars represent differences of plus or minus one standard deviation. See the text for details.

branches of the S-L-G line solid-supercritical gas equilibrium exists at all pressures. Consequently the critical end points define the region where solid-supercritical fluid extraction would take place. Also, solid solubilities are enhanced and very sensitive to changes in temperature and pressure near the end points (46), making them technologically interesting.

The naphthalene-CO_2 system has been previously investigated. Tsekhanskaya et al. (47) have measured solubilities of naphthalene in supercritical CO_2 at temperatures of 35, 45, and 55°C and pressures up to 324 bar, and McHugh and Paulaitis (48) have reported solubilities at 35, 55°C, and two higher temperatures to pressures of about 290 bar. However, neither study has clearly defined the solid naphthalene solubilities near the UCEP. In a later paper by McHugh and Yogan (49) in which the S-L-G phase lines were determined for various systems, it was pointed out that the higher temperature solubility data reported by McHugh and Paulaitis (48) was in fact for liquid naphthalene-supercritical fluid equilibria. The discrepancy can arise from the fact that the previous studies used techniques that require external sampling to determine the solubilities--important phase information can therefore be obscured.

The NMR method we have developed gives a direct, in situ determination of the solubility and also allows us to obtain phase data on the system. In this study we have measured the solubilities of solid naphthalene in supercritical carbon dioxide along three isotherms (50.0, 55.0, and 58.5°C) near the UCEP temperature over a pressure range of 120-500 bar. We have also determined the pressure-temperature trace of the S-L-G phase line that terminates with the UCEP for the binary mixture. Finally, we have performed an analysis of our data using a quantitative theory of solubility in supercritical fluids to help establish the location of the UCEP.

The experimental solubility data for solid naphthalene in supercritical carbon dioxide, given as moles naphthalene dissolved per liter, are shown in Figure 6. Qualitatively the three pressure-composition isotherms show characteristic behavior for a solid-supercritical fluid system. Each isotherm initially shows a large increase in solubility with increasing pressure, and then a limiting value is reached at higher pressures.

Diepen and Scheffer (46) were the first to show that near either the lower or upper critical end point the solubility of a solid in a supercritical fluid is enhanced and also very sensitive to changes in temperature and pressure; our solubility isotherms show this effect for both end points. First, the isotherms cross at about 140 bar so that the solubility at the lowest temperature (50.0°C) is largest at 120 bar. This is a result of approaching the lower critical end point region (which should be close to the critical point of pure CO_2 as previously mentioned). At temperatures and pressures near this LCEP the solubility enhancement results in lower temperature isotherms having the greater solubilities. The effect of the upper critical end point is also well shown by our data. The 58.5°C isotherm shows a large increase in solubility at about 235 bar; the slope of the isotherm is near zero. As Van Welie and Diepen

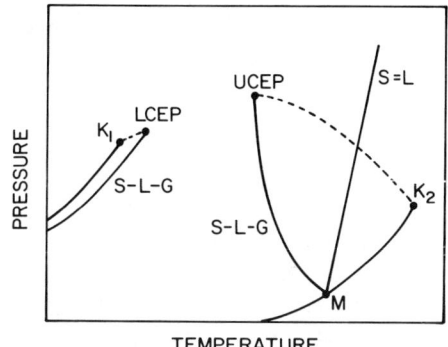

Figure 5. Pressure-temperature diagram for the naphthalene-CO_2 system. K_1 and K_2 are the critical points of pure carbon dioxide and naphthalene, respectively. The critical mixture line (---) is intersected at the lower (LCEP) and upper (UCEP) critical end points by the solid-liquid-gas (S-L-G) freezing point depression curve. Point M is the triple point of pure naphthalene, and the line S = L is the melting cure of the pure solid.

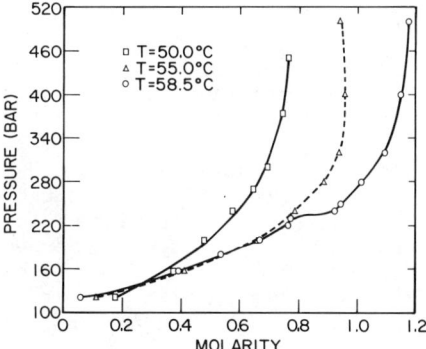

Figure 6. Experimental solubilities for solid naphthalene in supercritical carbon dioxide expressed in moles naphthalene dissolved per liter solution.

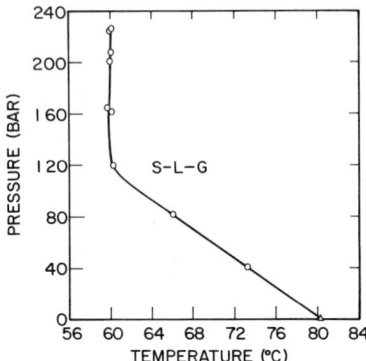

Figure 7. Experimental pressure-temperature trace of the three-phase solid-liquid gas line that terminates at the UCEP for the naphthalene-CO_2 system (O). The S-L-G line begins at the melting point of naphthalene, 80.3°C (Δ).

demonstrated in the naphthalene-supercritical ethylene system (50), exactly at the UCEP temperature a solubility isotherm should show a horizontal point of inflection at the UCEP pressure. Our 58.5°C isotherm inflection point therefore shows that we are operating in close proximity to the UCEP.

The location of the UCEP is quite important in these mixtures: this critical point gives the maximum temperature at which solid-gas equilibrium exists at all pressures. In order to obtain its location, we first used our NMR method to determine the S-L-G line. The results of our experimental determination of the pressure-temperature trace for the solid-liquid-gas equilibrium line that terminates at the UCEP for the naphthalene-carbon dioxide system is shown in Figure 7.

This phase line and the location of the UCEP have been previously determined by McHugh (51) and McHugh and Yogan (49). However, our data differ from the results of those studies. The previous studies have found a substantial temperature minimum in the phase line at about 57.5°C and 117 bar (i.e., as pressure increases, the S-L-G line begins with a negative slope (dP/dT), reaches the temperature minimum, and then proceeds with a positive slope to the UCEP). This would suggest, for example, that as pressure was increased along a 58.5°C isotherm the S-L-G line would be crossed and a liquid phase would be present. On the contrary, our solubility measurements along this isotherm showed no liquid phase formation. As can be seen in Figure 6, we obtained only a slight temperature minimum for the S-L-G line, and a 58.5°C isotherm does not intersect our S-L-G phase boundary. We thus believe our S-L-G line results to be more accurate since they better correlate with our solubility data.

The high pressure termination of the S-L-G line gives the pressure-temperature location of the upper critical end point. The value for the UCEP temperature shown by the high pressure portion of our S-L-G line is 60.1°C. This agrees with the value for the UCEP temperature reported by McHugh and Yogan (49). Next, in order to determine the UCEP pressure, we have used the theory of Gitterman and Procaccia (52). This analysis gave the UCEP pressure as P_c = 227 bar. In this way we have been able for the first time to determine the location of the UCEP for the naphthalene carbon dioxide system. These results show that our novel NMR method can yield both solubility data and phase information when studying equilibria in supercritical fluid mixtures.

Acknowledgments

This work was partially supported by the Department of Energy under Grant DE-FG-82PC50800.

Literature Cited

1. Irani C. A.; Funk, E. W. "Recent Developments in Separation Science"; N. N. Li, Ed., CRC Press, Cleveland, Ohio, 1977, v. 3A, p. 171.
2. Williams, D. F. Chem. Eng. Science 1981, 36, 1769.
3. Schneider, G. M. Angew. Chem. Int. Ed. 1978, 17, 716.
4. Brunner, G.; Peter S. Ger. Chem. Eng. 1982, 5, 181.
5. Jonas, J. Science 1982, 216, 1179.
6. Baker, E. S.; Brown, D. R.; Jonas, J. J. Phys. Chem. 1984, 88, 5425.
7. Baker, E. S.; Brown, D. R.; Lamb, D. M.; Jonas, J. J. of Chem. and Eng. Data 1985, 30, 141.
8. Lamb, D. M.; Jonas, J. J. Phys. Chem. in press.
9. Jonas, J. Rev. Sci. Instrum. 1972, 43, 643.
10. Cantor, D. M.; Jonas, J. Anal. Chem. 1976, 48, 1904.
11. Cantor, D. M.; Jonas, J. J. Mag. Res. 1977, 28, 157.
12. Finney, R. J.; Wolfe, M.; Jonas, J. J. Chem. Phys. 1977, 67, 4004.
13. Trappeniers, N. J.; Prins, K. O. Physica 1967, 33, 435.
14. DeFries, T. H.; Jonas, J. J. Mag. Res. 1979, 35, 111.
15. Lamb, W. J.; Jonas, J. J. Chem. Phys. 1981, 74, 913.
16. Lamb, W. J.; Hoffman, G. A.; Jonas, J. J. Chem. Phys. 1981, 74, 6875.
17. Brown, D. R.; Jonas, J. J. Polym. Sci. Phys. Ed. 1984, 22, 655.
18. Leyden, D. E.; Cox, R. M. "Analytical Applications of NMR"; John Wiley and Sons, New York, 1977.
19. Dick, W. F. L.; Hedley, A. G. M. "Thermodynamic Functions of Gases"; F. Din, Ed., Butterworths, London, 1962, v. 2, p. 88.
20. Michels, A.; Geldermans, M. Physics 1942, 9, 967.
21. Sass, A.; Dodge, B. F.; Bretton, R. H. J. Chem. Eng. Data 1967, 12, 158.
22. Hamann, H.; Richtering, H.; Zucker, U. Ber. Buns. Phys. Chem. 1966, 70, 1084.
23. Chapman, S.; Cowling, T. G. "Mathematical Theory of Non-Uniform Gases"; Cambridge Univ. Press, Cambridge, 1970, 3rd Ed.
24. Alder, B. J.; Gass. D. M.; Wainwright, T. E. J. Chem. Phys. 1970, 53, 3813.
25. Dymond, J. H. Physica 1974, 75, 100.

26. Dymond, J. H.; Alder, B. J.; J. Chem. Phys. 1966, 45, 2061.
27. Jonas, J. "NATO ASI on High Pressure Chemistry"; H. Kelm, Ed., D. Reidel Publ. Co., Dordrechi, Holland, 1978, p. 65.
28. Jonas, J. Rev. Phys. Chem. Japan 1980, 50, 19.
29. Chandler, D. J. Chem. Phys. 1974, 60, 3500, 3508.
30. Wolfe, M.; Arndt, E.; Jonas, J. J. Chem. Phys. 1977, 67, 4012.
31. Andersen, H. C.; Chandler, D.; Weeks, J. D. Adv. Chem. Phys. 1976, 34, 105.
32. Finney, R. J.; Wolfe, M.; Jonas, J. J. Chem. Phys. 1977, 67, 4012.
33. Whitehead, J. C.; Williams, D. E. J. Inst. Fuel 1975, 48, 397.
34. Berkowitz, N. "An Introduction to Coal Technology"; Academic Press: New York, 1979.
35. Worthy, W. Chem. Eng. News 1983, 61, 35.
36. Parkhurst, H. J.; Lee, Y.; Jonas, J. J. Chem. Phys. 1971, 55, 1368.
37. Wilbur, D. J.; Jonas, J. J. Chem. Phys. 1971, 55, 5840.
38. Wilbur, D. J.; Jonas, J. J. Chem. Phys. 1975, 62, 2800.
39. Monchick, L.; Mason, E. A. J. Chem. Phys. 1961, 35, 1676.
40. Knappwost, A.; Ruhe, F.; Raschti, M.; Wochnowski, H.; Ankara, U. Z. Phys. Chem. (Wiesbaden) 1980, 122, 143.
41. Slattery, J. C.; Bird, R. B. AICHE J. 1958, 4, 137.
42. Takahashi, S. J. Chem. Eng. Japan 1974, 7, 417.
43. Dawson, R.; Khoury, F.; Kobayashi, R. AICHE J. 1970, 16, 725.
44. Mathur, G. P.; Thodos, G. AICHE J. 1965, 11, 613.
45. Gubbins, K. E.; Shing, K. S.; Strett, W. B. J. Phys. Chem. 1983, 87, 4573.
46. Diepen, G. A. M.; Scheffer, F. E. C. J. Phys. Chem. 1953, 57, 575.
47. Tsekhanskaya, Yu. V.; Iomtev, M. B.; Mushkina, E. V. Russ. J. Phys. Chem. (Engl. Transl.) 1964, 38, 1173.
48. McHugh, M.; Paulaitis, M. E. J. Chem. Eng. Data 1980, 25, 326.
49. McHugh, M. A.; Yogan, T. J. J. Chem. Eng. Data 1984, 29, 112.
50. Van Welie, G. S. A.; Diepen, G. A. M. Rec. Trav. Chim. Pays-Bas 1961, 80, 673.
51. McHugh, M. A. Ph.D. Dissertation, University of Delaware, Newark, DE, 1981.
52. Gitterman, M.; Procaccia, I. J. Chem. Phys. 1983, 78, 2648.

RECEIVED June 24, 1986

Chapter 3

Application of Solvatochromic Probes to Supercritical and Mixed Fluid Solvents

S. L. Frye, C. R. Yonker, D. R. Kalkwarf, and R. D. Smith

Chemical Methods and Separations Group, Chemical Technology Department, Pacific Northwest Laboratory, Battelle Memorial Institute, Richland, WA 99352

> Changes of the polarity/polarizability of supercritical CO_2, N_2O, Freon-13, NH_3, and CO_2/methanol mixtures, were measured by observing the shift of the UV absorption of 2-nitroanisole as a function of temperature and pressure. The polarities of the supercritical fluid solvents are compared to those of liquid solvents through use of the Kamlet-Taft π^* polarity scale. In pure supercritical solvents the polarity was found to correlate primarily with density. The data for CO_2 indicate that a change in the cybotatic region occurs at a fluid density of ~0.3 g/cm^3. Adding methanol as a modifier to supercritical CO_2 gradually increases the polarity of the solvent and in solutions with over 10% methanol the composition of the cybotatic region appears to be enriched in methanol (compared to the bulk composition).

Supercritical fluids are increasingly being used as solvents in both analytical and process applications. The increasing interest in the use of supercritical fluid solvents is partially due to the dependence of fluid solvent properties on temperature and pressure. One is able to change to a "different" solvent simply by changing the fluid temperature or pressure. This variability in solvent properties adds an extra degree of freedom in solvent use. However, there are a variety of complex interactions between the solvent and solute which complicate understanding the behavior of supercritical solvents. Some of the interactions are nonspecific, occurring between any solvent-solute pair, while other interactions (such as hydrogen bonding) are dependent upon the specific solvent and solute. Because the strength of such interactions depends on intermolecular distances, the relative importance of each contribution is density dependent. A large body of work exists that studies the various solvent-solute interactions using liquid solvents; however, very few results have been reported for supercritical fluid solvents.

This study was aimed at investigating solvent-solute interactions in the supercritical fluid (dense gas) regime, particularly

0097-6156/87/0329-0029$06.00/0
© 1987 American Chemical Society

the dependence of specific solvent interactions on temperature and pressure. A spectroscopic technique has been used to obtain information about the solvent environment in the vicinity of a solute molecule and to monitor changes in that region as the fluid temperature and pressure are varied. The results show that the region surrounding a solute molecule can change more drastically than the bulk solvent density and provide a means for comparison of liquid and supercritical fluid solvents. This method should be particularly useful in exploring mixed fluid systems.

Approach

The wide variety of possible solvent-solute interactions requires that any scale used to quantify solvent properties will be complex. Unfortunately, no universally accepted scale of solvating power has been devised. It does not seem reasonable to develop an entirely new scale for supercritical fluid solvents, especially since it is desirable to compare the solvent behavior of supercritical fluids with that of liquid solvents.

Of the existing solvent scales, the Kamlet and Taft π^* scale (1) appears appropriate for use with supercritical fluid solvents. The basis of the Kamlet and Taft scale is the effect of solvent polarity and polarizability on the energy of the $\pi \rightarrow \pi^*$ electronic transition of a solute probe molecule. Solvent molecules in proximity to a solute (the cybotatic region) differentially affect the electronic energy levels of solutes (the solvatochromic effect) (2). In this manner, the electronic transitions of some molecules are sensitive probes of the solute environment. Since the probe molecules selected by Kamlet and Taft have π^* electronic states which are more polar than the ground state, a change in the polarity/polarizability of the solvent medium changes the electronic energy gap, and thus the position of the absorption band. Kamlet and Taft have developed an empirical relationship between measured solute absorption maxima in a solvent and the polarity/polarizability of that solvent:

$$\nu = \nu_0 + s\pi^*$$

ν = absorption maximum in test solvent
ν_0 = absorption maximum in reference solvent
s = solute dependent parameter
π^* = measured solvent polarity/polarizability parameter

In subsequent work Kamlet and Taft added additional terms to this relationship to account for hydrogen bonding and other types of solvent-solute interactions (3,4).

Since its introduction the Kamlet-Taft π^* scale has been applied to a large number of diverse liquid solvents (5). Several studies have shown that the π^* solvent scale is consistent with other popular measures of solvent strength (6). It has also been shown that there are good correlations between the empirical measurements of the Kamlet and Taft scale and several theoretical models of solvent behavior (7). Use of the solutes chosen by Kamlet and

Taft permits direct comparison between the results for supercritical fluid solvents and the known behavior of liquid solvents.

Experimental

The solvatochromic probe molecule chosen for this work was 2-nitroanisole (Aldrich Chemical Co.). The s value reported in the literature for 2-nitroanisole is -2.428 ± 0.195 (1). A known s value for the solute allows one to calculate the change in the supercritical fluid solvents' π^* value as temperature or pressure changes. The reference absorption maxima for 2-nitroanisole is 32.56×10^3 cm^{-1} (v_0) in cyclohexane (1).

Absorption spectra of 2-nitroanisole in supercritical CO_2, N_2O, Freon-13, ammonia and CO_2-methanol mixtures were obtained on a Cary model 1605 spectrophotometer operated in the dual beam mode. The gases used as supercritical solvents were of the highest purity available from the supplier (Matheson) and were further filtered prior to use. The mixed solvent system of CO_2-methanol was obtained from Scott Speciality Gases (15.4 wt% methanol), and other mixtures were made in the laboratory. Spectra of 2-nitroanisole in n-pentane, methanol, tetrahydrofuran and acetonitrile (Burdick & Jackson) were obtained using quartz cells with a 1-cm light path and with a pure solvent blank in the reference beam. Vapor phase and supercritical fluid spectra were obtained using an air reference.

The cylindrical high-pressure cell was constructed from stainless steel (SS 304) and had dimensions of 1 in. diameter by 2 in. length, with a 3/16-in.-diameter hole drilled along the axis. High-pressure 1/16-in. O.D. stainless steel inlet and outlet connections were silver soldered into the cell, which allowed for supercritical fluid to flow through and purge the cell. Each end of the cell had a seat for a 1/4-in. O.D., 1/4-in.-thick quartz window, and was threaded to accept a brass compression nut. A 1/4-in. O.D. Teflon o-ring was placed on each side of each window to provide cushioning and make a gas-tight seal. The optical path length of the assembled cell was approximately 1.5 cm with a total volume of 350 µℓ. The absorption cell was electrically heated and the temperature monitored with a thermocouple. For the mixed fluid solvent systems a similar, larger volume high-pressure cell was used. Either assembly could be placed in the sample compartment of the spectrophotometer, and a single mode temperature controller (West) provided temperature regulation to within ±0.5°C. The supercritical solvent was delivered to the cell by a high-pressure syringe pump (High Pressure Equipment) and was connected to the cell via a high-pressure liquid chromatographic sampling valve (Rheodyne Inc.). The sampling valve contained a 10 µℓ sample loop; this allowed for easy introduction of 2-nitroanisole into the absorption cell. A pressure transducer (Model 204, Setra Systems) provided pressure measurement (±10 psi).

The experiments were conducted as follows: the sample was loaded into the sample loop of the injection valve and a valve at the cell outlet was opened to allow fluid from the pump to flow through the cell. The sample loop was then switched in line and the absorbance monitored to detect the appearance of the sample in the cell. The solute (probe) concentration was diluted by introduction of additional solvent to obtain the desired concentration (and absorbance). The outlet valve was then closed and the absorption spectrum of the 2-nitroanisole was recorded. Subsequent scans at different pressures or temperatures were made after appropriate equilibration periods. In all cases the pure fluid spectra were recorded to ensure that there were no interferences.

Results and Discussion

Pure Fluids. Table I gives the critical properties for the four pure fluids studied. Table II lists the absorption maxima for the probe molecule (2-nitroanisole) in a variety of liquid and supercritical fluid solvents. Comparing the results for the liquids, it is obvious that increasing the polarity of the solvent causes the absorption to be shifted to longer wavelengths (i.e., red shifted). As expected, the vapor phase corresponds to the least polar or polarizable "solvent". The results for the supercritical fluid solvents at the reduced density of 1.5, given in Table II, show them to be comparable to liquid solvents in polarity, although not as polar as the subcritical liquids. The reduced polarity of the supercritical solvents is understandable when the much lower density of these fluids (at these experimental conditions) is taken into account. In fact, at a reduced density of 0.8 all the fluids give solvatochromic shifts less than that observed for pentane.

At more "liquid-like" densities, the solvatochromic shifts in supercritical fluids approach those observed in the corresponding liquids. Figures 1 and 2 depict the pressure dependence of the wavelength of the absorption maximum of 2-nitroanisole in supercritical CO_2, N_2O, $CClF_3$ (Freon-13), and NH_3. These measurements reflect the effect of pressure (fluid density) on the cybotatic region of these solvents. It is clear that the fluid density affects the cybotatic region, as evidenced by the shift in the absorption maximum with pressure; and it is also evident that the magnitude of the shift is fluid dependent.

A more direct comparison of the results from the different fluids is obtained by plotting the peak position vs reduced density, as shown in Figure 3. With Freon-13 as the solvent the peak maximum is at greater wave numbers, which is characteristic of a less-polar solvent. There is only a small shift in the peak position as the density is changed. CO_2 and N_2O exhibit quite similar peak positions and shifts at approximately equal reduced densities, which is consistent with their similar physicochemical behavior and critical points. Finally, the results for NH_3 indicate that it is the most

TABLE I. Critical properties for four pure fluids

Supercritical Fluid	$T_c(K)$	$P_c(atm)$	$\rho_c(g/cm^3)$
CO_2	304.2	72.8	0.46
N_2O	309.6	71.7	0.45
$CClF_3$	302.0	38.1	0.58
NH_3	405.6	111.3	0.24

TABLE II. π^* Solubility Scale Parameters and the Peak Position of Absorbance Maximum for 2-Nitroanisole

Solvent	T (K)	P (atm)	ρ (g/cm^3)	$\nu_{max} \times 10^{-3}$ Observed (cm^{-1})	$\nu_{max} \times 10^{-3}$ Calculated (cm^{-1})(a)	π^*
Vapor	--	--	--	34.97	35.13	-1.06(b)
CClF$_3$ (c)	303	90	0.87(f)	33.26 ± 0.07	--	-0.29 ± 0.03
CClF$_3$ (e)	296	270	1.15(f)	33.07 ± 0.10	--	-0.21 ± 0.04
N$_2$O (c)	323	180	0.68(f)	33.85 ± 0.09	--	-0.12 ± 0.03
Pentane	298	1	0.626	32.89	--	-0.08(d)
CO$_2$ (c)	323	180	0.69(f)	33.73 ± 0.10	--	-0.07 ± 0.04
N$_2$O (e)	296	270	0.95(f)	32.63 ± 0.05	--	-0.03 ± 0.02
CO$_2$ (e)	296	270	0.95(f)	32.47 ± 0.05	--	0.04 ± 0.03
NH$_3$ (c)	413	170	0.36(f)	31.75 ± 0.25	--	0.34 ± 0.07
Tetrahydrofuran	298	1	0.888	31.45	31.15	0.58(b)
Methanol	298	1	0.791	31.06	31.10	0.60(b)
Acetonitrile	298	1	0.777	30.86	30.74	0.75(b)
NH$_3$ (e)	296	270	0.62(f)	30.63 ± 0.04	--	0.80 ± 0.02

(a) ν_{max} values calculated based on ν_0 = 32.56 x 10^{-3} cm^{-1}.
(b) π^* values from reference 5.
(c) Supercritical fluid, ρ_r = 1.5.
(d) Estimate π^* value for pentane from reference 5.
(e) ν_{max} determined for subcritical liquid.
(f) Estimated from "Gas Encyclopedia"; Elsevier Scientific: Amsterdam, 1976.

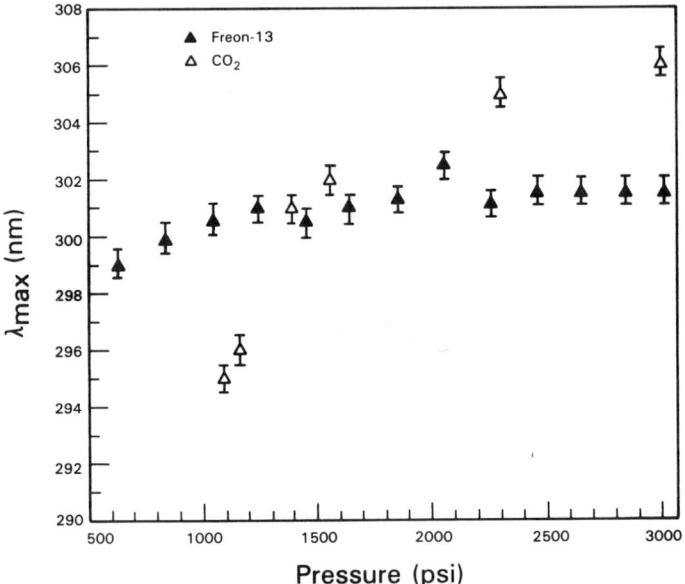

Figure 1. Wavelength of maximum absorbance (λ_{max}) versus pressure (psi) for 2-nitroanisole with CO_2 and Freon-13.

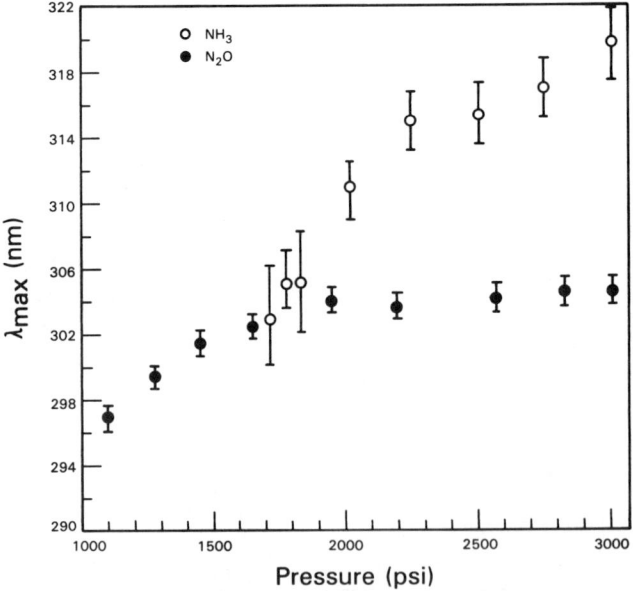

Figure 2. Wavelength of maximum absorbance (λ_{max}) versus pressure (psi) for 2-nitroanisole with N_2O and NH_3.

polar solvent examined, displaying the largest variation in the cybotactic region as the fluid density is changed.

From the position of the absorbance maximum in different fluids, the corresponding π^* parameter values are readily calculated, allowing comparison of supercritical fluids at various experimental conditions with conventional liquid solvents. In Figure 4 the calculated π^* values are plotted vs the reduced density for the various fluids. Two sets of points have been plotted for CO_2; one set of measurements were made by varying the pressure at constant temperature, the other is for various temperatures at constant pressure. The two data sets fall on a single curve (within experimental error), indicating that the fluid density is the prime factor in determining the value of π^*. The data for the other fluids was obtained under isothermal conditions.

The data indicate that as the reduced density of a fluid is increased the cybotactic region becomes more polar/polarizable, with an especially large effect for supercritical NH_3. The low π^* values for Freon-13 at these densities are similar to those of the perfluoro-alkanes (5). The values for supercritical CO_2 and N_2O range between those of the perfluoro-alkanes to that of the n-alkanes (n-heptane, etc.) (5). Ammonia gives π^* values which range from approximately that of n-heptane to that of ethyl acetate or tetrahydrofuran (5) over the density range studied.

Comparison of π^* values with solubility parameters for the various liquids and fluids, calculated as described by Giddings and coworkers (10), shows a general correlation for the less polar solvents. Ammonia and the polar liquid solvents diverge from this correlation, suggesting the operation of specific interactions which contribute to the greater magnitude of the shifts observed for ammonia.

It is possible to compare these results with those reported in the literature for supercritical CO_2. Hyatt (8) reports π^* values for CO_2 from -0.52 to -0.0 which compares very favorably with the present value of -0.55 at a reduced density of 0.43. Sigman et al. (9) used a number of solutes to determine π^* values for CO_2 at fluid densities of 0.86, 0.68, and 0.46 g/cm^3 of -0.12, -0.22, and -0.45, respectively. The present values at similar densities are -0.05, -0.10, and -0.25. The source of this discrepancy is unknown, although Sigman et al. attribute the variation in π^* values among their various indicators to specific solute effects.

The Kamlet-Taft π^* polarity/polarizability scale is based on a linear solvation energy relationship between the $\pi \rightarrow \pi^*$ transition energy of the solute and the solvent polarity (1). The Onsager reaction field theory (11) is applicable to this type of relationship for nonpolar solvents, and successful correlations have previously been demonstrated using conventional liquid solvents (7). The Onsager theory attempts to describe the interactions between a polar solute molecule and the polarizable solvent in the cybotactic region. The theory predicts that the stabilization of the solute should be proportional to the polarizability of the solvent, which can be estimated from the index of refraction. Since carbon dioxide is a nonpolar fluid it would be expected that a linear relationship

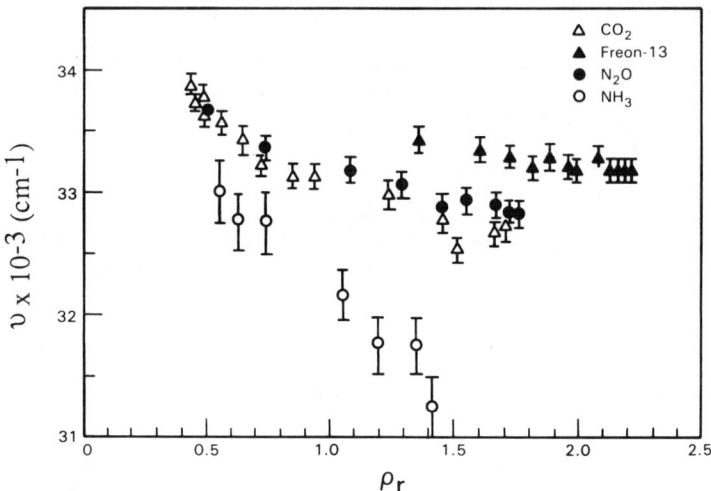

Figure 3. ν_{max} versus reduced density (ρ_r) for 2-nitroanisole with CO_2, N_2O, Freon-13 and NH_3.

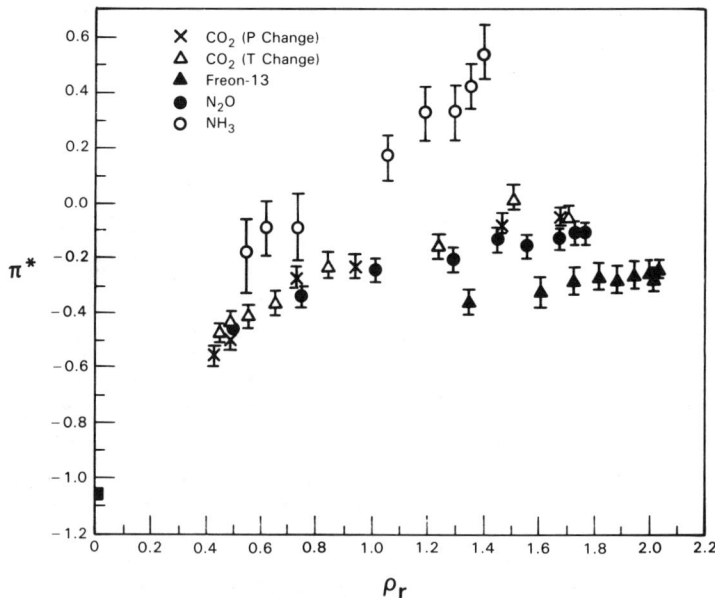

Figure 4. π^* versus reduced density (ρ_r) for 2-nitroanisole with CO_2, N_2O, Freon-13 and NH_3.

would be obtained between the measured π^* values at different densities and the Onsager reaction field function $L(n^2)$:

$$L(n^2) = (n^2 - 1) / (2n^2 + 1)$$

The Onsager function is dependent on the refractive index of the solvent (n). For supercritical CO_2 the refractive index is dependent on the fluid density and can be calculated from the Lorentz-Lorenz refraction equation:

$$\left(\frac{n^2 - 1}{n^2 + 2}\right)\left(\frac{1}{\rho}\right) = A_R + B_R(T)\rho + \ldots$$

where A_R and $B_R(T)$ are the first and second refractometric viral coefficients (12,13). Kholodov, Timoshenko, and Yaminov (12) and Bose and St. Arnaud (13) have reported values for these coefficients for supercritical CO_2, enabling calculation of the refractive index at various temperatures and pressures.

A plot of the Onsager function vs the measured π^* values for CO_2 is given in Figure 5. A change in the slope of the plot occurs at a density of ~0.3 g/cm^3 (reduced density of ~0.7). The nonlinearity of the plot suggests that a change in the character of the cybotatic region occurs as the density of CO_2 increases. At low fluid densities the solvating power of the solvent should be low, approaching the vapor phase limit as the density is decreased. A linear least squares fit to these data yield an intercept value of -1.02 for π^*, which compares favorably with the vapor phase limit of -1.06 ± 0.10 reported by Essfar, Guihenuef, and Abboud (14). At higher fluid densities it acts more like a liquid solvent as more complex multiple solvent-solute interactions become important. In fact, data for the subcritical liquid shows the π^* value for the fluid and liquid, at equal densities, to be identical within experimental limits. At some intermediate fluid density a transition occurs from the low pressure region to solvent behavior more characteristic of liquids, leading to the change in slope observed in this plot. It is not clear why the transition takes place at this density, however it is interesting to note that in this density region virial coefficients beyond the second (indicating multi-molecule interactions in the solvent) become significant and significant enhancements in solubility are typically observed. Further work is ongoing to elucidate this behavior.

Mixed Fluids. Adding additional components to supercritical fluid solvents, entrainers in supercritical fluid extraction systems and modifiers in supercritical fluid chromatography, can extend or significantly alter the fluid solvating properties (15,16). By varying the composition of the fluid mixture it is possible to extend the range of fluid solvent properties while retaining the advantages of supercritical solvents.

The nature of mixed fluid solvent systems and their role in various supercritical fluid applications is complicated by the increased complexity of the phase behavior. Several different types of mixed fluid phase behavior have been identified (17,18), some of them involving multiple liquid and vapor phases. Without knowledge of the specific phase behavior of a given mixed fluid it is not possible to interpret correctly solvent behavior over the range of temperatures and pressures of interest.

Mixtures of CO_2 and methanol were selected for the initial investigation of the solvatochromic behavior in supercritical fluid systems. This combination is of interest as it combines the low critical temperature and pressure of carbon dioxide with a polar, less volatile modifier. This system exhibits relatively simple Type I phase behavior and several groups have published measurements of mixture critical points (19-21). At intermediate compositions the critical pressure for this fluid is much higher than that of either pure CO_2 or pure methanol, reaching a maximum of approximately 2400 psi (20).

Figure 6 illustrates the effect of pressure on the position of the 2-nitroanisole absorption maximum for several different solvent compositions. It is apparent that adding methanol causes the absorption to be red-shifted, corresponding to an increase in solvent polarity, as expected. However, even 9.5%-methanol does not shift the absorption maximum to the value measured in pure methanol (see Table II). It is interesting to note that the 5.6%-methanol mixture appears to be affected by increasing pressure (similarly to pure carbon dioxide), but above 9.5%-methanol pressure no longer effects the peak position. It can be speculated that at the higher methanol concentration the cybotatic region is enriched in methanol relative to the bulk concentration and that this methanol rich environment is not highly sensitive to pressure effects, as would be the case for pure methanol solvent. Similar behavior is shown in Figure 7 in which peak position is plotted vs pressure at different temperatures for a solvent which is 15% methanol. Again, the pressure change results in no significant effects over the range studied, but peak position is dependent upon temperature. These thermochromic shifts are similar in magnitude to those reported by Suppan and Tsiamis (22).

It is possible to compare these spectroscopic measurements of a methanol modifier in supercritical carbon dioxide to results obtained via chromatographic separations with similar fluid systems. The spectroscopic results indicate that adding methanol does increase the fluid polarity, but small additions of methanol do not drastically alter the solvent behavior. A similar result has been found in this laboratory by measuring the retention times of a polar test mixture with capillary columns (23). In contrast, previous reports with packed columns found a large effect with very small (less than 1%) additions of methanol to CO_2 (16,24). A reasonable conclusion is that the methanol modifier has a measurable effect on the supercritical fluid mobile phase, but a much more significant effect in modifying the packed column stationary phase. These differences can be attributed to the typically large numbers of active sites on typical chromatographic packings compared to the highly inert and deactivated polymeric stationary phases used in our capillary supercritical fluid studies (25,26).

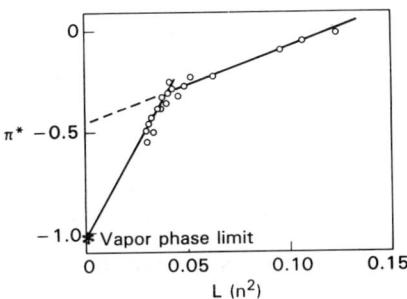

Figure 5. π^* versus Onsager reaction field function ($L(n^2)$) for CO_2, at 323 K.

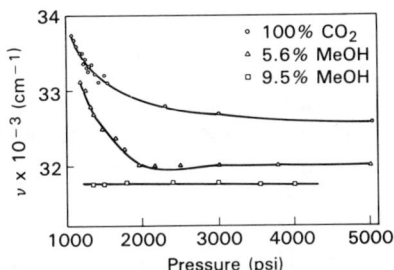

Figure 6. Absorption maxima versus pressure for 2-nitroanisole in CO_2-methanol systems at 50°C.

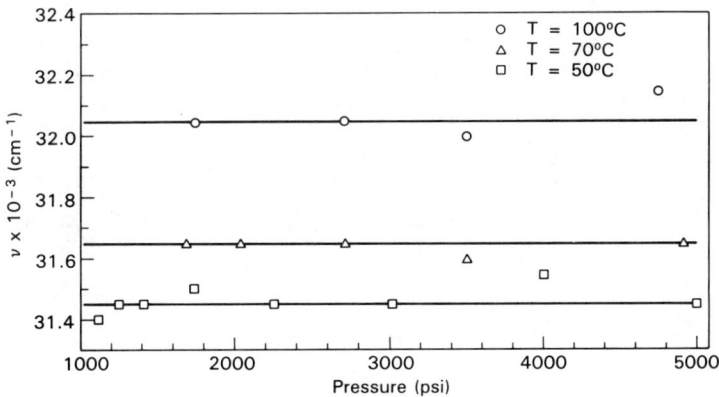

Figure 7. Absorption maxima versus pressure for 2-nitroanisole in 15.4% methanol-CO_2 at various temperatures.

Conclusions

This study demonstrates the feasibility of exploring the solvating properties of pure supercritical fluid solvents and solvent mixtures using solvatochromic measurements. The technique of measuring solvatochromic shifts is valuable for probing the specific interactions between solvent and solute molecules under varying conditions of temperature and pressure. By using the Kamlet and Taft π^* scale it is possible to directly compare the solvation regions for supercritical fluid and liquid solvents. It was demonstrated that the polarity/polarizability of the cybotatic region is dependent on both the fluid and the fluid density and that at equivalent densities similar shifts are observed for the supercritical fluid and the liquid. Using the Onsager reaction field function it is possible to also demonstrate that a qualitative change in the nature of the solvation region occurs as the fluid density increases from gas-like to liquid-like densities. By studying a mixed supercritical fluid solvent it was possible to verify that addition of a methanol modifier increased the polarity of the fluid, as was expected. This study also suggests that at higher bulk methanol concentrations the cybotatic region may have a different composition than the bulk solution. Extension of these studies is anticipated to increase our understanding of the solvent-solute interactions which occur in supercritical fluids.

Acknowledgment

This work has been supported by the U.S. Department of Energy, Office of Basic Energy Sciences, under Contract DE-AC06-76RLO 1830.

Literature Cited

1. Kamlet, M. J.; Abboud, J.-L. M.; Taft, R. W. J. Am. Chem. Soc. 1977, 99, 6027.
2. Bayliss, N. S.; MacRae, E. G. J. Phys. Chem. 1954, 58, 1002.
3. Taft, R. W.; Kamlet, M. J. J. Am. Chem. Soc. 1976, 98, 2886.
4. Kamlet, M. J.; Taft, R. W. J. Am. Chem. Soc. 1976, 98, 377.
5. Kamlet, M. J.; Abboud, J.-L. M.; Abraham, M. H.; Taft, R. W. J. Org. Chem. 1983, 48, 2877.
6. Kamlet, M. J.; Abboud, J.-L. M.; Taft, R. W. Prog. Phys. Org. Chem. 1981, 13, 485.
7. Brady, J. E.; Carr, P. W. J. Phys. Chem. 1985, 89, 1813.
8. Hyatt, J. A.; J. Org. Chem. 1984, 49, 5097.
9. Sigman, M. E.; Lindley, S. M.; Leffler, J. E. J. Am. Chem. Soc. 1985, 107, 1471.
10. Giddings, J. C.; Myers, M. N.; McLaren, L.; Keller, R. A. Science 1968, 162, 67.
11. Onsager, L. J. Am. Chem. Soc. 1936, 58, 1486.
12. Kholodov, E. P.; Timoshenko, N. I.; Yaminov, A. L. Thermal Engineering 1972, 19, 126.
13. Bose, T. K.; St. Arnaud, J. M. J. Chem. Phys. 1979, 71, 4951.
14. Essfar, M.; Guihenuef, G.; Abboud, J.-L. M. J. Am. Chem. Soc. 1982, 104, 6786.

15. Levy, J. M.; Ritchey, W. M. Proc. 6th Int. Symp. Cap. Chromat. 1985, p. 925.
16. Blilie, A. L.; Greibrokk, T. Anal. Chem. 1985, 57, 2239.
17. Van Konynenberg, P. H. Ph.D. Thesis, University of California at Los Angeles, 1968.
18. Van Konynenberg, P. H.; Scott, R. L. Phil. Trans. Roy. Soc. 1980, 298, 495.
19. Semenova, A. I.; Emel'yanova, E. A.; Tsimmerman, S. S.; Tsiklis, D. S. Russian J. Phys. Chem. 1979, 53, 1428.
20. Brunner, E. J. Chem. Thermodynamics 1985, 17, 671.
21. Robinson, D. B.; Peng, D.-Y.; Samuel, Y.-K. C. Fluid Phase Equil. 1985, 24, 25.
22. Suppan, P.; Tsiamis, C. J. Chem. Soc. Faraday Trans 2 1981, 77, 1553.
23. Wright, B. W.; Smith, R. D. J. Chromatogr. 1986, 355, 367.
24. Gere, D. R. Hewlett-Packard Application Note AN800-2, 1983.
25. Wright, B. W.; Peaden, P. A.; Lee, M. L.; Stark, T. J. J. Chromatogr. 1982, 248, 17.
26. Wright, B. W.; Kalinoski, H. T.; Smith, R. D. Anal. Chem. 1985, 57, 2823.

RECEIVED July 8, 1986

Chapter 4

Effects of Supercritical Solvents on the Rates of Homogeneous Chemical Reactions

Sunwook Kim and K. P. Johnston

Department of Chemical Engineering, University of Texas, Austin, TX 78712

Solvatochromic shift data have been obtained for phenol blue in supercritical fluid carbon dioxide both with and without a co-solvent over a wide range in temperature and pressure. At 45°C, SF CO_2 must be compressed to a pressure of over 2 kbar in order to obtain a transition energy, E_T, and likewise a polarizability per unit volume which is comparable to that of liquid n-hexane. The E_T data can be used to predict that the solvent effect on rate constants of certain reactions is extremely pronounced in the near critical region where the magnitude of the activation volume approaches several liters/mole.

It is ironic that the large growth in SF extraction at the Max Planck Institute in Germany in the 1960's was the result of a serendipitous discovery of the solvent power of supercritical ethylene during the "Aufbau" reaction of triethylaluminum with ethylene (1). A number of recent articles review supercritical fluid (SF) extraction (2,3,4); however, the literature contains relatively few examples where supercritical fluid solvents have been used to modify or control reaction rate constants (5,6,7). Liquid phase reactions have been studied over wide pressure ranges, e.g. 1-10 kbar, to determine activation volumes, i.e. the pressure derivative of the rate constant. These studies essentially ignore the highly compressible near supercritical region where activation volumes can become infinite, either positively or negatively. Simmons and Mason (7) observed an abrupt decrease in the activation volume near the critical conditions for the cyclic dimerization of C_2F_3Cl. Paulaitis et al. (6) measured activation volumes as low as -500 cc/mol for the Diels-Alder cycloaddition of isoprene and maleic anhydride in CO_2. These examples provide an indication of the effects of supercritical solvents on the rate constants of homogeneous chemical reactions although a much larger data base is needed.

Supercritical fluid solvents offer some potential advantages compared with liquids as an environment for chemical reactions. It

0097-6156/87/0329-0042$06.00/0
© 1987 American Chemical Society

will be demonstrated that reaction rate constants of certain reactions may be altered markedly by small modifications in the pressure of a SF solvent. Reaction products could be separated efficiently using supercritical fluid extraction. Industrial applications of reactions in SF solvents include: the detoxification of wastewater, ethylene polymerization, isomerization of n-paraffins, synfuels processing, and the reaction of ethylene with triethylaluminum (5).

Solvent strength scales based on solvatochromism, i.e., shifts in the absorption wavelength of indicator dyes caused by the solvent, are used commonly to correlate and to predict rate constants for liquid phase reactions (8-13). Linear solvation energy relationships based on solvatochromic parameters have been used to correlate and predict solubility phenomena, absorption maxima in IR, NMR, ESR, and UV-visible spectroscopy, solvent effects on reaction rate constants and equilibrium constants, free energies and enthalpies of formation of acid-base complexes, retention indices in gas-liquid chromatography and high-performance-liquid chromatography, and finally, physiological and toxicological quantitative structure-activity relationships (14). Although solvatochromic "solvent strength" scales have such widespread application, they are just beginning to become available for supercritical fluid solvents (15,16) for very limited pressure and temperature ranges. Solvatochromic data are presented in this symposium in the article by Frye et al.

Solvatochromic data, specifically absorption or transition energies (E_T's), have been obtained for the dye phenol blue in supercritical fluids as a function of both temperature and pressure. These data will be used to compare the "solvent strength" of these fluids with liquid solvents. We will use the terms "solvent strength" and "E_T" synonymously in this paper such that they include the magnitude of the polarizability/volume as well as the dipole moment. The "solvent strength" has been characterized by the spectroscopic solvatochromic parameter, E_T, for numerous liquid solvents (9,11,17,18).

Reaction rate constants and activation volumes will be predicted as a function of pressure in the highly compressible near critical region and in the highly dense, less compressible region using E_T. In addition, the magnitude of the compression of a supercritical fluid about a solute in the highly compressible region will be determined qualitatively. These spectroscopic data, which describe interactions at the molecular level, will benefit significantly the understanding of both phase equilibria and solvent effects on reaction rates in the supercritical fluid state.

Experimental

Phenol blue (benzoquinone N-[(4-dimethylamino) phenyl] imine, Aldrich >97%) was purified by recrystallization and chromatographic separation (30). The purity was checked by the melting point (161-162°C), and the absorption maximum in acetone (582 nm) and in CCl_4 (565 nm). The dye was of chromatographic purity as determined by thin-layer chromatography on silica gel.

The volume of the cylindrical high pressure cell was 14 cm^3, and the pressure and temperature ranges were 0 to 6000 psi and -30 to 250°C, respectively. The inside diameter was 1.75 cm, and the wall thickness was 1.67 cm. Because of the large 6 cm path length of the cell, the spectral shift can be measured for concentrations as low as 10^{-6}M at 0.1 Absorbance units. Each 1 cm-thick by 2.5 cm in diameter sapphire window was flat to one wavelength of yellow light, and the optical axis was perpendicular to the face. A teflon o-ring was inserted between the window and a flat surface on the 316 stainless steel vessel. The cell was thermostated using a ¼" copper heat exchange coil which was jacketed with fiberglass insulation.

The temperature was indicated and controlled to ±0.1°C with a platinum resistance probe which extended 1 mm inside the inner surface of the cell. The pressure was adjusted using a 100 cc Ruska syringe pump and was measured to within ±0.1% with a 710A Heise digital pressure gauge which is traceable to an NBS standard. The pressure varied less than 0.15 bar (2 psi) during a spectral scan. The wavelength accuracy was ±0.2 nm for the Varian (Cary) 2290 spectrophotometer.

The cell was loaded with 10^{-5}g phenol blue, evacuated, and pressurized with the solvent. A spectrum was obtained at a given pressure after the temperature and absorbance equilibrated. The absorption band was scanned over a range of 450-600 nm 2-3 times to obtain the average λ_{max}. The reproducibility in λ_{max} was ±0.2 nm which corresponds to a precision in E_T of ±0.02 kcal/mol.

Results and Discussion

Solvent strength in the critical region. All of the experiments were performed with the dye phenol blue which has been well-characterized both experimentally and theoretically in liquid solvents ([20],[21],[22]). Since the dipole moment of phenol blue increases 2.5 debye upon electronic excitation ([8]), it is a sensitive probe of the local solvent environment. For example the absorption maxima occur at 550 and 608 nm in n-hexane and methanol, respectively. The excited state is stabilized to a greater extent than the ground state as the "solvent strength" is increased, which is designated as a red shift.

The transition energy, E_T, of phenol blue in CO_2 is plotted versus pressure in Figure 1 along the saturation curve and at supercritical conditions. The behavior of the E_T versus pressure plot is similar to that of density versus pressure, such that E_T is fairly linear in density as shown in Figure 2. The effect of temperature is relatively small at a given density. In the highly compressible near critical region, the E_T is extremely sensitive with respect to pressure along an isotherm, and likewise with respect to temperature along an isobar (see Figure 1). At high reduced pressures where the fluid is relatively incompressible, E_T is only a weak function of pressure and temperature as is the case for liquids near the triple point. Supercritical fluid solvents exhibit an interesting property which is not observed for pure liquid solvents in that the density and thus the "solvent strength" (E_T) may be adjusted over a continuum using small changes in temperature and pressure.

In Table I, several SF solvents are compared with liquid n-hexane at a reduced temperature ($T_r = T/T_c$) near 1.05. For each

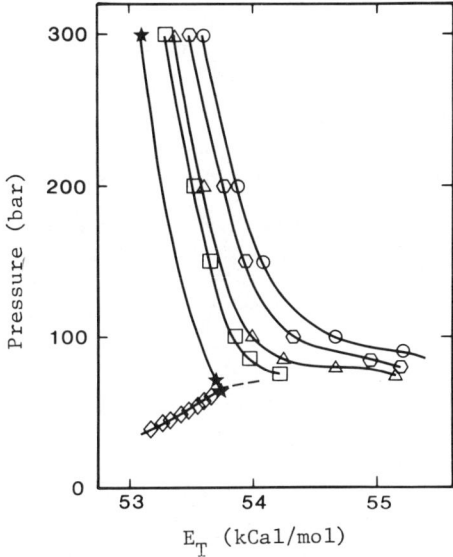

Figure 1. Transition energy (E_T) of phenol blue in CO_2 ($\Diamond$ = saturation curve, ★ = 25 °C, □ = 31 °C, △ = 35 °C, ⬯ = 40 °C, and ○ = 45 °C).

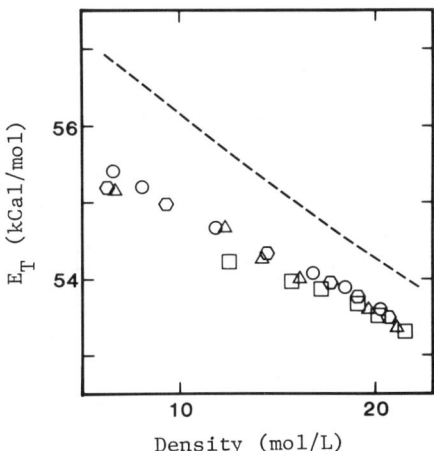

Figure 2. E_T of phenol blue in CO_2 vs. density (□ = 31 °C, △ = 35 °C, ⬯ = 40 °C, ○ = 45 °C, and --- is calculated E_T using Equation 5).

fluid, the pressure is chosen to match the E_T with the value for n-hexane, so that the comparison is made at constant "solvent strength". Many investigators have implied that the solvent strength of a supercritical fluid approaches that of a liquid at the point where the density approaches that of a liquid. This is clearly a misconception. At 45°C ($T_r = 1.05$), the molar density of CO_2 approaches that of n-hexane at 130 bar, yet it is a weaker solvent for aliphatic and aromatic hydrocarbons at these conditions (2,3). The reason for this difference is that the average polarizability per molecule of CO_2 is only 26.5×10^{-25} cm^3 whereas it is 123×10^{-25} cm^3 for hexane. At a given molar density, the polarizability per unit volume is much greater for hexane. It is also greater for hexane at a given mass density despite the difference in the molecular weights.

Table I. Comparison of SF Solvents with n-Hexane at constant E_T

Solvent	T (°c)	$T_r =$ T/T_c	P (bar)	$P_r =$ P/P_c	ρ (g/cc)	δ [1] (cal/cc)$^{\frac{1}{2}}$	α/v [2]
C_2H_4 [3]	25	1.056	1700	33.8	0.57	7.0	0.052
CO_2	45	1.046	2800	37.9	1.32	10.3	0.048
CF_3Cl [3]	40	1.037	1300	33.2	1.95	6.9	0.051
$n-C_6H_{14}$	25	-	1	-	0.66	7.3	0.057

[1] Hildebrand solubility parameter
[2] polarizability per volume
[3] reference (19)

The E_T's of the nonpolar solvents, CF_3Cl and C_2H_4, become equal to that of n-hexane at a pressure in the range of 1-2 kilobar. Notice that the Hildebrand solubility parameters of these three solvents are roughly equivalent at this condition of constant E_T. The same result is also observed for the polarizabilities/volume of these solvents. Again, the molar densities of these supercritical fluids are considerably higher than that of n-hexane at this equivalence point in solvent strength, since the polarizabilities/molecule are lower.

The E_T of CO_2 at 45°C reaches that of n-hexane at 2.8 kbar. At this pressure, the polarizability/volume of SF CO_2 is a little less than that of n-hexane, which suggests that there are other molecular interactions between CO_2 and phenol blue in addition to dispersion and induction. The likely possibilities include electron donor-acceptor forces and dipole-quadrupole interactions.

The E_T's of CO_2, C_2H_4, and CHF_3 are compared versus reduced density at constant reduced temperature in Figure 3. The curves for CO_2 and ethylene coincide even though the polarizability/ molecule of ethylene is 1.6 times that of CO_2. The primary reason for this is that the molar density of CO_2 is 1.4 times that of ethylene at constant reduced density and temperature. The secondary factor is that CO_2 is slightly acidic and has a quadrupole moment otherwise it would give a smaller red shift. For nonpolar solutes that do not exhibit dipolar and acid-base interactions, the solvent strength of CO_2 would be less than that of ethylene at constant reduced temperature and density. In addition, at a given reduced temperature and density, the reduced pressures are similar (e.g., see Figure 3 and Table 1) thus the actual pressure for CO_2 is about 1.5 times that for C_2H_4.

The solubility data for naphthalene in ethylene and in CO_2 are consistent with the E_T data in Figure 3. The proper way to make the comparison is to use the enhancement factor instead of the solubility. The enhancement factor equals y_2P/P_2^S, which is simply the actual solubility divided by the solubility in an ideal gas. The enhancement factor removes the effect of vapor pressure which is useful for comparing fluids at constant reduced temperature but at different actual temperatures. In terms of the fugacity coefficient of the solute, ϕ_2, the enhancement factor is given by

$$E = \exp(v_2^S P/RT)/\phi_2 \tag{1}$$

where v_2^S is the molar volume of the solid. If a comparison is made at constant reduced temperature and reduced density but at different pressures, the enhancement factor removes the Poynting effect of pressure on the condensed phase. In summary, the enhancement factor is a measure of the solvent strength in the SF phase. The enhancement factor for naphthalene in ethylene at 25°C (T_r = 1.05) is about 3 times that in CO_2 at 45°C (T_r = 1.05) for a wide range in reduced density. The larger solubility in ethylene is consistent with the above discussion concerning Figure 3, since naphthalene is not a very strong Lewis base compared with phenol blue.

Solvent effect on rate constants. In this section, the rate constant will be predicted qualitatively in CO_2 for the Diels-Alder cycloaddition of isoprene and maleic anhydride, a reaction which has been well-characterized in the liquid state (23,24). In a previous paper, we used E_T data for phenol blue in ethylene to predict the rate constant of the Menschutkin reaction of tripropylamine and methyliodide (19). The reaction mechanisms are quite different, yet the solvent effect on the rate constant of both reactions can be correlated with E_T of phenol blue in liquid solvents. The dipole moment increases in the Menschutkin reaction going from the reactant state to the transition state and in phenol blue during electronic excitation, so that the two phenomena are correlated. In the above Diels-Alder reaction, the reaction coordinate is isopolar with a negative activation volume (8,23),

$$\Delta v^{\ddagger} = \bar{v}_M - \bar{v}_A - \bar{v}_B = -RT(d\ln k_x/dP) \tag{2}$$

where M is the transition state, A and B are the reactants and k_x is the rate constant based on mole fraction units. Using Regular Solution Theory, it was demonstrated that the rate constant increases with the cohesive energy density (δ^2) of the solvent for this reaction as was observed experimentally (23). Since there is little difference in the dipole moment of the transition state versus the reactants, the solvent effect on the rate constant is small compared with the Menschutkin reaction.

The rate constant data for the Diels-Alder reaction of isoprene and maleic anhydride (23,24) may be correlated with E_T of phenol blue at 35°C by

$$\ln k_x = -a\, E_T + b \tag{3}$$

where the constants a and b are -1.053 and 43.70, respectively. Using Eqs. 2 and 3, the activation volume may be expressed as

$$\Delta v^{\ddagger} = +aRT\rho k_T \, (\partial E_T/\partial \rho)_T \tag{4}$$

where k_T is the isothermal compressibility, $1/\rho(\partial \rho/\partial P)_T$, and it is assumed that a and b are pressure independent. The density derivative of E_T is relatively constant compared to ρk_T, especially in the highly compressible region, and is similar for both liquids and supercritical fluids. Since phenol blue exhibits a red shift, $(\partial E_T/\partial \rho)_T$ is negative, as is $\Delta v^{\ddagger}$ which reaches a minimum at the same density where the compressibility is a maximum.

The logarithm of the rate constant of the above Diels-Alder reaction is predicted versus pressure in Figure 4 at 35°C. The rate constant increases by a factor of ten for this pressure range which corresponds to a density range of 6.7 to 21 mol/ℓ. The negative of the slope of this plot, or the activation volume becomes largely negative in the highly compressible near critical region as shown in Table II. For example, $\Delta v^{\ddagger}$ reaches -4000 at 75 bar. At 35°C and 300 bar, $\Delta v^{\ddagger}$ is only moderately negative, i.e., -55 cc/mol compared with -37.4 cc/mol in liquid ethylacetate at 35°C. At each pressure in Figure 4, the activation volume is negative since the magnitude of $\bar{v}_M$ is less than that of the sum of $\bar{v}_A$ and $\bar{v}_B$ for this cycloaddition reaction (see Eq. 2). In the highly compressible region, the $\bar{v}_i$'s become largely negative since the solute causes a compression of the fluid solvent as was explained in detail both experimentally and theoretically (26,27,28). As a result, the activation volume is also an extremely large negative number.

Table II. Prediction of the Activation volume of the Diels-Alder reaction between isoprene and maleic anhydride in supercritical carbon dioxide at 35°C

Pressure (bar)	75	80	85	100	200	300
$\Delta v^{\ddagger}$ (cc/mol)	-4000	-2000	-950	-225	-70	-55

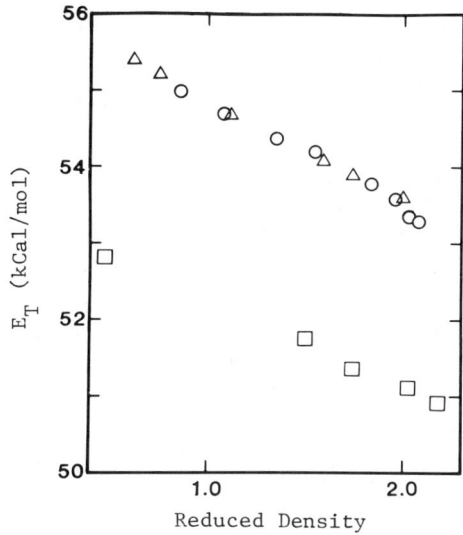

Figure 3. E_T of phenol blue in several fluids vs. reduced density ($\triangle$ = CO_2 - 45 °C, $\bigcirc$ = C_2H_4 - 25 °C (19), and $\square$ = CHF_3 - 40 °C (19).

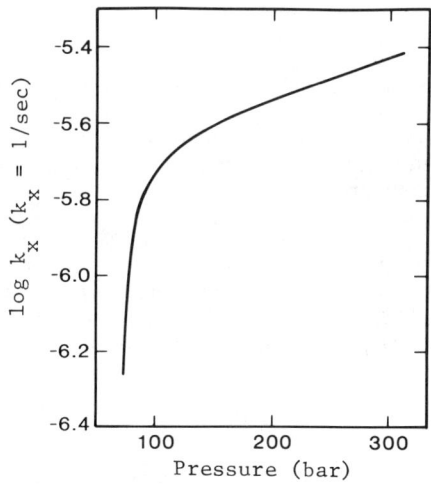

Figure 4. Predicted rate constant for the Diels-Alder reaction of isoprene and maleic anhydride in CO_2 at 35 °C.

The extremely pronounced solvent effect in the near critical region can also be explained using Eq. 4. As mentioned above, $(\partial E_T/\partial \rho)_T$ is similar for both liquids and supercritical fluids. However, the magnitude of activation volume can be much larger in supercritical fluids because of the much higher compressibility. This analysis can also be applied to the Menschutkin reaction of tripropylamine and methyliodide, which has an activation volume of -65 cc/mol in liquid CCl_4 at 30°C. In ethylene at 66 bar and 25°C, $\Delta v^\ddagger$ is predicted to be -5000 cc/mol. Both of these reactions exhibit an extremely pronounced solvent effect in the near critical region since the reactions have relatively large (in magnitude) activation volumes even in liquids. The solvent effect on the Menschutkin reaction would be even more pronounced in a supercritical fluid with a dipole moment where the "solvent strength" would be even more sensitive to density. These predictions suggest that future measurements of rate constants in supercritical fluids could potentially demonstrate a solvent effect of orders of magnitude.

<u>Local solvent compression</u>. The next application of the solvatochromic data will be to determine the magnitude of the local compression of a supercritical fluid solvent in the immediate environment of the solute. The E_T of a dye such as phenol blue can be predicted in liquids where no specific interactions are present by treating the solvent as a homogeneous polarizable dielectric (<u>22,29</u>). The intrinsic "solvent strength", E_T°, describes dispersion, induction, and dipole-dipole forces and is given by (22).

$$E_T^\circ = A\left(\frac{n^2-1}{2n^2+1}\right) + B\left(\frac{D-1}{D+2} - \frac{n^2-1}{n^2+2}\right) + C \qquad (5)$$

where n is the refractive index and D is the dielectric constant. The constants A, B and C are functions of the properties of the dye such as the dipole moment and oscillator strength of the ground and excited states and the cavity radius. The first term in Eq. 5 includes dispersion and solute permanent dipole-solvent induced dipole forces, thus it is a function of the refractive index of the solvent. The second term describes the effects of permanent dipole-dipole forces (orientation). The E_T° of phenol blue was correlated successfully with Eq. 5 for 21 non-hydrogen bonding liquid solvents (<u>20,22,29</u>).

The measured transition energy, E_T, includes the intrinsic "solvent strength" given by Eq. 5 and the specific "solvent strength", E_T^S, which describes an exaggerated solvent effect such as hydrogen bonding such that

$$E_T \text{ (experimental)} = E_T^O + E_T^S \qquad (6)$$

For phenol blue, the E_T^S is zero for ethylene and CF_3Cl, but non-zero for the Lewis acids CF_3H and CO_2. One of the attractive features of solvatochromic scales is that the non-specific and specific interactions may be separated since the former can be calculated straightforwardly using Eq. 5.

The E_T is not influenced by factors such as repulsive forces between the solute and solvent since they are identical in the ground and excited states. The locations of the molecules do not change during the time frame of electronic excitation. The theory for solubility phenomena is much more complicated; for example, it includes repulsive forces. As a result, solubility data give less information about the properties of the solvent, or the local solvent environment about the solute.

In order to analyze the behavior of phenol blue in CO_2, it is useful to review the results for ethylene since it has a specific "solvent strength" of zero (19, see Figure 5). The dashed line was calculated using Eq. 5 where the constants A and B were obtained from the literature (20,30) and C was chosen to force agreement with experiment in the dense incompressible region. The dashed line fits the data at high density where ethylene is relatively incompressible. At a lower density, e.g. 7 mol/l, ethylene is highly compressible so that it "clusters" about the solute due to attractive forces. Eckert et al (27,28) measured solute partial molar volumes on the order of -10 l/mol in CO_2 and in ethylene in the highly compressible region. This clustering or local compression phenomenon which increases the number of solvent molecules near the solute causes an additional red shift of 1 kcal/mol at a density of 7 mol/l. The additional red shift is the difference between the experimental point and that which is predicted using the theory for a homogeneous liquid shown by the dashed line. This difference becomes small as the isothermal compressibility diminishes as explained below.

The local density of solvent about the solute may be determined by comparing the experimental and calculated curves. Consider points A and B in Figure 5 at a constant value of E_T, i.e., 55 kcal/mol. A hypothetical homogeneous fluid at point B gives the same "solvent strength" as the actual fluid at point A. The local density about the solute ρ_i^B exceeds the bulk density ρ_i^A due to compression, such that

$$\rho_i^B = \frac{\rho_i^A}{V_{ij}} \int_0^{r_{ij}} g_{ij}(\underline{r}) \, d\underline{r} \tag{7}$$

where $g_{ij}(r)$ is the unlike pair radial distribution function. Using Kirkwood-Buff solution theory (31), it was shown that the difference between the local and bulk densities is (19)

$$\rho_{12} - \rho = \frac{\rho k_B T k_T}{V_{12}} (1 - \bar{v}_2^\infty / k_B T k_T) \tag{8}$$

where V_{12} is a measure of the volume of the first coordination shell. Based on previous results (Eckert et al., 27, 28), it is reasonable to assume

$$\bar{v}_2^\infty = a k_T + b \tag{9}$$

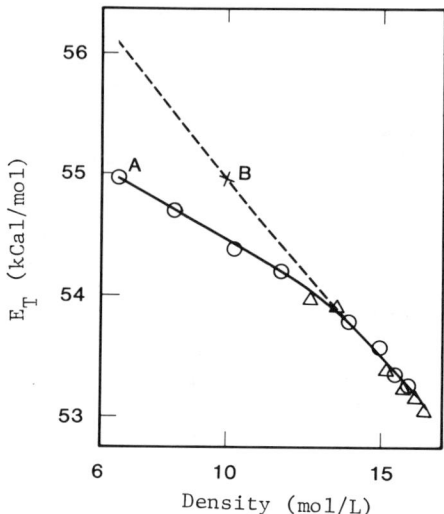

Figure 5. E_T of phenol blue in ethylene vs. density (O = 25 °C, △ = 10 °C, and --- is calculated E_T using Equation 5).

Using Eqs. 8 and 9, the final result is

$$(\rho_{12}^{1} - \rho)/\rho = a'k_T + b' \qquad (10)$$

where a' and b' are functions of temperature and V_{12}. The linearity suggested by Eq. 10 has been obtained from the data in Figure 5 (<u>19</u>).

The CO_2 data in Figure 2 are more difficult to analyze because of the acid-base interactions. The dashed line is the calculated result for $E_T°$ of a homogeneous polarizable dielectric (Eq. 5) using the same values of the constants A, B, and C as obtained above for ethylene. At high density, where the local compression effect disappears, the specific "solvent strength" which is the difference between the data and $E_T°$ is about 0.7 kcal/mol. To put this in perspective, the specific "solvent strengths" are 6.3, 1.6, 1.3 for phenol blue in the liquid solvents m-cresol (strong Lewis acid), methanol, and chloroform, respectively (<u>20</u>,<u>30</u>)). In the highly compressible region, the red shift exceeds the dashed line due to two coupled effects, acid-base forces, and local solvent compression.

The final set of solvatochromic data are shown in Figure 6 for phenol blue in SF CO_2 doped with various amounts of the co-solvent or entrainer, methanol. Consider a pressure of 100 bar where the E_T of phenol blue in CO_2 is 54 kcal/mol. The red shift is increased more by the addition of 3.5 mole percent methanol at constant pressure than by an increase in the pressure of pure CO_2 of over 200 bar. The large specific "solvent strength" of methanol causes this behavior. The red shift caused by the co-solvent is in

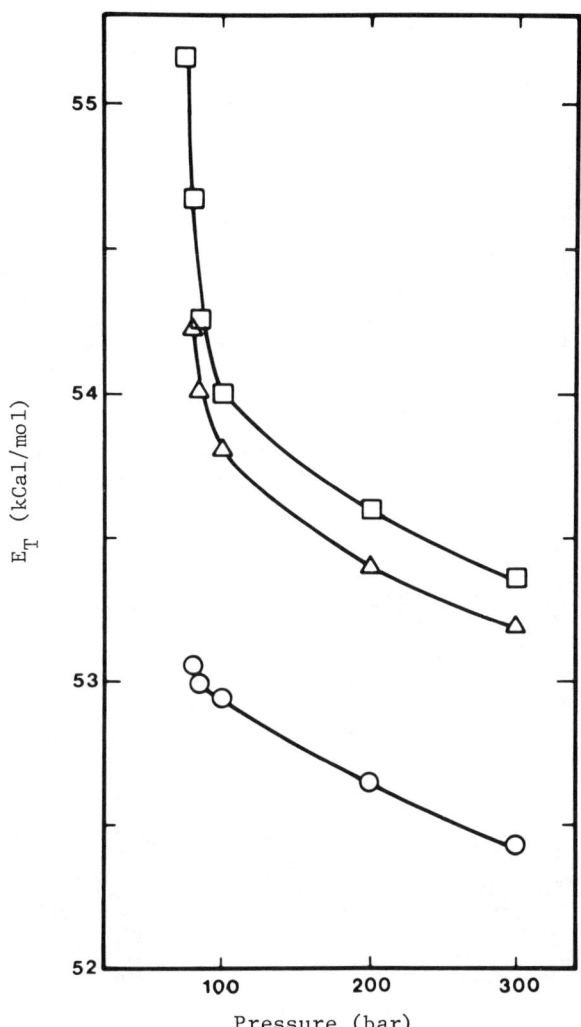

Figure 6. E_T of phenol blue in CO_2-methanol mixtures (□ = pure CO_2, △ = CO_2 - 1% CH_3OH, ○ = CO_2 - 3.5% CH_3OH).

excess of the amount calculated if it is assumed that the mixture is random. This indicates that the dye is solvated preferentially by methanol such that methanol's local concentration exceeds its bulk concentration. A large number of solubilities of solids were measured in CO_2 with and without co-solvent by Johnston and co-workers (32,33). In many of these systems, small amounts of co-solvents produced a greater increase in the solubility than pressure increases of hundreds of atmospheres. The relationship between the spectroscopic data and the solubility data is the subject of ongoing research.

Acknowledgments

This material is based on work supported by the National Science Foundation under Grant No. CPE-8306327. Any opinions, findings, and conclusions or recommendations expressed in this publication do not necessarily reflect the views of the National Science Foundation.

Acknowledgment is made to the Donors of the Petroleum Research Fund, administered by the American Chemical Society, for partial support of this work. Further support is acknowledged from the Dow Chemical Company Foundation and the Separations Research Program at The University of Texas.

Literature Cited

1. Zosel, K. Angew. Chem. Int. Ed. 1978, 17, 702.
2. Paulaitis, M. E.; Krukonis, V. J.; Kurnik, R. T.; Reid, R. C. Rev. in Chem. Eng. 1983 1(2), 179.
3. Johnston, K. P. "Kirk-Othmer Encyclopedia of Chemical Technology," John Wiley & Sons: New York, 1984.
4. McHugh, M. A. "Extraction with Supercritical Fluids," in "Recent Development in Separation Science," Li, N. N. and Carlo, J. M., Eds., CRC Press: Boca Raton, 1984; Vol. IX.
5. Randall, L. G. Sep. Sci. Technol. 1982, 17(1), 1.
6. Alexander, G.; Paulaitis, M. E. AIChE Annual Meeting, #140d, San Francisco, 1984.
7. Simmons, G. M.; Mason, D. M. Chem. Eng. Sci. 1972, 27, 89.
8. Reichardt, C. "Solvent Effects in Organic Chemistry," Verlag Chemie, Weinheim, New York, 1979.
9. Dack, M.R.J., "Solutions and Solubilities Part II," John Wiley & Sons: New York, 1976.
10. Tamura, K.; Imoto, T. Bull. Chem. Soc. of Japan 1975, 48(2), 369.
11. Reichardt, C. Angew. Chem. Int. Ed. 1979, 18, 98.
12. Kamlet, M. J.; Hall, T. N.; Taft, R. W. J. Org. Chem. 1979, 44, 2599.
13. Kamlet, M. J.; Abboud, J. L.; Taft, R. W. J. Amer. Chem. Soc. 1981, 103(5), 1080.
14. Taft, R. W.; Abraham, M. C.; Doherty, R. M.; Kamlet, M. J. Nature 1985, 313(31), 384.
15. Hyatt, J. A. J. Org. Chem. 1984, 49, 5097.
16. Sigman, M. E.; Lindley, S. M.; Leffler, J. E. J. Amer. Chem. Soc. 1985, 107, 1471.
17. Kamlet, M. J.; Abboud, J. L.; Taft, R. W. J. Amer. Chem. Soc. 1977, 99(18), 6027.

18. Kamlet, M. J.; Abboud, J. L.; Abraham, M. H.; Taft, R. W. J. Org. Chem. 1983, 48, 2877.
19. Kim, S.; Johnston, K. P. "Molecular Interactions in Dilute Supercritical Fluid Solutions," Submitted to Ind. Eng. Chem. Fund. 1985.
20. Figueras, J. J. Amer. Chem. Soc. 1972, 93(13), 3255.
21. Kolling, O. W. Anal. Chem. 1981, 53, 54.
22. McRae, E. G. J. Phys. Chem. 1957, 61, 562.
23. Grieger, R. A.; Eckert, C. A. Trans. Faraday Soc. 1970, 66, 2579.
24. Eckert, C. A.; Hsieh, C. R.; McCabe, J. R. AIChE J. 1974, 20(1), 20.
25. Silber, E. Ph.D. Thesis, Texas Tech. Univ, 1971.
26. Eckert, C. A.; Paulaitis, M. E.; Johnston, K. P. J. Phys. Chem. 1981, 85, 1770.
27. Eckert, C. A.; Ziger, D. H.; Johnston, K. P.; Ellison, T. K. Fluid Phase Equilib. 1983, 14, 167.
28. Eckert, C. A.; Ziger, D. H.; Johnston, K. P.; Kim, S. J. Phys. Chem. 1986 (in press).
29. Mataga, N.; Kubota, T. "Molecular Interactions and Electronic Specta," Marcel Dekker, Inc.: New York, 1970.
30. Kolling, O. W.; Goodnight, J. L. Anal. Chem. 1973, 45(1), 160.
31. Kirkwood, J. C.; Buff, F. P. J. Chem. Phys. 1951, 19(6), 774.
32. Johnston, K. P.; Dobbs, J. M.; Wong, J. M.; Lahiere, R. J. Ind. Eng. Chem. Fund. 1986 (in press).
33. Johnston, K. P.; Dobbs, J. M.; Wong, J. M. J. Chem. Eng. Data 1986 (in press).

RECEIVED June 25, 1986

CHEMICAL REACTIONS

Chapter 5

Organic Chemistry in Supercritical Fluid Solvents: Photoisomerization of *trans*-Stilbene

Tetsuo Aida and Thomas G. Squires[1]

Applied Organic Chemistry, Energy & Minerals Resources Research Institute, Iowa State University, Ames, IA 50011

> The wide range of supercritical fluid properties accessible through modest manipulations of pressure and temperature affords a unique approach to investigating and, perhaps, controlling chemical processes. In contrast to liquid phase behavior, the photostationary cis/trans ratio obtained from photoisomerization of trans-stilbene in supercritical CO_2 is dependent on system pressure. This behavior is probably attributable to significant changes in solvent viscosity in this pressure range. Techniques for investigating reactions under supercritical conditions are described, and an approach to assessing the "supercritical fluid effect" is suggested.

The unusual solvent properties of supercritical fluids (SCFs) have been known for over a century (1). Just above the critical temperature, T_c, forces of molecular attraction are balanced by kinetic energy; and fluid properties, including solvent power, exhibit a substantial pressure dependence. Many complex organic materials are soluble at moderate pressures (80 to 100 atmospheres); and SCF solvent power increases dramatically when the pressure is increased to 300 atmospheres. The pressure responsive range of solvent properties thus attainable provides a tool for investigating the fundamental nature of molecular interactions and is also being exploited in important areas of applied research (2,3).

In the latter activity, supercritical fluids have been utilized extensively in the thermochemical solubilization of coal (4), for selective extraction of naturally occurring materials (5,6), and in various separation techniques including destraction (7) and supercritical fluid chromatography (8,9). The physicochemical principles underlying these applications have also been investigated (10). In view of the high level of interest in manipulating complex organic mixtures with supercritical fluids, it is surprising that these fluids have seen little use as solvents for organic reactions

---

[1]Current address: Associated Western Universities, Inc., 142 East 200 South, Salt Lake City, UT 84111

0097-6156/87/0329-0058$06.00/0
© 1987 American Chemical Society

(11,12). The pivotal role of solvent properties in controlling the course and rate of chemical reactions is well established (13); and, thus, the same kind of pressure responsive solvent-solute interactions which control extractions and separations in SCFs can be expected to direct chemical pathways and influence reaction rates.

Much of our understanding of the nature of chemical reactions has been derived by observing the responses of these processes to changes in solvent properties. While investigations of this type have been highly productive, interpretation of the results has often been blurred by uncertainties inherent in the experimental approach. Heretofore, liquid solvents have been used in these investigations, and solvent properties have been changed by adjusting the solvent composition. Unfortunately, no quantitative relationship between bulk solvent properties and chemical phenomena has emerged, presumably due to selective solvent-solute interactions which result in differences between bulk solvent composition and microscopic solvent composition (13).

In contrast to liquid solvents, the properties of a single SCF solvent can be altered over a wide range through modest manipulations of pressure and/or temperature. At constant reduced temperature, $(T_R=T/T_c)$, between 1.0 and 1.1, solvent properties are extremely responsive to very accessible changes in pressure (800 to 4000 psi) (14). SCFs thus provide an unprecedented opportunity to investigate the effects of solvent properties on chemical reactions without changing the solvent composition.

This application of SCFs, while highly touted, has received scant experimental attention (15). Investigations have been limited to Diels-Alder reactions (11,16), electrochemical reactions (17,18), polymerizations (19,20), and high temperature processes (21,22). Recent semiempirical treatments of SCF solvent properties (23,24) have provided a basis for interpreting solvent effects in SCFs.

Investigations of organic reactions in supercritical solvents are subject to several constraints, one attributable to supercritical fluid properties and others imposed for interpretive and experimental simplicity. Because supercritical fluid properties are affected by changes in temperature, a reaction should be selected which does not require heat for initiation and is not highly exothermic. Additionally, for experimental simplicity and clarity of interpretation, a clean, well-understood reaction should be chosen; and one should expect an experimentally observable response to changes in pressure. Finally, a unimolecular reaction which produces a single product obviates the complication of controlling the concentrations of two reactants and simplifies product analysis. The photoisomerization of trans-stilbene meets these requirements.

The photoisomerization of stilbene is one of the most extensively studied photoreactions (25). Solvent effects have been thoroughly investigated for both the direct and photosensitized isomerizations, and a model has been developed which attributes these effects to solvent viscosity (26). Increased viscosity inhibits direct photoisomerization of the cis isomer, but facilitates that of trans-stilbene. As a result, the cis/trans ratio of the photostationary state increases with increasing solvent viscosity. The wide range of viscosities which are attainable by pressure manipulation of supercritical carbon dioxide provides an excellent opportunity to probe the effect of viscosity on stilbene photochemistry in the same solvent.

We chose carbon dioxide for our initial investigations because the properties of supercritical carbon dioxide are well known and the solubility of a relatively large number of organic materials are known for this medium. Although carbon dioxide is interactive with solute molecules, it is generally unreactive. Furthermore, the supercritical working range ($1.0 < T_R < 1.1$), 31.5 to 60°C, is experimentally convenient; and many organic reactions have been extensively investigated in this temperature range.

Experimental

Photoisomerizations were carried out using a 450 watt Hanovia medium pressure mercury source, and each reaction was followed to the photostationary state by gas chromatographic determination of the cis/trans ratios. Twice recrystallized trans-stilbene (Aldrich), spectrophotometric grade cyclohexane, and dry, high purity carbon dioxide were used in these experiments.

A schematic diagram of our supercritical fluid reactor is shown in Figure 1. Both the pumping and pressure control devices are adaptations of those utilized by other investigators (27). The photochemical reactor was constructed from thick-walled quartz capillary tubing (2mm i.d.) and was incorporated into the supercritical system by affixing stainless steel fittings to the tubing with epoxy resin. Except for an occasional faulty joint, this section of the apparatus performed flawlessly.

Flow rates through the system were controlled by quartz capillaries connected in parallel and were measured with a gas buret connected to the capillary tips. Using this method, a wide range of flow rates at various system pressures were attained. To obtain a particular flow rate, all valves except V_1' and V_3' in the reactant addition system (see Figure 2) were closed; and the appropriate combination of capillary tips (in the exit manifold) were valved on-line.

The concentration of trans-stilbene flowing into the photoreactor was controlled by the reactant addition system illustrated in Figure 2. trans-Stilbene was coated onto glass beads and loaded into the feeder tube. After immersing the feeder tube in a constant temperature bath, the system was sealed and pressurized. As shown in Figure 3, temperature was an effective means of controlling the concentration of trans-stilbene exiting the feeder tube.

If necessary, concentrations to the reactor were further adjusted through manipulation of the valve system, e.g. opening valves V_1' and V_2' to dilute the solution entering the photoreactor. This component was very effective for delivering a wide range of constant concentrations to the photoreactor. Furthermore, downstream concentrations of stilbene were considerably below the saturation level, and plugging of the system was not a problem.

Temperature in the photoreactor was controlled to ±0.1°C by immersion in a constant temperature Dewar flask. For experiments using cyclohexane as the solvent, the reactor was filled with a solution of trans-stilbene, loosely stoppered, and, after reaching the desired temperature, irradiated. Periodically, aliquots were withdrawn and analyzed by gas chromatography.

Experiments using CO_2 were conducted in the flow mode utilizir

5. AIDA AND SQUIRES *Photoisomerization of* trans-*Stilbene* 61

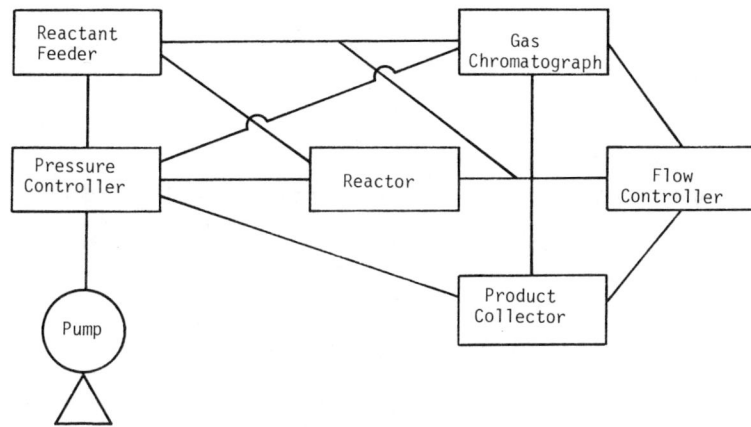

Figure 1. Schematic diagram of supercritical fluid reactor.

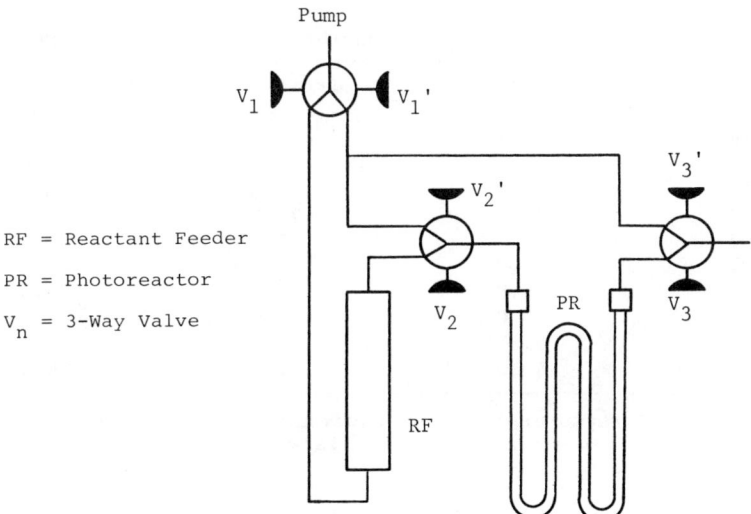

RF = Reactant Feeder
PR = Photoreactor
V_n = 3-Way Valve

Figure 2. Supercritical fluid reactant addition system.

techniques described above to control flow rates and concentrations of trans-stilbene supplied to the reactor. The product stream from the reactor was directed to either a product isolation loop or an on-line gas chromatographic analysis loop, and the products were analyzed to determine the ratio of cis-to trans-stilbene and to detect other photoproducts. A constant photostationary ratio was achieved in one to four minutes, well within the residence time in the irradiation zone.

Results and Discussion

A widely used method for assessing supercritical fluid phenomena consists of comparing physical and chemical behavior above the critical point with corresponding behavior in the subcritical liquid. Because this approach (unrealistically) seeks to observe discontinuous behavior between states, the results of such experiments are often ambiguous. In the present study, we have compared the photoisomerization of trans-stilbene in subcritical and supercritical CO_2; and, to model liquid behavior, we have also carried out these isomerizations in cyclohexane. In all three systems, the effects of temperature and concentration on the cis/trans ratio were compared; and, for the CO_2 systems, the effect of pressure on this photostationary ratio was also probed. The results from these experiments are shown in Tables I and II and are plotted in Figures 4 through 6.

Table I. Photoisomerization of trans-Stilbene in Cyclohexane[a]

Initial Concentration (mg/ml)[b]	Temperature (°C)	Irradiation Time (Minutes)	Photostationary State (cis/trans)
2.14	21.7	70.0	6.1
2.14	40.0	70.0	6.1
1.07	21.7	20.0	6.9
1.07	40.0	30.0	6.9
0.50	21.7	10.0	6.4
0.50	40.0	15.0	6.7
0.25	21.7	10.0	6.4
0.25	40.0	5.0	6.6

a. Atmospheric pressure; direct irradiation with a Hanovia 450 watt medium pressure mercury vapor lamp.
b. Pure trans-stilbene was used.

From the results obtained in cyclohexane (Table I), it is clear that, in the condensed phase, we can expect very little effect on the photostationary state due to changes in temperature or concentration. On the other hand, these factors cause large changes when the reaction is carried out in liquid or (especially) SCF CO_2. The contrast in behavior is even more apparent from Figures 4 and 5.

The responses of the cis/trans ratio to changes in pressure for liquid (25°C) and supercritical fluid (40°C) CO_2 are plotted in Figure 6. There is a noticeable effect in liquid CO_2, but the photo-

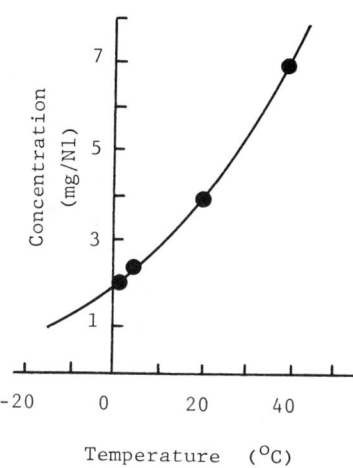

Figure 3. Effect of temperature on CO_2 solubility of trans-stilbene.

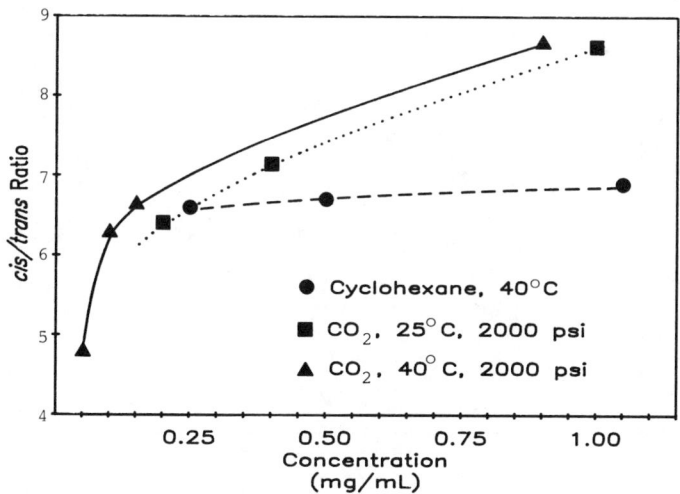

Figure 4. Concentration effects in the photoisomerization of trans-stilbene.

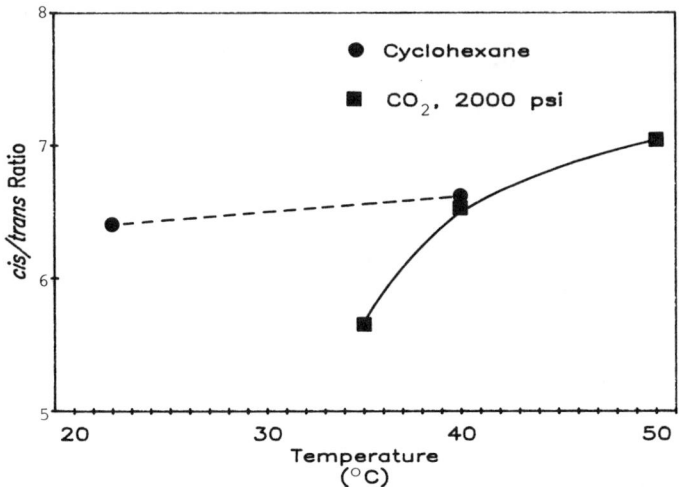

Figure 5. Temperature effects in the photoisomerization of trans-stilbene.

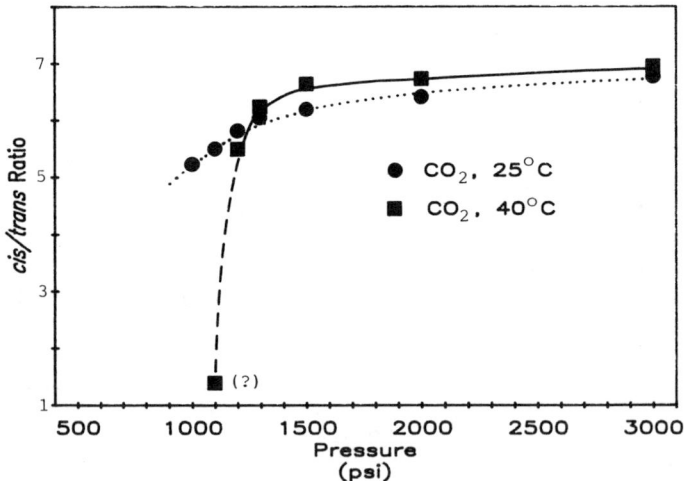

Figure 6. Pressure effects in the photoisomerization of trans-stilbene.

stationary ratio is clearly more responsive to pressure manipulations in the supercritical fluid state. The observed behavior is consistent with the model developed by Saltiel and D'Agostino (26) which attributes increases in the cis/trans ratio to increases in solvent viscosity. As the pressure ranges from 1100 to 3000 psi, there is a slight increase in the viscosity of liquid CO_2, but there is a three-fold increase in the viscosity of supercritical CO_2.

Table II. Photoisomerization of trans-Stilbene in Carbon Dioxide[a]

Initial Concentration (mg/NL)[b]	Reactor Concentration (mg/ml)[c]	Temperature (°C)	Pressure (psi)	Photostationary State (cis/trans)
2.00	0.98	25	2000	8.63
2.00	0.87	40	2000	8.68
0.80	0.39	25	2000	7.15
0.40	0.16	25	1000	5.23
0.40	0.16	25	1100	5.51
0.40	0.05	40	1116	1.38[d]
0.40	0.17	25	1200	5.82
0.40	0.07	40	1200	5.51
0.40	0.18	25	1300	6.06
0.40	0.11	40	1300	6.24
0.40	0.18	25	1500	6.19
0.40	0.15	40	1500	6.65
0.40	0.20	25	2000	6.42
0.40	0.17	40	2000	6.73
0.40	0.21	25	3000	6.80
0.40	0.19	40	3000	6.91
0.28	0.13	35	2000	5.66
0.28	0.12	40	2000	6.54
0.28	0.11	50	2000	7.04
0.24	0.10	40	2000	6.29
0.12	0.052	40	2000	4.80

a. Irradiation in a flow reactor with a 450 watt Hanovia medium pressure mercury vapor lamp.
b. Pure trans-stilbene was used. Normal liter (NL): the quantity of CO_2 which has a volume of 1 liter at 25°C and 1 atmosphere.
c. Calculated from the reported density of pure CO_2.
d. Value approximate due to pressure and flow fluctuations under these conditions.

While the implications for manipulating chemical pathways, rates, and equilibria in supercritical fluids are exciting, the experiments reported here provide no direct evidence about the nature of the interactions responsible for this behavior, and this will be the focus of future investigations.

Acknowledgments

We wish to thank Professor R. S. Hansen and Dr. C. G. Venier for helpful discussions. We gratefully acknowledge the support of Iowa State University through the Energy & Minerals Resources Research Institute.

Literature Cited

1. Hannay, J.B.; Hogarth, J. J. Proc. Royal Soc. (London) Ser. A **1879**, *29*, 324.
2. Paulaitis, M.E.; Krukonis, V.J.; Kurnik, R.T.; Reid, R.C. Rev. in Chem. Eng., **1983**, *1(2)*, 179.
3. Johnston, K.P. In "Kirk-Othmer Encyclopedia of Chemical Technology"; John Wiley & Sons: New York, 1984.
4. Maddocks, R.R.; Gibson, J.J. J. Chem. Eng. Prog. **1977**, *73(6)*, 59.
5. Caragay, A.B. Perfumer Flavorist, **1981**, *6(4)*, 43.
6. Hubert, P.; Vitzthun, O.G. Angew. Chem., Int. Ed. Engl. **1978**, *17*, 710.
7. Zosel, K. Angew. Chem., Int. Ed. Engl. **1978**, *17*, 702.
8. Meyers, M.N.; Giddings, J.C. Prog. Sep. Purif., **1970**, *3*, 133.
9. van Wasen, U.; Swaid, I.; Schneider, G.M. Angew. Chem., Int. Ed. Engl. **1980**, *19*, 575.
10. Schneider, G.M., ibid., **1978**, *17*, 716.
11. Hyatt, J.A. J. Org. Chem. **1984**, *49*, 5097.
12. Squires, T.G.; Venier, C.G.; Aida, T. Fluid Phase Equil. **1983**, *10*, 261.
13. Reichardt, C. J. Pure & Appl. Chem. **1982**, *54*, 1867.
14. Bartmann, D.; Schneider, G.M. J. Chromatogr. **1973**, *83*, 135.
15. McHugh, M.A.; Krukonis, V.J. "Supercritical Fluid Extraction: Principles and Applications"; Butterworths: Boston, 1986; Chap. 11.
16. Alexander, G.; Paulaitis, M.E. paper presented at the A.I.Ch.E. Annual Meeting at San Francisco, CA, November, 1984.
17. Crooks, R.M.; Fan, F.-R.F.; Bard, A.J. J. Am. Chem. Soc. **1984**, *106*, 6851.
18. Silverstri, G.; Gambino, S.; Filardo, G.; Cuccia, C.; Guarino, E. Angew. Chem., Int. Ed. Engl. **1981**, *20*, 101.
19. Cottle, J.E. Supercritical polymerization of olefins. U.S. Patent 3,294,772. 1966.
20. Hagiwara, M.; Mitsui, H.; Machi, S.; Kagiya, T. J. Polym. Sci., Part A **1968**, *6*, 603.
21. Koll, P.; Metzger, J. Angew. Chem. Int. Ed. Engl. **1978**, *17*, 754.
22. Metzger, J.O.; Hartmans, J.; Malwitz, D.; Koll, P. In "Chemical engineering at supercritical fluid conditions"; Paulaitis, M.E.; Penninger, J.M.L.; Gray, R.D.; Davidson, P., Eds.; Ann Arbor Science: Ann Arbor, MI, 1983; Chap. 14.
23. Frye, S.L.; Yonker, C.R.; Kalkwarf, D.R.; Smith, R.D. Prepr. Pap.-Am. Chem. Soc., Div. Fuel Chem. **1985**, *30(3)*, 7.
24. Sigman, M.E.; Lindley, S.M.; Leffler, J.E. J. Am. Chem. Soc. **1985**, *107*, 1471-1472.
25. Saltiel, J.; Charlton, J.L. In "Rearrangements in Ground and Excited States, Vol. 3"; Academic Press: New York, 1980, pp. 25-89.
26. Saltiel, J.; D'Agostino, J.T. J. Am. Chem. Soc. **1972**, *94*, 6445.
27. Swaid, I.; Schneider, G.M. Ber. Bunsenges. Phys. Chem. **1979**, *83*, 969.

RECEIVED August 29, 1986

Chapter 6

Solvent Effects During the Reaction of Coal Model Compounds

Martin A. Abraham and Michael T. Klein

Department of Chemical Engineering, University of Delaware, Newark, DE 19716

> The reaction of benzylphenylamine in supercritical water and supercritical methanol is shown to involve parallel pyrolysis and solvolysis reaction pathways. Reactant conversion passed through a minimum as the solvent loading increased while the product spectrum shifted from pyrolysis-like to solvolysis-like. The observed kinetics were consistent with two mechanistic interpretations. The first involves parallel pyrolysis and solvolysis steps, each having pressure-dependent rate constants. The second assumes the formation of a solvent cage, which suppresses the pyrolysis kinetics as the solvent loading increases from zero to above its critical density.

The chemical reactions that accompany the extraction of volatiles (1) from hydrocarbon resources are frequently obscured by the complexities of the reaction system. In contrast, the comparative simplicity of model compound structures and product spectra permit resolution of reaction fundamentals (2) and subsequent inference of the factors that control real reacting systems. Herein is described the use of model compounds to probe the kinetics of pyrolysis and solvolysis reactions that likely occur during the extraction of volatiles from coals and lignins.

Previous studies of the reactions of guaiacol (orthomethoxyphenol) (3), dibenzyl ether (4), and benzylphenylamine (5) in dense water elucidated parallel hydrolysis and pyrolysis pathways, the selectivity to the latter increasing with water density. Reactant decomposition kinetics were interestingly nonlinear in water density and consistent with two mechanistic interpretations. The first involved "cage" effects, as described for reactions in liquid solutions (6). The second led to parallel pyrolysis and solvolysis reaction pathways wherein associated rate constants were dependent upon pressure. These two schemes are probed herein through the reactions of benzylphenylamine (BPA) in water and methanol.

0097-6156/87/0329-0067$06.00/0
© 1987 American Chemical Society

Experimental

Table I summarizes the experimental conditions of reactant's concentration, solvent loading, temperature, and holding time. Measured amounts of the commercially available (Aldrich) substrate benzylphenylamine (BPA), the solvent (tetralin, water or methanol), and the inert internal standard biphenyl were loaded at room temperature into "tubing bomb" microreactors that have been described elsewhere (5).

Table I. Experimental Conditions for Reaction of BPA

Solvent	BPA Conc. (mol/L)	Reaction Temp. (°C)	Solvent Loading (mol/L)	Holding Time (min)
None (Neat)	0.64	386	--	5 - 50
Tetralin	0.59	386	2.95	5 - 60
Water	0.59	386	0.4 - 21.5	5 - 50
Methanol	0.59	340, 386	0.85 - 13.6	5 - 60

Sealed reactors were immersed into a fluidized sand bath held constant at the desired reaction temperature, which was attained by the reactors in about 2 min; this heat-up period was small compared to ultimate reaction times (up to 60 min) and was, in any case, identical for all runs. Products were identified by GC-MS and quantitated by GC as described elsewhere (4,5).

Results

Table II summarizes the major reaction products. Neat pyrolysis of BPA led to toluene, aniline, and benzalaniline as major products with minor products including 1,2-diphenylethane, diphenylmethane, and 2-benzylaniline. BPA conversion (x) was nearly 1.0 in 30 min. Yields ($y_i = n_i/n_{BPA_0}$) of toluene and aniline were both approximately 0.5. Benzalaniline, a dehydrogenated BPA analog, was observed in a yield near 0.2, approximately one-half the yield of the major fragmentation products toluene and aniline.

Table II. Selectivity to Major Products of BPA Reactions at 386°C and 25 min

Product	Reactants			
	BPA/Neat	BPA/Tetralin	BPA/Water $\rho_r = 0.81$	BPA/MeOH $\rho_r = 1.0$
Toluene	0.5	0.63	0.31	0.55
Aniline	0.5	0.63	0.62	0.08
Benzalaniline	0.2	0.1	0.02	0.02
Benzyl Alcohol	--	--	0.25	0.10
Benzaldehyde	--	--	0.21	0.06
N-methylaniline	--	--	--	0.20

Thermolysis of BPA in tetralin was slower than neat pyrolysis, as a conversion of only 0.8 was reached after 60 min. The yields of toluene and aniline asymptotically approached 0.5; benzalaniline yield approached 0.1 after 60 min. Selectivity (y_i/x) to toluene, aniline, and benzalaniline was, respectively, 0.5, 0.5, and 0.2 for neat pyrolysis and 0.63, 0.63, and 0.13 for thermolysis in tetralin. The increased selectivity to toluene and aniline and the decreased selectivity to benzalaniline during reaction in tetralin suggests that this hydrogen donor competed with BPA as the hydrogen source.

The reaction of BPA in water at 386°C and ρ_r = 0.8 is summarized in Figure 1 and led to benzyl alcohol and benzaldehyde as well as the neat pyrolysis products. BPA conversion was approximately 0.8 after 40 min. Aniline was the major product with a yield of 0.5 after 40 min. Toluene, a primary and major product of neat pyrolysis, had a yield near 0.2. This was roughly equal to the yield of benzyl alcohol, which had a maximum value of 0.15 at 30 min; the yield of benzyl alcohol decreased after 30 min. The reduced yield of toluene relative to neat pyrolysis and the presence of benzyl alcohol indicate that water reacted with BPA.

The effect of water density on BPA conversion and product selectivity, for a constant reaction time of 10 min, is shown in Figure 2. BPA conversion passed through a minimum at a reduced water density of 0.1. The selectivity to aniline was relatively unaffected by the solvent density, but the selectivity to toluene and benzalaniline decreased and the selectivity to benzyl alcohol increased as the solvent loading increased.

The reaction of BPA in supercritical methanol at 340°C and ρ_r = 0.5 is summarized in Figure 3 and led to N-methylaniline as a major product. Benzaldehyde and benzyl alcohol were observed in yields less than 0.1. The yield of toluene was 0.56, roughly equivalent to that observed from neat pyrolysis, whereas the aniline yield was approximately 0.1, substantially reduced from that observed in the neat case. The yield of N-methylaniline surpassed that of aniline, as a value of 0.4 was attained after 120 min. Reaction in ^{13}C-labeled methanol showed the label to reside on the methyl group in the N-methylaniline product, definitively indicating that methanol reacted with BPA.

The effect of methanol density on BPA conversion and product selectivity, at a constant reaction time of 60 min and a temperature of 340°C, was qualitatively similar to that observed for water. This is illustrated in Figure 4, where the minimum conversion occurs at a reduced methanol density of 0.6. The yield of toluene was relatively unaffected by changes in the methanol density, whereas the yield of aniline decreased and that of N-methylaniline increased as the solvent loading increased.

Discussion

BPA reaction in water or methanol yielded solvation products in addition to those observed from pyrolysis neat and thermolysis in the hydrogen donor solvent, tetralin. BPA conversion passed through a minimum at a reduced solvent density of 0.6 and 0.1 for reaction in methanol and water, respectively. Results qualitatively similar to those observed in Figures 2 and 4 have been noted previously for

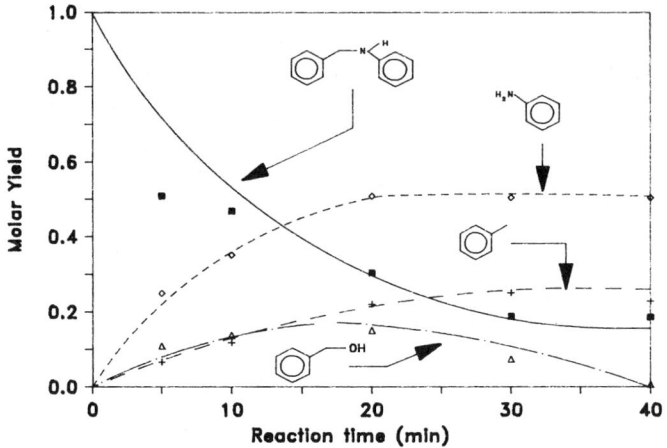

Figure 1. The reaction of BPA in supercritical water at 386°C and $\rho_r = 0.81$.

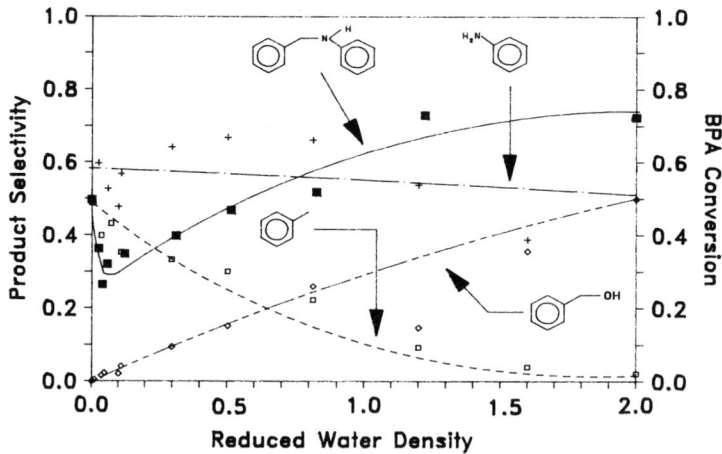

Figure 2. Effect of water density on BPA conversion and product selectivity for reaction at 386°C after 10 min.

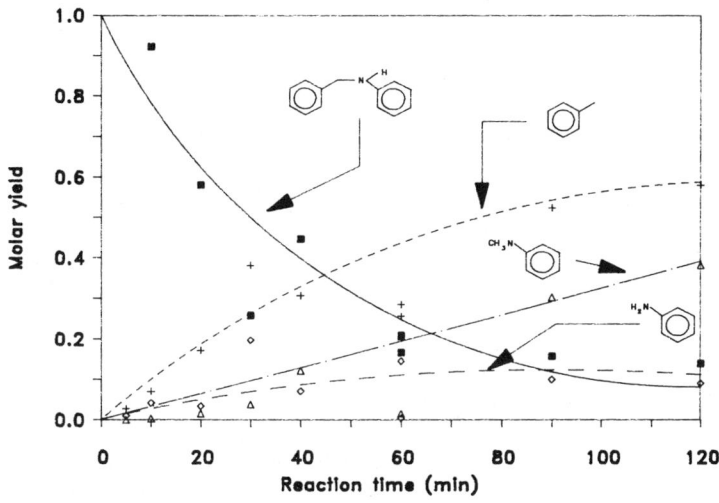

Figure 3. The reaction of BPA in methanol at 340°C and ρ_r = 0.5.

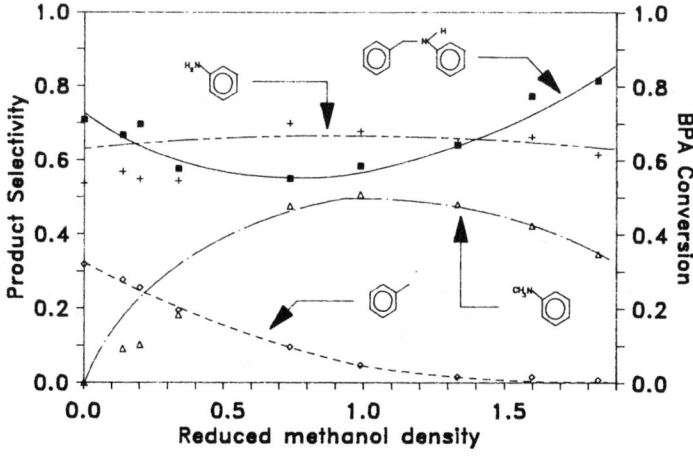

Figure 4. Effect of methanol density on BPA conversion and product selectivity for reaction at 340°C after 60 min.

reactions in solution with networks containing pressure-dependent rate constants (7). The minimum in BPA conversion is explained by the simultaneous decrease in the rate of the pyrolysis reaction and increase in the rate of the solvolysis reaction.

These results are consistent with the reaction pathways shown in Figure 5. The major neat pyrolysis pathway, illustrated in Figure 5a, requires two molar equivalents of BPA for the formation of one mole each of toluene, aniline, and carbon-rich products (CRP) which include the observed benzalaniline. The network for thermolysis in tetralin is a combination of the neat pyrolysis pathway (Figure 5a) and a pathway wherein tetralin and BPA react to one mole each of toluene and aniline, and 0.5 molar equivalents of naphthalene. Figure 5b depicts a direct solvolysis pathway for water and methanol solvents. Here the BPA can either (i) react by the neat pyrolysis pathway with the consumption of another BPA, to give one mole each of toluene, aniline, and benzalaniline or (ii) proceed through the solvolysis pathway to give oxygenated products and aniline (or N-methylaniline for reaction in methanol). The selectivity to oxygenated products would increase as the solvent loading increases.

The minima in BPA conversion observed for reaction of BPA in water and methanol can be explained by allowing the rate constants of Figure 5b to be dependent on pressure. For each solvent loading (and thus pressure) studied, the pseudo-first-order rate constants for the pathways of Figure 5b are shown in Table III. These were calculated using a sequential simplex search based on the Complex Method of Box (8) where the objective function was the square of the deviations between predicted and experimental values. The pressure generated by water was estimated from PVT data (9) and the methanol pressure was estimated using a Peng-Robinson equation of state.

Transition state theory explicitly provides the pressure dependence of an nth order rate constant for an elementary step through the volume of activation, $\Delta V^{\ddagger}$, as indicated in Equation 1.

$$\frac{\partial \ln k_c}{\partial P} = -\frac{\Delta V^{\ddagger}}{RT} + (\Delta n^{\ddagger}) \frac{\partial \ln c_T}{\partial P} \qquad (1)$$

Note that $\Delta n^{\ddagger}$ is the difference between the number of moles in the transition state and the molecularity of the elementary step. In Equation 3, k_c is the observed rate constant based on concentration, and c_T is the total concentration of the reaction mixture (10). The volume of activation is the difference between the partial molar volumes of the activated state and the reactants.

For reactions in solution, pressures in the range of kilobars are generally required before the rate constant is affected significantly. This is because typical values of the activation volume range from -20 to +20 cc/mol (11) and, hence, ln (k/k_0), which is equal to $-\Delta V^{\ddagger} \Delta P/RT$, does not become significant (e.g. ~1) until large changes in pressure (e.g., $\Delta P \cong 2.74$ kbar for $\Delta V^{\ddagger} = -20$cc/mol) occur. However, Simmons and Mason (12) have shown that near the critical point of a fluid the volume of activation may approach $\pm \infty$. Moreover, as can be seen from Equation 1, the effect of pressure on the rate constant is attributable to not only the intrinsic volume of activation but also the compressibility of the fluid which may be large in a fluid near its critical point. Hence, the net "apparent" volume of

Table III. Apparent Rate Constants for BPA Reaction in SC Solvents

ρ_r	Estimated Pressure (atm)	$k_1 \times 10^3$ (min^{-1})	$k_2 \times 10^3$ (L(mol)$^{-1}$(min)$^{-1}$)
\multicolumn{4}{c}{Reaction in water at 386°C}			
0.0	0	69	--
0.025	22	26	17
0.05	45	12	5.3
0.1	87	13	5.6
0.3	184	14	2.2
0.5	227	15	1.6
0.8	249	17	2.0
1.2	264	18	1.8
\multicolumn{4}{c}{Reaction in methanol at 386°C}			
0.1	43.1	22	6.4
0.3	117.3	8.1	2.6
0.5	183.0	8.6	1.5
0.75	264.7	7.7	1.8
1.0	358.9	6.1	1.3
1.25	480.7	2.7	1.0
1.6	739.2	3.2	1.0
\multicolumn{4}{c}{Reaction in methanol at 340°C}			
0.0	0.0	22	--
0.1	39.2	5.8	1.7
0.3	101.8	4.0	0.93
0.5	151.6	2.9	1.4
0.75	208.1	2.7	0.87
1.0	271.0	1.3	0.84
1.3	375.9	1.0	0.72
1.6	547.0	0.38	0.50

activation is the sum of an intrinsic term and a term arising from the solution compressibility, both of which may be unusually large. Thus, the effects of pressure on the reaction rate constant can be observed at pressures significantly lower than previously noted in the liquid-phase literature.

The effects of solvent density on the observed rates can also be interpreted through the onset of cage effects. As the solvent loading increases, the system density increases from a gas-like to a liquid-like value. In dense fluids, an addition reaction such as $A + B \rightarrow AB$ is considered to follow, in series, the formation of an ecountered pair AB^* (6). Similarly, the fragmentation reaction $AB \rightarrow A + B$ can be limited by the rate of breaking apart the encountered pair AB^*. For the pathways of Figure 6, the encountered species is BPA^*. Rate constants k_1, k_{-1}, and k_3 are intrinsic, whereas k_2 is the inverse of the duration of an encounter. Rate constant k_2 is proportional to the diffusion coefficient of the encountered species, which will be high in the gas phase and several orders of magnitude lower in the liquid phase. Allowing BPA^* to exist in a pseudo-steady-state permits derivation of a rate expression for the pathways of Figure 6 as:

$$-\frac{dBPA}{dt} = BPA \left[\frac{2k_1 k_2}{k_{-1} + k_2} + k_3 S \right] \quad (2)$$

Both terms on the right hand side of equation 2 depend on solvent loading. In the first term, k_2 is inversely proportional to the time of encounter, τ_{AB}, which in turn is inversely proportional to the diffusivity. The diffusivity and thus k_2 decrease with increases in the density as:

$$k_2 \sim 1/\tau_{AB} \sim D \sim 1/\rho^a \quad (3)$$

The second of the terms of the right hand side of Equation 2, $k_3 S$, increases with solvent loading. The overall rate of BPA conversion may thus initially decrease as the solvent loading is increased due to the decrease in rate constant k_2; as the solvent loading is increased further, the rate of disappearance of BPA should increase because of the increase in the term $k_3 S$.

Conclusions

1. Supercritical solvents react with BPA to yield products that are not observed from neat pyrolysis or thermolysis in a hydrogen donor.
2. Increasing the solvent loading shifts the product spectrum from pyrolysis-like to solvolysis-like. In addition, a minimum in reactant conversion, arising from a decrease in the rate of the pyrolysis pathway, is observed.
3. The observed kinetics are consistent with two mechanistic interpretations. The first involves pressure-dependent rate constants as viewed through transition-state theory. The volume of activation may be unusually large due to the unique properties of a

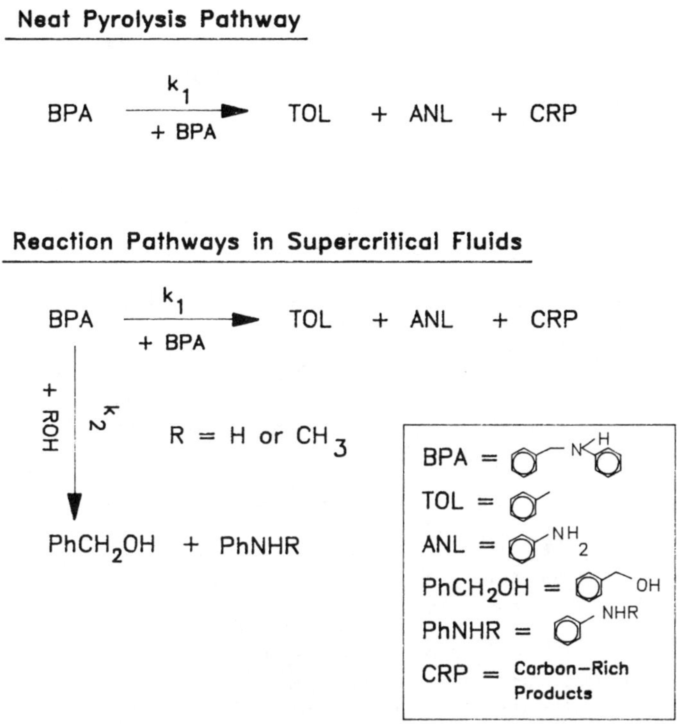

Neat Pyrolysis Pathway

$$BPA \xrightarrow[+ BPA]{k_1} TOL + ANL + CRP$$

Reaction Pathways in Supercritical Fluids

$$BPA \xrightarrow[+ BPA]{k_1} TOL + ANL + CRP$$

$$\downarrow {}^{+ ROH}_{k_2}$$

R = H or CH_3

$PhCH_2OH + PhNHR$

BPA = (phenyl)-NH-(phenyl)
TOL = (phenyl)-CH₃
ANL = (phenyl)-NH_2
$PhCH_2OH$ = (phenyl)-CH_2OH
PhNHR = (phenyl)-NHR
CRP = Carbon-Rich Products

Figure 5. Global pathways for the reaction of BPA neat and with supercritical fluid solvents.

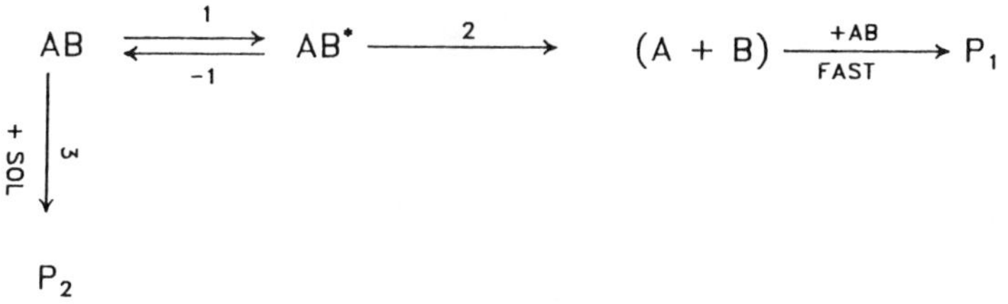

$$AB \underset{-1}{\overset{1}{\rightleftharpoons}} AB^* \xrightarrow{2} (A + B) \xrightarrow[FAST]{+AB} P_1$$

$$\downarrow {}^{+ SOL}_{3}$$

$$P_2$$

$$\frac{d \ln (AB)}{dt} = \frac{2k_1 k_2}{k_{-1}+k_2} + k_3 S$$

$$k_2 \propto \frac{1}{\tau_{AB}} \quad \propto D \quad \propto \frac{1}{\rho^a}$$

Figure 6. Model of parallel BPA solvolysis and pyrolysis through caged radical pair.

fluid near its critical point. The second allows solvent cage effects to suppress pyrolysis, such that at increasing solvent loading, the pyrolysis reaction becomes diffusion limited.

Literature Cited

1. Squires, T. G.; Aida, T.; Chen, Y.; Smith, B. F. ACS Div. Fuel Chem. Preprints 1983, 28(4), 228.
2. Simmons, M. B.; Klein, M. T. Ind. Eng. Chem. Fundam. 1985, 24, 55.
3. Lawson, J. R.; Klein, M. T. Ind. Eng. Chem. Fundam. 1985, 24, 203.
4. Townsend, S. H.; Klein, M. T. Fuel 1985, 64, 635.
5. Abraham, M. A.; Klein, M. T. Ind. Eng. Chem. Prod. Res. and Dev. 1985, 24, 300.
6. Moore, J. W.; Pearson, R. G. "Kinetics and Mechanism"; John Wiley & Sons: New York, 1981.
7. Kohnstam, G. In "Progress of Reaction Kinetics"; Porter, G,; Jennings, K. R.; Suppan, P.; Eds.; Pergamon Press: New York, 1965, Vol. V, p. 335.
8. Beveridge, G. S.; Schechter, R. S. "Optimization: Theory and Practice"; McGraw-Hill Book Co.: New York, 1970.
9. Keenan, J. H.; Keyes, F. G. "Thermodynamic Properties of Steam"; John Wiley & Sons: New York, 1986.
10. Simmons, G. M.; Mason, D. M. Chem. Eng. Sci. 1972, 27, 2307.
11. Eckert, C. A. Ann. Rev. Phys. Chem. 1972, 23, 239.
12. Simmons, G. M.; Mason, D. M. Chem. Eng. Sci. 1972, 27, 89.

RECEIVED August 27, 1986

Chapter 7

Heterolysis and Homolysis in Supercritical Water

Michael Jerry Antal, Jr., Andrew Brittain, Carlos DeAlmeida, Sundaresh Ramayya, and Jiben C. Roy[1]

Department of Mechanical Engineering, University of Hawaii, Honolulu, HI 96822

The reaction chemistry of simple organic molecules in supercritical (SC) water can be described by heterolytic (ionic) mechanisms when the ion product K_w of the SC water exceeds 10^{-14} and by homolytic (free radical) mechanisms when $K_w \ll 10^{-14}$. For example, in SC water with $K_w > 10^{-14}$ ethanol undergoes rapid dehydration to ethylene in the presence of dilute Arrhenius acids, such as 0.01M sulfuric acid and 1.0M acetic acid. Similarly, 1,3 dioxolane undergoes very rapid and selective hydration in SC water, producing ethylene glycol and formaldehyde without catalysts. In SC methanol the decomposition of 1,3 dioxolane yields 2 methoxyethanol, illustrating the role of the solvent medium in the heterolytic reaction mechanism. Under conditions where $K_w \ll 10^{-14}$ the dehydration of ethanol to ethylene is not catalyzed by Arrhenius acids. Instead, the decomposition products include a variety of hydrocarbons and carbon oxides.

Efficient thermochemical processes underlie the conversion of crude oil and natural gas liquids to higher value chemicals and fuels. Unfortunately, attempts to develop similar conversion processes for biomass feedstocks (such as wood chips and bagasse) have been frustrated by the non-specificity of high temperature pyrolysis reactions involving biopolymer substrates (1,2). For example, the pyrolysis of bagasse yields a liquid mixture of carbohydrate sirups and phenolic tars, a gas composed primarily of carbon oxides and hydrogen, and a solid charcoal. Variations of the conventional engineering parameters (temperature, heating rate, residence time and pressure) do not provide a good control over the complex set of concurrent and consecutive pyrolysis reactions. Thus it has not been possible to engineer the pyrolysis reactions to produce a few high value products from biomass materials.

Recent advances in materials technology, high pressure pumps and other high performance liquid chromatography (HPLC) equipment, have created new opportunities (3) for fundamental studies of

[1]Current address: Gonoshathaya Pharmaceuticals Limited, P.O. Nayarhat via Dhmrai, Dhaka, Bangladesh

chemical reactions in solvents at very high pressures (>30 MPa) and temperatures (>400° C). For sufficiently dilute solute-solvent mixtures at pressures $P > P_c$, no liquid-vapor phase transition occurs as the mixture is heated. Furthermore, when the temperature T is near the solvent's critical temperature T_c and $P > P_c$ many solvents manifest extraordinary properties (4,5). These unusual properties offer opportunities for the control of chemical reactions involving biopolymers and other substrates.

As discussed in the following sections, when the density of supercritical (SC) water exceeds 0.4 g/cm^3 the fluid retains its ionic properties (high dielectric constant and ion product) - even at high temperatures. These properties provide new opportunities to catalyze a variety of heterolytic bond cleavages with a high degree of specificity. Examples discussed in this paper include the dehydration of ethanol to ethylene and the hydration of 1,3 dioxolane to glycol and formaldehyde. In each of these examples the specificity of the heterolytic bond cleavage is high; whereas the conventional, higher temperature, free radical reactions offer lower yields of the desired products. In the case of ethanol dehydration, findings reported here may have exciting implications for the production of ethylene from ethanol.

Prior Work

Interest in the use of SC solvents as a reaction media is founded upon recent advances in our understanding of their unique thermophysical and chemical properties. Worthy of special note are those thermophysical properties (6) which can be manipulated as parameters to selectively direct the progress of desirable chemical reactions. These properties include the solvent's dielectric constant (7), ion product (8,9), electrolyte solvent power (10,11), transport properties [viscosity (12), diffusion coefficients (13) and ion mobilities (14)], hydrogen bonding characteristics (15), and solute-solvent "enhancement factors" (6). All these properties are strongly influenced by the solvent's density ρ in the supercritical state.

For example, SC water with ρ =0.47 g/cm^3 at 400° C (P = 35. MPa) enjoys a dielectric constant of about 10 (comparable to a polar organic liquid under normal conditions), an ion product of 7×10^{-14} (vs 10^{-14} at room conditions) and a dynamic viscosity μ =0.57 millipoise (vs 10 at room conditions). Under these conditions SC water behaves as a water-like fluid with strong electrolytic solvent power, high diffusion coefficients and ion mobilities, and considerable hydrogen bonding. These properties favor chemical reactions involving heterolytic (ionic) bond cleavages which can be catalyzed by the presence of acids or bases.

Dramatic changes occur when the temperature of the SC water is raised to 500° C at constant pressure (ρ=0.144 g/cm^3). Decreases in the dielectric constant to a value of 2 and ion product to 2.1×10^{-20} cause the fluid to lose its water-like characteristics and behave as a high temperature gas. Under these conditions homolytic (free radical) bond cleavages are expected to dominate the reaction chemistry. Thus by using the engineering parameters of temperature and pressure one can dramatically change the chemical properties of the solvent (dielectric constant and ion product) to favor heterolytic or homolytic bond cleavages. This paper emphasizes the manipulation of these parameters as a means for engineering the reaction chemistry of biopolymer materials.

Apparatus and Experimental Procedures

Figure 1 is a schematic of the SC flow reactor used in this work. Prior to the initiation of flow, the system is brought up to pressure by an air compressor. Afterwards, an HPLC pump forces a pure solvent into the reactant accumulator at a measured rate of flow. This flow displaces the solvent/solute reactant mixture out of the accumulator, through the reactor and a 10 port valve dual loop sampling system, and into the product accumulator. The flow of products into the second accumulator displaces air through a back-pressure regulator and into a water displacement apparatus, which measures the rate of air flow at ambient conditions. The reactant flow is rapidly heated to reaction temperature by the entry heat guard, and maintained at isothermal conditions by a Transtemp Infrared furnace and an exit heat guard. Samples captured in 5.54 ml sample loops are released into sealed, evacuated test tubes for quantitative analysis by GC, GC-MS, and HPLC instruments within the laboratory. The outer annulus of the reactor is a 4.6 mm ID Hastelloy C-276 tube, and the inner annulus is a 3.2 mm OD sintered alumina tube, giving the reactor an effective hydraulic diameter of 1.4 mm. The alumina tube accommodates a movable type K thermocouple along the reactor's axis, which provides for the measurement of axial temperature gradients along the reactor's functional length. Radial temperature gradients are measured as differences between the centerline temperature and temperatures measured at 10 fixed positions along the outer wall of the reactor using type K thermocouples. The location of the movable thermocouple within the reactor is measured electronically to within 0.01 mm by a TRAK digital position read out system. The entire reactor and sampling system is housed in a "bullet proof" enclosure which can be purged of air (oxygen) during studies involving flammable solvents (such as methanol).

The following representative nondimensional numbers characterize the reactor operating at 400°C: Re = 420, Pr = 1.86, Sc = 0.86, Pe_m = 358, Pe_h = 776 and Da = 0.4. Figure 2 displays a typical temperature profile of the reactor during operation. Because the thermal diffusivity of SC water is comparable to that of many high quality insulation materials, gross radial temperature gradients can easily exist in a flow reactor. As shown in Figure 2, radial temperature gradients within the annular flow reactor are negligible. A computer program, which accurately accounts for the effects of the various fluid (solvent, solvent and solute, air) compressibilities on flow measurements, calculates mass and elemental balances for each experiment. A typical experiment evidences mass and elemental balances of 1.00±0.05.

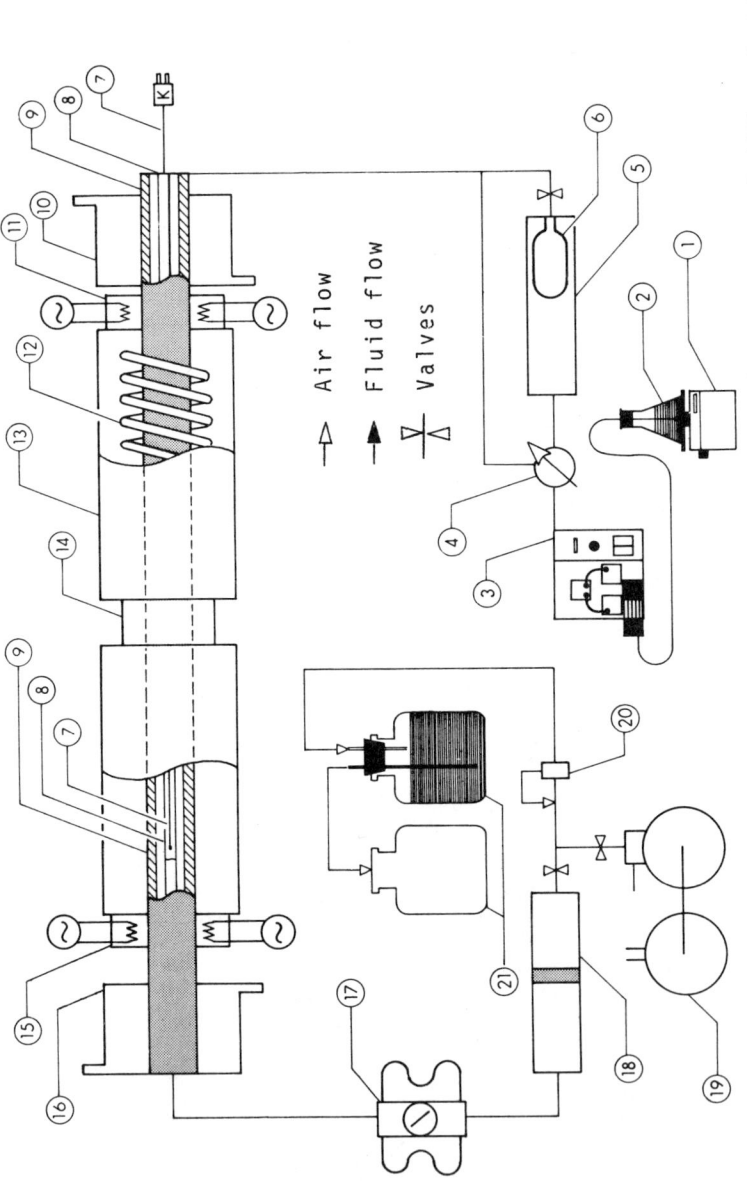

Figure 1. Supercritical flow reactor. Key: (1) Mettler balance; (2) flask with H$_2$O (filtered and deaerated); (3) HPLC pump; (4) bypass (three-way) valve; (5) feed cylinder; (6) weather balloon with feed solution; (7) probe thermocouple (type K); (8) ceramic annulus; (9) Hastelloy C-276 tube; (10) entrance cooling jacket; (11) entrance heater; (12) furnace coils; (13) quartz gold-plated IR mirror; (14) window (no coils); (15) guard heater; (16) outlet cooling jacket; (17) ten-port dual-loop sampling valve; (18) product accumulator; (19) air compressor; (20) back-pressure regulator; and (21) outflow measuring assembly.

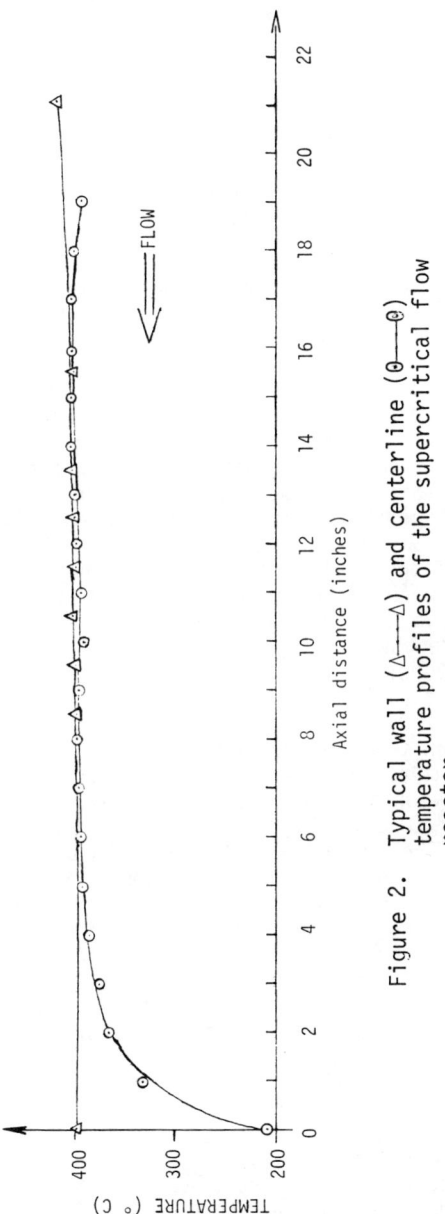

Figure 2. Typical wall (△——△) and centerline (⊙——⊙) temperature profiles of the supercritical flow reactor.

Results

Results of experiments probing the dehydration chemistry of ethanol in SC water (P = 34.5 MPa) are summarized in Table 1. If equilibrium were established, the conversion of ethanol to ethylene at 400° C and 34.5 MPa would be 74%. Unfortunately, the uncatalyzed reaction is very slow and little ethylene is formed. Moreover, many Arrhenius acid catalysts either decompose or react with the ethanol at these high temperatures, forming unwanted byproducts. For example, nitric acid completely decomposes and leads to the formation of CO_2, CO and other byproducts. Formic acid also decomposes in SC water under these conditions, and does not catalyze the dehydration reaction. At high concentrations (0.1M) sulfuric acid suffers some decomposition and reacts with the ethanol, as well as catalyzing the dehydration reaction. However, at lower concentrations, the acid exerts a strong and selective catalytic influence on the dehydration of ethanol. As shown in Table 1, in the presence of 0.01 M sulfuric acid, ethanol undergoes 21% decomposition in 13s at 400°C, yielding 82% ethylene.

Unlike formic acid, acetic acid is absolutely stable in SC water to 500° C at 34.5 MPa. At 400° C the presence of acetic acid (1.0M) triples the rate of formation of ethylene from ethanol; however, the formation of the liquid byproduct ethyl acetate accounts for 46% of the ethanol which reacted. Normally the presence of a strong acid (such as H_2SO_4) is required to catalyze the formation of the acetate from ethanol and acetic acid. Consequently, the formation of the acetate (along with diethylether as a minor, <1%, additional product) surprised us. This result seems to indicate the ability of SC water to act as an acid catalyst at high temperatures when $K_w > 10^{-14}$.

As mentioned earlier, at 500° C and 34.5 MPa supercritical water has a small dielectric constant, a very low ion product, and behaves as a high temperature gas. These properties would be expected to minimize the role of heterolysis in the dehydration chemistry. As shown in Table 1, the conversion of ethanol to ethylene at 500° C is small, even in the presence of 0.01M sulfuric acid catalyst. The appearance of the byproducts CO, CO_2, CH_4 and C_2H_6 points to the onset of nonselective, free radical reactions in the decomposition chemistry, as would be expected in the high temperature gas phase thermolysis of ethanol.

Similar experiments involving the hydration of 1,3 dioxolane in water (see Figure 3) at 350° C and 34.5 MPa evidenced the complete decomposition of the reactant and produced a yield of 99.8% ethylene glycol and formaldehyde (formed in equimolar amounts) with a residence time of 100s. An experiment involving 1,3 dioxolane in SC methanol at 450° C and 13.8 MPa formed the expected product 2-methoxyethanol, substantiating the role of the solvent medium in the reaction chemistry as indicated in Figure 3. The extraordinary specificity and rate of hydration reaction rendered the use of catalysts unnecessary. These results contrast with earlier findings (18) concerning the high temperature (>600° C), free radical decomposition of 1,3 dioxolane in steam at 0.1 MPa, where a wide variety of gaseous products (including H_2, CO, CO_2, CH_4, C_2H_4 and C_2H_6) were observed and no specificity in the reaction chemistry could be realized.

TABLE I. Ethanol (1M) Dehydration in SC Water

Catalyst	Catalyst Conc.(M)	Temp. (°C)	Press. (MPa)	Res. Time (s)	Conversion (%)	C_2H_4	C_2H_6	Gaseous Product Yield (%)* CO_2	H_2	CO	CH_4	Carbon Balance
-	-	400	31.7	49	2.6	17	1.6	0.0	15	0.0	0.0	0.98
H_2SO_4	0.01	400	34.5	13	21	82	0.0	0.0	0.0	0.0	0.0	0.96
HNO_3	0.01	400	31.7	48	12	1.2	0.1	57	15	14	0.8	0.93
CH_3COOH	1.0	400	31.7	52	9.5	12	0.2	0.1	2.3	0.0	0.0	0.98
-	-	500	31.7	66	2.0	25	25	3.0	218	3.5	7.8	1.00
H_2SO_4	0.01	500	34.5	8.5	17	40	0.2	1.1	42.	1.8	2.9	0.91

*Yield = (mole product)/(mole reactant consumed)

+A low background concentration of H_2 is observed, even in blank experiments with pure water as a reactant.

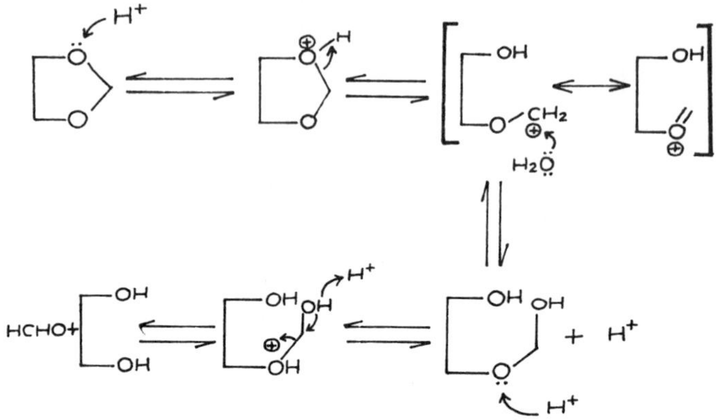

Figure 3. Heterolytic formation of glycol and formaldehyde from 1,3 dioxolane in SCW (in collaboration with M. Jones).

Discussion

Heterolytic reactions can be distinguished from homolytic (free radical) reactions by the following criteria:
1) heterolytic reactions usually can be catalyzed by Arrhenius acids and bases.
2) heterolytic reactions usually are influenced by the nature (polar vs non-polar) of the solvent medium.
3) heterolytic reactions are often quite specific.
4) homolytic reactions are not influenced by Arrhenius acid or base catalysts.
5) because of termination steps, homolytic reactions are often non-specific.

Findings reported in this paper include:
1) the catalysis of ethanol dehydration by two Arrhenius acids in SC water at 400°C where $K_w > 10^{-14}$
2) the influence of the solvent (water vs methanol) on the products and rates of the uncatalyzed decomposition of 1,3 dioxolane.
3) the extraordinary specificity of these two reactions when $K_w > 10^{-14}$.
4) the ineffectiveness of H_2SO_4 as a catalyst for ethanol dehydration in SC water at 500°C where $K_w \ll 10^{-14}$.
5) the loss of specificity of the ethanol dehydration reaction at 500°C where $K_w \ll 10^{-14}$.

These findings cause us to conclude that aqueous phase chemical reactions, usually observed at much lower temperatures, can be conducted in SC water providing $K_w > 10^{-14}$. When $K_w \ll 10^{-14}$ SC water behaves as a high temperature gas and free radical reactions predominate.

The formation of the ethyl acetate ester from ethanol and acetic acid also points to the role of carbocation chemistry in SC water when $K_w > 10^{-14}$. In addition, this reaction suggests the possible role of SC water as a catalytic medium (due to the "high" concentrations of H+ and OH-) at sufficiently high temperatures and pressures.

The catalytic dehydration of ethanol to ethylene in SC water may be commercially important (16). Although high quality commercial alumina catalysts exist for the vapor phase dehydration of ethanol, the commercial processes require the ethanol feedstock to be relatively free of water. Hence the ethanol must be distilled from the ethanol-water mixture which is the product of fermentation processes. By avoiding this distillation step, and securing phase separation of the ethylene product from the ethanol-water reactant, SC dehydration of ethanol could enjoy advantages over existing commercial technologies.

Because of the potential commercial significance of this work, we are presently developing kinetic expressions for the rate of ethylene formation in the SC water environment. We are also measuring the rate of ethanol dehydration in the vicinity of the critical point of water to determine if the properties of the fluid near the critical point have any influence on the reaction rate. In the near future we plan to begin studies of the reaction chemistry of glucose and related model compounds (levulinic acid) in SC water.

Conclusions

The significance of this work is its identification of SC water as a medium which supports and enhances aqueous phase chemistry ordinarily observed at much lower temperatures. Fundamental studies of the reaction chemistry of biopolymer related model compounds described in this paper offer insights into the details of reaction mechanisms, and facilitate the choice of reaction conditions which enhance the yields of valuable products. Chemical reaction engineering in supercritical solvents, based on the ability to choose between heterolytic and homolytic reaction mechanisms with foreknowledge of results, holds much promise as a new means to improve our utilization of the vast biopolymer resource.

Acknowledgments

This research was supported by the National Science Foundation under grant CPE 8304381, the Department of Planning and Economic Development of the State of Hawaii, and the Coral Industries Endowment. The authors thank Dr. Maria Burka (NSF), Kent Keith and Dr. Tak Yoshihara (DPED), and David Chalmers (Coral Industries) for their interest in this work. The authors also thank William Mok for his continuing assistance with the flow reactor, Dr. Ali Tabatabaie-Raissi, and Professor Maitland Jones (Princeton) for many stimulating discussions. The comments of the reviewers were appreciated.

Literature Cited

1. Antal, M.J. "Biomass Pyrolysis. A Review of the Literature - Part 1: Carbohydrate Pyrolysis" in Advances in Solar Energy, (Eds. K.W. Boer and J.A. Duffie) Vol. 1, (American Solar Energy Society, Boulder, CO) 1983.

2. Antal, M.J. "Biomass Pyrolysis. A Review of the Literature - Part 2: Lignocellulose Pyrolysis" in Advances in Solar Energy, 2, (Plenum Press, N.Y.) 1985.
3. Paulaitis, M.E., Penninger, J.M.L., Gray, R.D, Jr. and Davidson, P. "Chemical Engineering at Supercritical Fluid Conditions" Ann Arbor Science, Ann Arbor, 1983.
4. Franck, E.U. "Supercritical Water" Endeavor, 1968, **27**, 55-59.
5. Marshall, W.L. "Water at High Temperatures and Pressures" Chemistry, 1975, **48**, 6-12.
6. Franck, E.U. "Thermophysical Properties of Supercritical Fluids with Special Consideration of Aqueous Systems", Fluid Phase Equilib. 1983, 10(2-3), 211-22.
7. Franck, E.U. "Water and Aqueous Solutions at High Pressures and Temperatures".
8. Marshall, W.L and Franck, E.U., J. Phys. Chem. Ref. Data, 1981, **10**, 295-304.
9. Quist, A.S., Marshall, W.L., and Jolley, H.R. J. Phys. Chem., 1965, **69**, 2726-2735.
10. Marshall, W.L., Rev. Pure. Appl. Chem., 1968, **18**, 167-186.
11. Quist, A.S., J. Phys. Chem., 1970, **74**, 3396-3402.
12. Bruges, E.A. and Gibson, M.R., J. Mech. Eng. Sci., 1969, **11**, 189-205.
13. Lamb, W.J., Hoffman, G.A., Jonas, J., J. Chem. Phys., 1981, **74**(12), 6875- 80.
14. Franck, E.U., Hartmann, D., and Hensel, F., J. Phys. Chem., 1968, **72**, 200-206.
15. Gorbatyi, Y.E., Probl. Fix.-Khim. Petrol., (Eds. Zharikov, V.A., Fonarev, V.I., Korikovskii, S.P.) 1979, 2, 14-24.
16. Luchi, N.R. and Trindade, S.C., Hydrocarbon Proc., 1982, 179-183.
17. Knozinger, H. The Chemistry of the Hydroxyl Group, Part 2, (Ed. S. Patai), Interscience, London 1971.
18. Cutler, A., Ph.D. thesis, Princeton University, Princeton, 1984.

RECEIVED October 7, 1986

PHASE EQUILIBRIA

Chapter 8

A Statistical Mechanics Based Lattice Model Equation of State
Applications to Mixtures with Supercritical Fluids

Sanat K. Kumar, R. C. Reid, and U. W. Suter

Department of Chemical Engineering, Massachusetts Institute of Technology, Cambridge, MA 02139

> A Statistical-Mechanics based Lattice-Model Equation of state (EOS) for modelling the phase behaviour of polymer-supercritical fluid mixtures is presented. The EOS can reproduce qualitatively all experimental trends observed, using a single, adjustable mixture parameter and in this aspect is better than classical cubic EOS. Simple mixtures of small molecules can also be quantitatively modelled, in most cases, with the use of a single, temperature independent adjustable parameter.

During the last decade, increasing emphasis has been placed on the use of the equation of state (EOS) approach to model and correlate high-pressure phase equilibrium behaviour. More successful applications have employed some form of cubic EOS (1-3) although others (e.g., 4) have been proposed. However, as the types of systems studied have become more complex, the inherent weaknesses of a cubic EOS have become apparent. We, in particular, are interested in studying phase behaviour of systems comprising polymer molecules in the presence of a supercritical fluid. Here the size disparity of the component molecules can be large. One approach would have been to adopt the modified perturbed hard chain theory (5,6) which has been adapted for mixtures of large and small hydrocarbon molecules. We, however, elected to study whether lattice theory models could be of value for systems of our interest. Studies based on this approach have been attempted for different systems (7-15), and an interesting model has been proposed by Panayiotou and Vera (16). Our approach is similar in many respects to the last reference although significant differences appear in treating mixtures.

Pure Components

Theory. Molecules are assumed to "sit" on a lattice of coordination number z and of cell size v_H. Each molecule (species 1) is assumed to occupy r_1 sites (where r_1 can be fractional), and the lattice has empty sites called holes. There are N_0 holes and N_1 molecules. To account for the connectivity of the segments of a molecule, an effective chain length q_1 is defined as,

$$zq_1 = zr_1 - 2r_1 + 2 \qquad 1)$$

wherein it has been assumed that chains are not cyclic. zq_1 now represents the effective number of external contacts per molecule. The interaction energy between segments of molecules is denoted by $-\epsilon_{11}$, while the interaction energy of any species with a hole is zero. Only nearest neighbour interactions are considered, and pairwise additivity is assumed. The canonical partition function for this ensemble can be formally represented as

$$\Omega = \sum_{\substack{\text{all} \\ \text{states } \{n\}}} \exp(-\beta E_{\{n\}}) \qquad 2)$$

where $\beta=1/kT$. On the assumption of random mixing of holes and molecules, and following the approach of Panayiotou and Vera (16), we obtain an expression for Ω which is valid outside the critical region of the pure component, i.e.,

$$\Omega = \left(\frac{\delta}{\sigma}\right)^{N_1} \frac{(N_0 + r_1 N_1)!}{N_0! \, N_1!} \left(\frac{(N_0 + N_1 q_1)!}{(N_0 + N_1 r_1)!}\right)^{z/2} \exp\left(\frac{\beta}{2} z N_1 q_1 \, \epsilon_{11} \frac{N_1 q_1}{N_0 + N_1 q_1}\right) \qquad 3)$$

where δ is the number of internal arrangements of a molecule and σ a symmetry factor. Using the following reducing parameters

$$(z/2)\epsilon_{11} = P^* v_H = RT^* \qquad 4)$$

and defining $\underline{V}$, the total volume of the system,

$$\underline{V} = v_H(N_0 + r_1 N_1) \qquad 5)$$

an EOS that defines the pure component is obtained, i.e.,

$$\frac{\tilde{P}}{\tilde{T}} = \ln\left(\frac{\tilde{v}}{\tilde{v}-1}\right) + \frac{z}{2} \ln\left(\frac{\tilde{v} + (q/r) - 1}{\tilde{v}}\right) - \frac{\vartheta^2}{\tilde{T}} \qquad 6)$$

Here ϑ is the effective surface fraction of molecules and the tilde (~) denotes reduced variables. All quantities, except v in the EOS, are reduced by the parameters in Equation 4. The specific volume v, is reduced by v^*, the molecular hard-core volume,

$$v^* = N_1 r_1 v_H \qquad 7)$$

Expressions for the chemical potential of a pure component can also be derived from Equation 3 and standard thermodynamics (17).

Determination of pure component parameters. In order to use the EOS to model real substances one needs to obtain ϵ_{11} and v^*. For a pure component below its critical point, a technique suggested by Joffe et al. (18) was used. This involves the matching of chemical potentials of each component in the liquid and the vapour phases at the vapour pressure of the substance. Also, the actual and predicted saturated liquid densities were matched. The set of equations so obtained was solved by the use of a standard Newton's method to yield the pure component parameters. Values of ϵ_{11} and v^* for ethanol and water at several temperatures are shown in Table 1. In this calculation v_H and z were set to 9.75×10^{-6} m^3 mole^{-1} and 10, respectively (16). The capability of the lattice EOS to fit pure component VLE was found to be quite insensitive to variations in z ($6<z<26$) and v_H (1.0×10^{-7} m^{3}mole$^{-1} < v_H < 1.5 \times 10^{-5}$ m^{3}mole^{-1}).

For a supercritical fluid (SCF) component, the pure component parameters were obtained by fitting P-v data on isotherms (300-380K). Preliminary data for these substances suggest that although the computed v^* is a weak function of temperature, ϵ_{11} is a constant within regression error.

Table 1: Pure Component parameters for ethanol and water at several temperatures ($z=10$, $v_H=9.75 \times 10^{-6}$ m^{3}mole^{-1})

T (K)	Ethanol		Water	
	ϵ_{11}(N-m/mole)	v^*(cm^{3}g^{-1})	ϵ_{11}(N-m/mole)	v^*(cm^{3}g^{-1})
283	1357.59	1.2018	3596.56	0.9602
293	1355.47	1.2016	3516.20	0.9685
303	1314.34	1.2193	3438.16	0.9767
313	1294.18	1.2276	3362.54	0.9588
323	1274.90	1.2358	3289.36	0.9943
333	1256.49	1.2437	3218.53	1.0030
343	1238.87	1.2515	3150.14	1.0123
353	1222.04	1.2589	3083.99	1.0216
363	1205.89	1.2661	3020.02	1.0310
373	1190.43	1.2731	2958.16	1.0406
393	1161.42	1.2864	2840.43	1.0602
413	1134.68	1.2989	2730.13	1.0804
433	1109.83	1.3109	2626.57	1.1011

Discussion. In order to understand qualitatively the behaviour of the lattice EOS, it was examined in the limit of small molecules (q,r $\longrightarrow$ 1). In this case Equation 6 simplifies to the form,

$$\frac{\tilde{P}}{\tilde{T}} = \ln\left(\frac{\tilde{v}}{\tilde{v}-1}\right) - \frac{1}{\tilde{T}\tilde{v}^2} \qquad 8)$$

The first term can be identified with a "hard-sphere" repulsion term, while the second accounts for attractive forces. The second term can be rewritten as,

$$Z_a \equiv \frac{P_a v}{RT} = - \frac{(P^* v^{*2})}{v^2} \equiv - \frac{a}{v^2} \qquad 9)$$

Thus, the attractive term represented in Equation 9 has the same form as the attractive term in the van der Waals EOS (19). On examining the data in Table 1 and computing the parameter a in Equation 9, it was found that for a temperature change of 150K, the change in the value of this parameter was always less than 3%, although the computed values of v^* and ϵ_{11} themselves showed a 7% variation. In the limit of small molecules, therefore, the lattice EOS has a term that approximates the van der Waals type attractive term closely.

The behaviour of the repulsive term of the lattice EOS is more complicated and will not be discussed in detail. At liquid-like densities this repulsion term is a better approximation to the hard spheres repulsion than the van der Waals repulsion term. At gas-like densities, the repulsion term of the lattice model and the van der Waals EOS have the same functional form.

Binary Mixtures

<u>Theory</u>. Consider a mixture of N_0 holes, N_1 molecules of species 1 and N_2 molecules of species 2. Following Panayiotou and Vera (16) the following mixing rules are assumed for the mixture parameters r_M, q_M and v_M^*.

$$r_M = \sum x_i r_i \qquad 10)$$

$$q_M = \sum x_i q_i \qquad 11)$$

$$v_M^* = \sum x_i v^*_i \qquad 12)$$

Lattice coordination numbers (z) and the cell volumes (v_H) for both the pure components and mixture lattices are assumed to have the same value. The partition function for this ensemble can be formulated following Equation 2. It is assumed now that the partition function, far from the binary critical point can be approximated by its largest term. Since molecule segments and holes can distribute themselves non-randomly, the partition function must incorporate terms to account for this effect. The nonrandomness correction Γ_{ij}

allows for distribution of the segments of species i about the segments of species j over the random values of such contacts. It is defined through the equation

$$N_{ij} = N^0_{ij} \; \Gamma_{ij} \qquad \qquad 13)$$

where N_{ij} is the actual number of i-j contacts and N^0_{ij} is the number of i-j contacts in the completely random case. Expressions for the non-randomness correction must be obtained through the solution of the "quasichemical" equations (20). These equations can be solved in a closed analytic form only in the case of a two-component system (including holes). In order to ensure the mathematical tractability of the binary results, it is therefore assumed that holes distribute randomly while molecules do not.

The solution for the quasichemical expressions for the pseudo two-component system yields an expression for the nonrandomness correction Γ_{ij}, which can be represented mathematically as,

$$\Gamma_{ij} = \frac{2}{1 + [1 - 4\bar{\vartheta}_i \bar{\vartheta}_j (1 - g)]^{1/2}} \qquad (i \neq j) \qquad 14)$$

where,

$$g = \exp(\vartheta \Delta \epsilon / kT), \qquad \qquad 15)$$

$$\Delta \epsilon = \epsilon_{11} + \epsilon_{22} - 2\epsilon_{12} \qquad \qquad 16)$$

$\bar{\vartheta}_i$ is the surface fraction of i molecule segments on a hole-free basis and ϑ is the total surface area fraction of molecule segments. Equation 16 immediately suggests a combining rule for ϵ_{12} as a measure of the departure of the mixture from randomness, i.e.,

$$\epsilon_{ij} = \begin{vmatrix} \epsilon_{ii}, & i=j \\ 0.5 \, (\epsilon_{ii} + \epsilon_{jj}) \, (1 - k_{ij}), & i \neq j \end{vmatrix} \qquad 17)$$

The mixing rule for ϵ arises naturally through the formulation of the canonical partition function, i.e.,

$$\epsilon_M = \sum_i \sum_j \bar{\vartheta}_i \bar{\vartheta}_j \Gamma_{ij} \epsilon_{ij} \qquad \qquad 18)$$

An EOS for the mixture and chemical potentials for component i in the mixture can now be derived using standard thermodynamics.

$$\frac{\tilde{P}}{\tilde{T}} = \ln\left(\frac{\tilde{v}}{\tilde{v}-1}\right) + \frac{z}{2}\ln\left[\frac{\tilde{v}+(q/r)-1}{\tilde{v}}\right] - \frac{\vartheta^2}{\tilde{T}} +$$

$$+ 0.5zg\frac{\Delta\epsilon}{kT}\frac{\Delta\epsilon}{\epsilon^*}\frac{\left[\bar{\vartheta}_1\bar{\vartheta}_2\vartheta\,\Gamma_{12}\right]^2}{\left[1-4\bar{\vartheta}_1\bar{\vartheta}_2(1-g)\right]^{(1/2)}}\ln\left[\frac{\Gamma_{11}\,\Gamma_{22}}{\Gamma_{12}^2}\right] \qquad 19)$$

$$-\frac{\mu_i}{kT} = \lambda(T) + \ln q_i - \ln \bar{\vartheta}_i\vartheta + r_i(0.5z-1)\ln\left[\frac{\tilde{v}+(q/r)-1}{\tilde{v}}\right]$$

$$-\frac{q_i\vartheta^2}{\tilde{T}} + \frac{q\,\vartheta_i}{\tilde{T}\,\epsilon^*}\Big\{ 2\bar{\vartheta}_i\epsilon_{ii}\Gamma_{ii} + 2\bar{\vartheta}_j\epsilon_{ij}\Gamma_{ij} + \bar{\vartheta}_1\epsilon_{11} + \bar{\vartheta}_2\epsilon_{22} - \bar{\vartheta}_1\epsilon_{11}\Gamma_{1i}$$

$$- \bar{\vartheta}_2\epsilon_{22}\,\Gamma_{2i}\Big\} + \left(\frac{\vartheta\,\Delta\epsilon}{\tilde{T}\,\epsilon^*} - \ln\frac{\Gamma_{11}\,\Gamma_{22}}{\Gamma_{12}^2}\right) \times \frac{\bar{\vartheta}_1\bar{\vartheta}_2\,\Gamma_{12}^2\,q_i}{\left[1-4\bar{\vartheta}_1\bar{\vartheta}_2(1-g)\right]^{(1/2)}}$$

$$[\bar{\vartheta}_j(\bar{\vartheta}_j - \bar{\vartheta}_i)(1-g) - \bar{\vartheta}_1\bar{\vartheta}_2 g\frac{\Delta\epsilon}{kT}\vartheta(1-\vartheta)] - \ln\Gamma_{12} - 0.5zq_i[\bar{\vartheta}_1(1-\Gamma_{1i})$$

$$\ln\frac{\bar{\vartheta}_1\,\Gamma_{11}}{\bar{\vartheta}_2\,\Gamma_{12}} + \bar{\vartheta}_2(1-\Gamma_{i2})\ln\frac{\bar{\vartheta}_2\,\Gamma_{22}}{\bar{\vartheta}_1\,\Gamma_{12}}] + 0.5zq_i[\ln\frac{\bar{\vartheta}_1}{\bar{\vartheta}_2} - \Gamma_{11}\ln\frac{\bar{\vartheta}_1\Gamma_{11}}{\bar{\vartheta}_2\Gamma_{12}}]$$

$$\bar{\vartheta}_1(2\delta_{1i}-\bar{\vartheta}_1) + 0.5zq_i[\ln\frac{\bar{\vartheta}_2}{\bar{\vartheta}_1} - \Gamma_{22}\ln\frac{\bar{\vartheta}_2\,\Gamma_{22}}{\bar{\vartheta}_1\,\Gamma_{12}}]\bar{\vartheta}_2(2\delta_{2i}-\bar{\vartheta}_2) \qquad 20)$$

where μ_i represents the chemical potential of component i in a binary mixture, $j=3-i$ and $\lambda(T)$ is some universal temperature function. δ_{ij} is the Kronecker-delta function. Parameters used for obtaining these equations in a dimensionless form are defined in manner analogous to Equation 4.

The mixture EOS [Equation 19] has the three terms that are present in the pure component EOS. Also, it has an additional term which accounts for the non-randomness corrections that have been incorporated into the partition function expression. This last term, for all cases tested, is always at least 4 orders of magnitude smaller than the other three terms and can effectively be neglected. However, it is retained for the sake of mathematical consistency.

Results and Discussion

The expressions derived for the EOS and the chemical potential of component i in a binary mixture were used to model the phase equilibria of binary mixtures. A set of non-linear equations was obtained and solved by the use of a Newton's method.

Mixtures of small molecules (acetone-benzene, ethanol-water) were considered first. In Figures 1 and 2, a comparison is made between the predicted and experimental low-pressure VLE data ([21],[22]) for these systems. An excellent fit to the data is obtained in both cases, with the use of one apparently temperature <u>independent</u> parameter (k_{ij}) per binary.

The interesting aspect of this modelling is the temperature independence of k_{ij}. It was shown earlier that, if the pure components were small molecules, the lattice EOS has an attractive term with an essentially temperature independent $\underline{a}$. Extending this argument to mixtures results in the prediction of the temperature independence of k_{ij} for binary mixtures of small molecules.

In examining the sensitivity of the model to the assumed value of z, it was found for the ethanol-water system that the goodness of fit of the model was insensitive to the z value in the vicinity of z=10. For large values of z (z>15), however, it was found that the model was incapable of even qualitatively prediciting mixture VLE behaviour.

The applicability of the lattice EOS in the modelling of the VLE of mixtures of molecules of different sizes was examined next. The results for the H_2S-$\underline{n}$-heptane system at 310K and 352K are shown in Figure 3 ([23]). For the temperatures modelled, it is seen that there is a good agreement between the fitted and the experimental data, again with the use of one temperature independent binary interaction parameter.

The lattice model thus provides the capability to obtain good, quantitative fits to experimental VLE data for binary mixtures of molecules below their critical point. Its value lies in the fact that it performs equally well regardless of the size difference between the component molecules.

The model was then extended to the phase equilibrium modelling of solid-supercritical fluid (SCF) binaries. These mixtures are compositionally highly assymmetric, since the solid is crystalline and essentially pure. The phases in equilibrium are thus very different in composition. In Figure 4, the model behaviour is compared to experimental data for the naphthalene-carbon dioxide binary at 308 and 318K. Outside the critical region, the lattice EOS provides a good fit to the measured data ([24]). In the critical region, however, at one pressure, the equation of state predicts the presence of a three phase region. This is the cause for the discontinuity of the curves in the region of the sharp rise in solubility. Outside of the critical region (approximately 7% away from T_c or P_c) the model

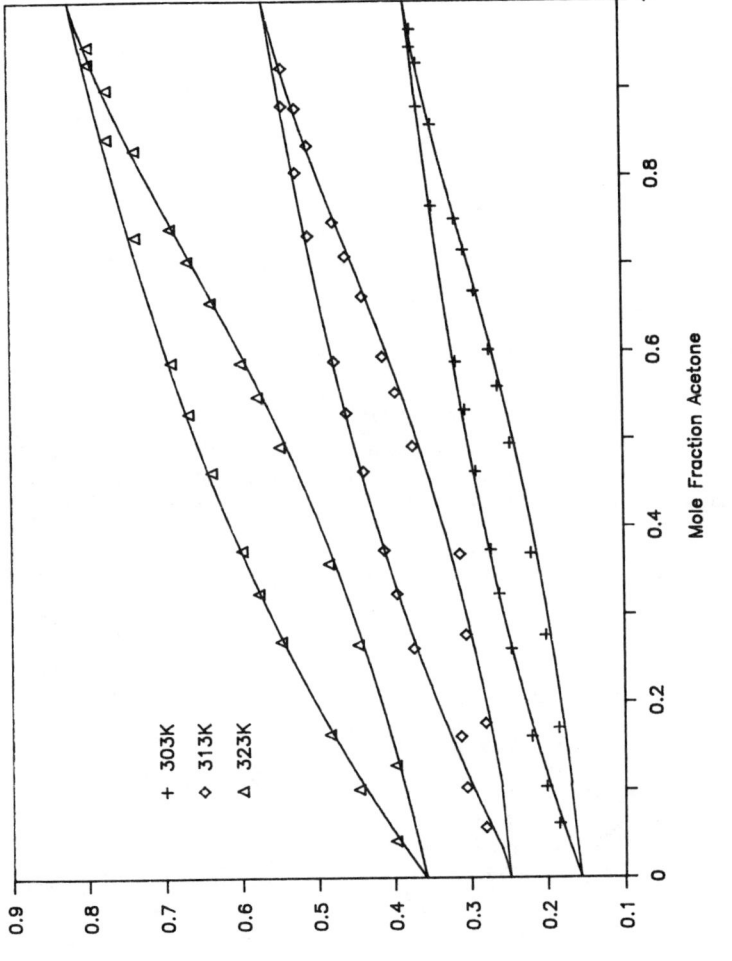

Figure 1. Comparison of lattice model prediction and experimental data of Weisphart (21) for the acetone-benzene binary at 303, 313 and 323 K ($z=10$, $v_H=9.75 \times 10^{-6}$ m^{3}mole^{-1}, $k_{ij}=0.022$).

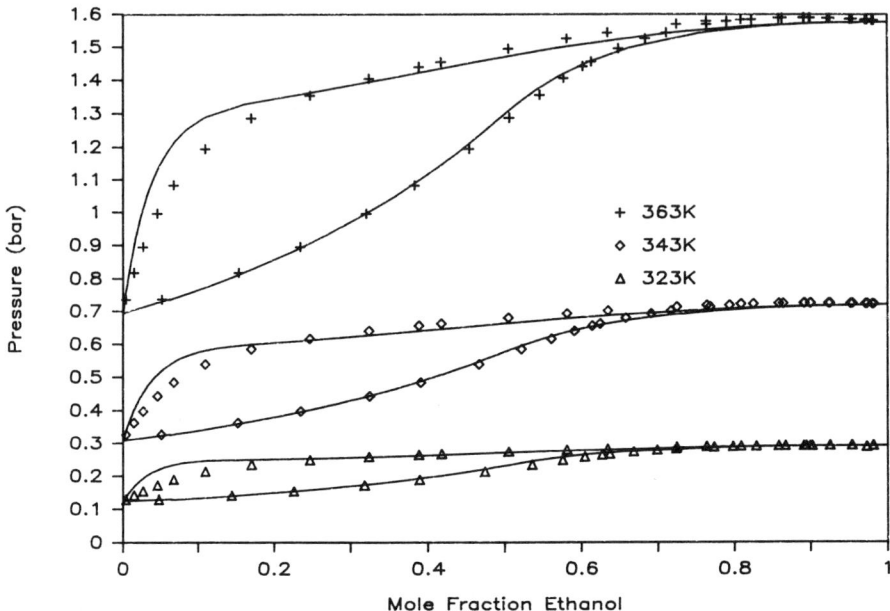

Figure 2. Comparison of the experimental data of Pemberton and Mash (22) for the ethanol-water binary at 323, 343, and 363 K with the lattice model predictions with z = 10, v_H = 9.75 x 10^{-6} $m^3 mole^{-1}$ and k_{ij} = 0.085.

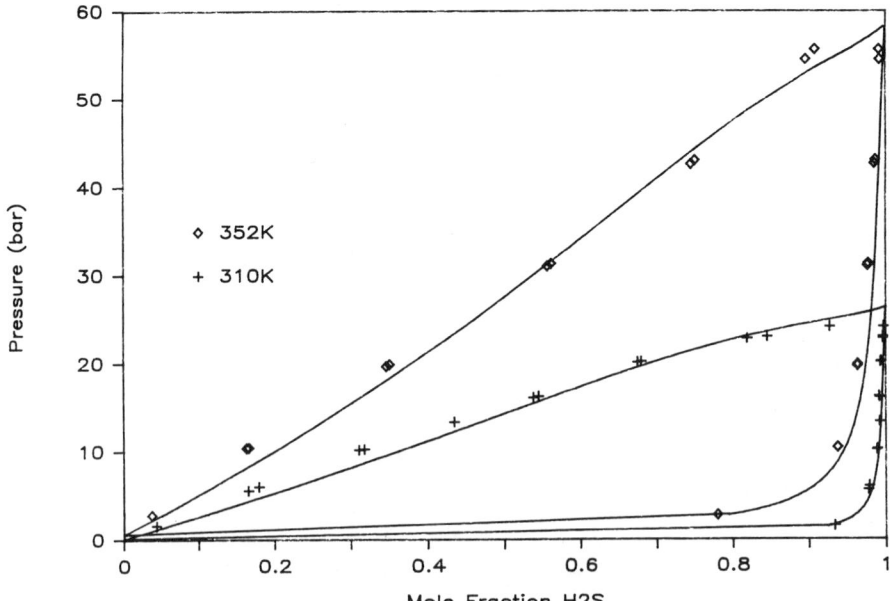

Figure 3. Comparison of the lattice model predictions and experimental data of Ng et al. (23) for the H_2S-n-heptane system at 310 and 352 K (z = 10, v_H = 9.75 x 10^{-6} $m^3 mole^{-1}$ and k_{ij} = 0.09).

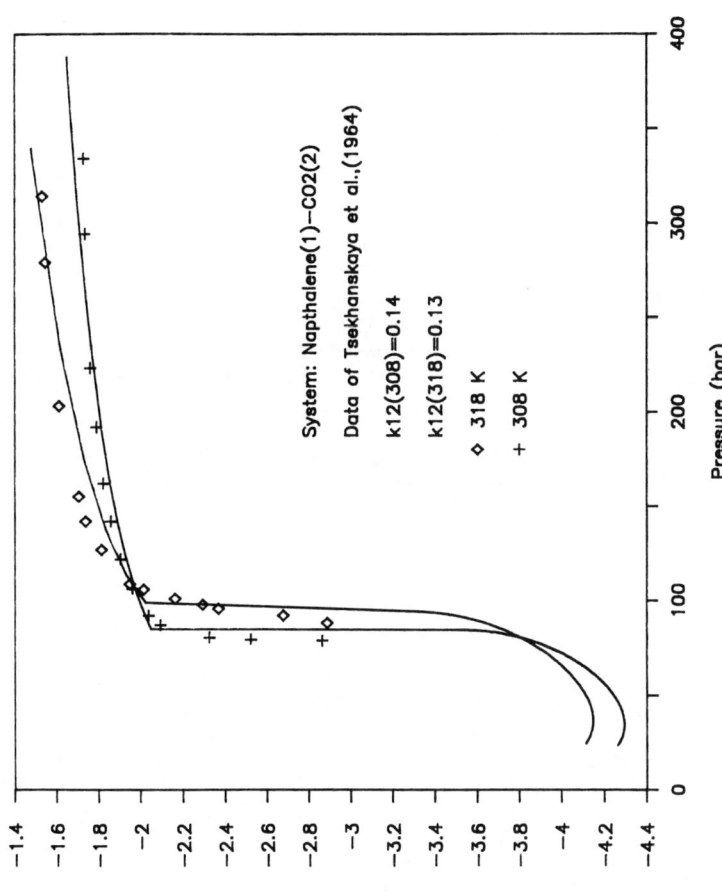

Figure 4. Comparison of the model predictions for the CO_2-naphthalene binary at 308 and 318 K with the experimental data of Tsekhanskaya et al. (24) ($z = 10$, $v_H^* = 9.75 \times 10^{-6}$ m^{3}mole^{-1}, $k_{ij}(308\ K) = 0.14$, and $k_{ij}(318\ K) = 0.135$).

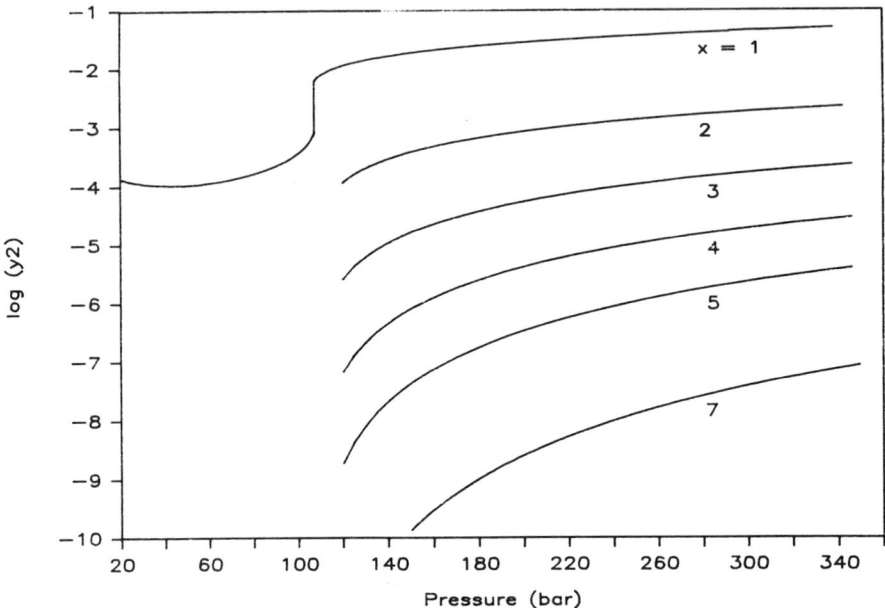

Figure 5. Lattice model predictions for the equilibrium fluid phase composition for a CO_2-polymer system at 328 K. Molecular weight of a monomeric unit is 100, while the degree of polymerization, ξ, varies between 1 and 7 ($z = 10$, $v_H = 9.75 \times 10^{-6}$ $m^3 mole^{-1}$, $k_{ij} = 0.10$).

predicts results within the accuracy of the experimental measurements.

The model behaviour in the critical region is not satisfactory. The main reason for this lies in one of the fundamental assumptions made in deriving the model; the one where the canonical partition function (Equation 2) is approximated by the largest term in the summation. Clearly, in the critical region this assumption is invalid, since there is no one dominating term in the summation for the partition function. The model must, therefore, be used with the full understanding of its shortcomings in the critical region.

In order to test the applicability of the model to polymer-SCF systems, a hypothetical system of CO_2 and a monodisperse ξ-mer with a monomeric unit molecular weight of 100 was simulated. Pure component parameters for the polymer, polystyrene, were obtained from Panayiotou and Vera (16). Constant values of k_{ij} were used for the polymer system, where the degree of polymerization, ξ, varied between 1 and 7. It was assumed that all chains had the same ϵ, and v^* scaled as the molecular weight of the chain. Figure 5 shows the results of the predicted mole fraction of the ξ-mer in the SCF phase.

The model simulates an experimentally observed trend (25) that the solubility of chains in a SCF shows a strong inverse dependence on the molecular mass of the polymer. Figure 5 shows that changing the molecular weight of the chain molecule from 100 to 700 causes a reduction in solubility of nearly 6 orders of magnitude. The model also shows that all the solubility plots tend to flatten out around 300 bar, as observed in experiments (25). Classical EOS like a modified cubic EOS (26), when applied to such systems, produce solubility curves which tend to show a sharp maximum around 200 bar. For polymer-SCF systems, therefore, the lattice EOS is believed to be superior to modified cubic EOS.

Conclusions

A statistical-mechanics based model for mixtures of molecules of disparate sizes has been presented in this paper. Results obtained to date demonstrate that the lattice EOS is probably more suited for modelling polymer-SCF equilibria than a modified cubic EOS, while for the other systems, outside the critical region, it performs as well as classically employed techniques.

The temperature independence of the a parameter [Equation 9] and the binary interaction parameter (k_{ij}) for systems of small molecules could lead to predictive models that merit closer examination.

Acknowledgement

The authors gratefully acknowledge support from the National Science Foundation, Division of Chemical, Biochemical and Thermal Engineering under Grant No. CBT 85-09945.

Literature Cited

1. Peng D.Y., Robinson D.R., "A new two constant equation of state", Ind. Eng Chem. Fundam., 1976, **15**(1),59.
2. Soave G., "Equilibrium constants from a modified Redlich-Kwong equation of state", Chem. Eng. Sci., 1972, **27**, 1197.
3. Carnahan N.F., Starling K.E., "Intermolecular forces and the equation of state for fluids", AIChEJ, 1972, **18**(6),1184.
4. Dieters U.K., "A new semi empirical equation of state for fluids I", Chem. Eng. Sci., 1981, **36**, 1139.
5. Donohue M.D., Prausnitz J.M., "Perturbed hard chain theory for fluid mixtures: Thermodynamic properties for mixtures in natural gas and petroleum technology", AIChEJ, 1975, **24**(5), 850.
6. Vimalchand, Donohue M.D., "Thermodynamics of quadropolar molecules: The Perturbed Anisotropic Chain Theory", Ind. Eng. Chem. Fundam, 1985, **24**, 849.
7. Vezetti D.J., "Solubilities of solids in supercritical fluids", J. Chem Phys.,1980, **77**, 1512.
8. Vezetti D.J., "Solubilities of solids in supercritical gases II. Extension to molecules of differing sizes", J. Chem. Phys., 1984, **80** (2),868-71.

9. Simha R., Somcynysky T., "On the statistical thermodynamics of spherical and chain molecule fluids", Macromolecules, 1969, 2(4), 342-350.
10. Jain R.K., Simha R., "On the equation of state of argon and organic liquids", J. Chem. Phys., 1980, 72(9), 4909-12.
11. Hirshfelder J.O., Curtiss C.O., Bird R.B., "Molecular theory of gases and liquids", John Wiley and Sons.,pp275-320, 1954
12. Kleintjens L.A., Koningsveld R., "Mean-Field Lattice-Gas Description of the System CO_2/H_2O", Sepn. Sci. and Tech., 1982, 17(1),215-233.
13. Kleintjens L.A., "Mean-Field Lattice Gas Description of Vapour-Liquid and Supercritical Equilibria", Fluid Ph. Equil., 1983, 10, 183-190.
14. Sanchez I.C., R.H. Lacombe, "An Elementary Molecular Theory of Classical Fluids. Pure Fluids", J. Phys. Chem., 1976, 80(21),2352-2362.
15. Sanchez I.C., R.H.Lacombe., "An Elementary Equation of State for Polymer Liquids", J. Polym. Sci. Pol. Lett. Ed., 1977, 15,71-75.
16. Panayiotou C., Vera J.H., "Statistical thermodynamics of r-mer fluids and their mixtures", Polymer Journal, 1982, 14(9),681-692.
17. Modell M., Reid R.C., "Thermodynamics and its applications",2 ed., Prentice Hall, 1983.
18. Joffe J., Schroeder G.M., Zudkevitch D., "Vapor-liquid Equilibria with the Redlich-Kwong Equation of State", AIChEJ, 1970, 48,261-266.
19. van der Waals J.D., Doctoral thesis, Leiden,1873.
20. Guggenheim E.A., "Mixtures", Clarendon Press, Oxford, 1954.
21. Weisphart J., "Thermodynamic equilibria of boiling mixtures", Springer-Verlag, Berlin,1975.
22. Pemberton R.C., Mash C.J., "Thermodynamic properties of aqueous nonelectrolyte mixtures II. Vapour pressures and excess Gibbs energies for water + ethanol at 303.15 K to 363.15 K determined by an accurate static method"., J. Chem. Therm., 1978, 10, 867-88.
23. Ng H-J., Kalra H., Robinson D.B., Kubota H., "Equilibrium phase properties of the toluene-hydrogen sulfide and n-heptane-hydrogen sulfide binary systems", J. Chem. Eng. Data., 1980, 25, 51-55.
24. Tsekhanskaya Yu. V., Iomtev M.B., Muskina E.V., "Solubility of naphthalene in ethylene and carbon dioxide under pressure", Russ. J. Phy. Chem., 1964, 38(9), 1173.
25. Schroeder E., Arndt K-F., "Loeslichkeitverhalten von Makromolekuelen in komprimierten Gasen", Faserforschung und Textiltechnik, 1976, 247(3), 135.
26. Panagiotopoulos A.Z., Reid R.C., "A new mixing rule for cubic equations of state for highly polar, asymmetric systems", in K.C.Chao and R.L.Robinson (Ed.), "Equations of State- Theories and Applications", ACS Symposium Series, 300,571-582(1986)

RECEIVED August 27, 1986

Chapter 9

Van der Waals Mixing Rules for Cubic Equations of State

E. H. Benmekki, T. Y. Kwak, and G. A. Mansoori

Department of Chemical Engineering, University of Illinois, Box 4348, Chicago, IL 60680

A new concept for the development of mixing rules for cubic equations of state consistent with statistical mechanical theory of the van der Waals mixing rules is introduced. Utility of this concept is illustrated by its application to the Redlich-Kwong (RK) and Peng-Robinson (PR) equations of state. The resulting mixing rules for the Redlich-Kwong and the Peng-Robinson equations of state are tested through prediction of solubility of heavy solids in supercritical fluids and prediction of phase behavior of binary mixtures of hydrocarbons and nonhydrocarbons.

There has been extensive progress made in recent years in research towards the development of analytic statistical mechanical equations of state applicable for process design calculations [1,2]. However, cubic equations of state are still widely used in chemical engineering practice for calculation and prediction of properties of fluids and fluid mixtures [3]. These equations of state are generally modifications of the van der Waals equation of state [4,5],

$$P = \frac{RT}{v - b} - \frac{a}{v^2} \qquad (1)$$

which was proposed by van der Waals [4] in 1873. According to van der Waals, for the extension of this equation to mixtures, it is necessary to replace a and b with the following composition-dependent expressions:

$$a = \sum_{i}^{n} \sum_{j}^{n} x_i x_j a_{ij} \qquad (2)$$

$$b = \sum_{i}^{n} \sum_{j}^{n} x_i x_j b_{ij} \qquad (3)$$

Equations 2 and 3 are called the van der Waals mixing rules. In these equations, a_{ij} and b_{ij} ($i=j$) are parameters corresponding to pure component (i) while a_{ij} and b_{ij} ($i \neq j$) are called the unlike-interaction parameters. It is customary to relate the unlike-interaction parameters to the pure-component parameters by the following expressions:

$$a_{ij} = (1 - k_{ij})(a_{ii} a_{jj})^{1/2} \qquad (4)$$

$$b_{ij} = (b_{ii} + b_{jj})/2 \qquad (5)$$

In Equation 4, k_{ij} is a fitting parameter which is known as the coupling parameter. With Equation 5 placed in Equation 3, the expression for b will reduce to the following one-summation form:

$$b = \sum_{i}^{n} x_i b_{ii} \qquad (3.1)$$

The Redlich-Kwong equation of state (6),

$$P = \frac{RT}{v-b} - \frac{a}{T^{1/2} v(v-b)} \qquad (6)$$

and the Peng-Robinson equation of state (7),

$$P = \frac{RT}{v-b} - \frac{a(T)}{v(v+b)+b(v-b)} \qquad (7)$$

$$a(T) = a(T_c) \{1 + \kappa(1 - T_r^{1/2})\}^2 \qquad (8)$$

$$a(T_c) = 0.45724 \, R^2 \, T_c^2 / P_c \qquad (8.1)$$

$$\kappa = 0.37464 + 1.54226\omega - 0.26992\omega^2 \qquad (9)$$

$$b = 0.0778 \, RT_c/P_c \qquad (10)$$

are widely used for thermodynamic property calculations.

The Theory of the Van Der Waals Mixing Rules

Leland and Co-workers (8-10) have been able to re-derive the van der Waals mixing rules with the use of statistical mechanical theory of radial distribution functions. According to these investigators, for a fluid mixture with a pair intermolecular potential energy function,

$$u_{ij}(r) = \epsilon_{ij} f(r/\sigma_{ij}) \qquad (11)$$

the following mixing rules will be derived:

$$\sigma^3 = \sum_i^n \sum_j^n x_i x_j \sigma_{ij}^3 \qquad (12)$$

$$\epsilon\sigma^3 = \sum_i^n \sum_j^n x_i x_j \epsilon_{ij}\sigma_{ij}^3 \qquad (13)$$

In these equations, ϵ_{ij} is the interaction energy parameter between molecule i and j while σ_{ij} is the intermolecular interaction distance between the two molecules. Knowing that coefficients a and b of the van der Waals equation of state are proportional to ϵ and σ according to the following expressions:

$$a = 1.1250 \, RT_c v_c \propto N_0 \, \epsilon\sigma^3 \qquad (14)$$

$$b = 0.3333 \, v_c \propto N_0 \, \sigma^3 \qquad (15)$$

We can see that Equations 12 and 13 are identical with Equations 3 and 2, respectively. Statistical mechanical arguments which are used in deriving Equations 12 and 13 dictate the following guidelines in using the van der Waals mixing rules:

(1) The van der Waals mixing rules are for constants of an equation of state.

(2) Equation 12 is a mixing rule for (molecular length)^3, and Equation 13 is a mixing rule for (molecular length)^3.(molecular energy). It happens that b and a of the van der Waals equation of state are proportional to (molecular length)^3 and (molecular length)^3.(molecular energy), respectively, and as a result, these mixing rules are used in the form which was proposed by van der Waals.

(3) Knowing that σ_{ij} (for $i \neq j$), the unlike-interaction diameters, for spherical molecules is equal to

$$\sigma_{ij} = (\sigma_{ii} + \sigma_{jj})/2 \qquad (16)$$

This will make the expression for b_{ij} for spherical molecules to be

$$b_{ij} = [\,(b_{ii}^{1/3} + b_{jj}^{1/3})/2\,]^3 \qquad (17)$$

Then for non-spherical molecules expression for b_{ij} will be

$$b_{ij} = (1-l_{ij})[\,(b_{ii}^{1/3} + b_{jj}^{1/3})\,]^3 \qquad (18)$$

With the use of these guidelines, we now derive the van der Waals mixing rules for the Redlich-Kwong and the Peng-Robinson equations of state. Similar procedure can be used for deriving the van der Waals mixing rules for other equations of state.

Mixing Rules for the Redlich-Kwong Equation of State. The Redlich-Kwong equation of state, Equation 6, can be written in the following form:

$$Z = \frac{PV}{RT} = \frac{v}{v-b} - \frac{a}{RT^{1.5}(v+b)} \qquad (19)$$

In this equation of state, b has the dimension of (molecular length)^3,

$$b = 0.26 v_c \propto N_0 \sigma^3$$

Then the mixing rule for b will be the same as the one for the first van der Waals mixing rules, Equation 3. However, mixing rule for a will be different from the second van der Waals mixing rule, Equation 2. Parameter a appearing in the Redlich-Kwong equation of state has dimension of $R^{-\frac{1}{2}}$. (molecular energy)$^{3/2}$ (molecular length)^3, that is

$$(a = 1.2828 RT_c^{1.5} v_c \propto N_0 (\epsilon/k)^{1.5} \sigma^3)$$

As a result the second van der Waals mixing rules, Equation 13, cannot be used directly for parameter a of the Redlich-Kwong equation of state. However, since $[(R^{\frac{1}{2}}ab^{\frac{1}{2}})]^{2/3}$ has the dimension of (molecular energy).(molecular length)^3, the second van der Waals mixing rules, Equation 13, can be written for this term. Finally the van der Waals mixing rules for the Redlich-Kwong equation of state will be in the following form:

$$a = \{\sum_i^n \sum_j^n x_i x_j a_{ij}^{2/3} b_{ij}^{1/3}\}^{1.5} / (\sum_i^n \sum_j^n x_i x_j b_{ij})^{1/2} \qquad (20)$$

$$b = \sum_i^n \sum_j^n x_i x_j b_{ij} \qquad (3)$$

These mixing rules, when joined with the Redlich-Kwong equation of state, will constitute the Redlich-Kwong equation of state for mixtures that is consistent with the statistical mechanical basis of the van der Waals mixing rules.

Mixing Rules for the Peng-Robinson Equation of State. In order to separate thermodynamic variables from constants of the Peng-Robinson equation of state, we will insert Equations 8 and 9 in Equation 7 and we will write it in the following form:

$$Z = \frac{v}{v-b} - \frac{c/RT + d - 2\sqrt{(cd/RT)}}{(v+b) + (b/v)(v-b)} \qquad (21)$$

where $c = a(T_c)(1 + \kappa)^2$ and $d = a(T_c) \kappa^2 / RT_c$

This form of the Peng-Robinson equation of state suggests that there exist three independent constants which are b, c, and d. Now, following the prescribed guidelines for the van der Waals mixing rules, mixing rules for b, c, and d of the Peng-Robinson equation of state will be

$$c = \sum_i^n \sum_j^n x_i x_j c_{ij} \qquad (22)$$

$$b = \sum_i^n \sum_j^n x_i x_j b_{ij} \qquad (23)$$

$$d = \sum_i^n \sum_j^n x_i x_j d_{ij} \qquad (24)$$

with the following combining rules:

$$c_{ij} = (1-k_{ij})(c_{ii} c_{jj})^{1/2} \qquad (25)$$

$$b_{ij} = (1-l_{ij})\{(b_{ii}^{1/3} + b_{jj}^{1/3})/2\}^3 \qquad (26)$$

$$d_{ij} = (1-m_{ij})\{(d_{ii}^{1/3} + d_{jj}^{1/3})/2\}^3 \qquad (27)$$

Applications for Supercritical Fluid Extraction Modeling

A serious test of mixture equations of state is shown to be their application for prediction of solubility of solutes in supercritical fluids (11). In the present report, we apply the Redlich-Kwong and the Peng-Robinson equations of state for supercritical fluid extraction of solids and study the effect of choosing different mixing rules on prediction of solubility of solids in supercritical fluids -Figures 1-5.

Thermodynamic Model. Solubility of a condensed phase, y_2, in a vapor phase at supercritical conditions (12) can be defined as:

$$y_2 = (P_2^{sat}/P)(1/\phi_2) \phi_2^{sat} \text{Exp}\{ \int_{P_2^{sat}}^{P} (v_2^{solid}/RT) dP\} \qquad (28)$$

where ϕ_2^{sat} is the fugacity coefficient at pressure P. Provided we assume v_2^{solid} is independent of pressure and for small values of P_2^{sat} the above expression will be converted to the following form:

$$y_2 = (P_2^{sat}/P)(1/\phi_2) \text{Exp}\{v_2^{solid}(P-P_2^{sat})/RT\} \qquad (29)$$

In order to calculate solubility from Equation 29 we need to choose an expression for the fugacity coefficient. Generally for calculation of fugacity coefficient an equation of state with appropriate mixing rules is used (12) in the following expression:

$$RT \ln \phi_i = \int_V^\infty [(\partial P/\partial n_i)_{T,V,n_{j \ne i}} - (RT/V)] dV - RT \ln Z \qquad (30)$$

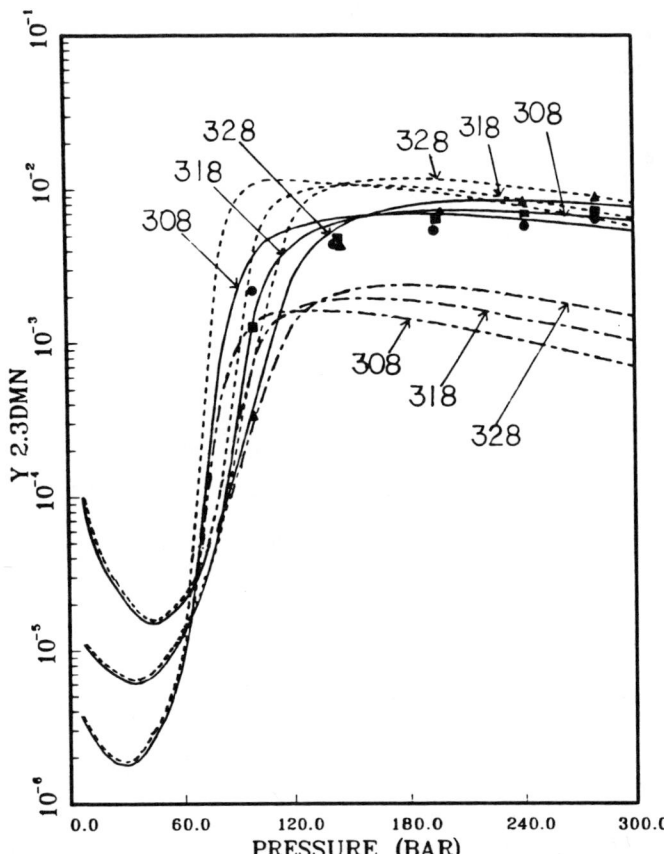

Figure 1. Solubility of 2,3 dimethyl naphthalene (DMN) in supercritical carbon dioxide using the Redlich-Kwong equation of state. —-—, original mixing rules with $k_{ij}=0$; - - -, original mixing rules with fitted k_{ij}; ——, new mixing rules with $k_{ij}=0$; •, ■ and ▲, experimental data (13).

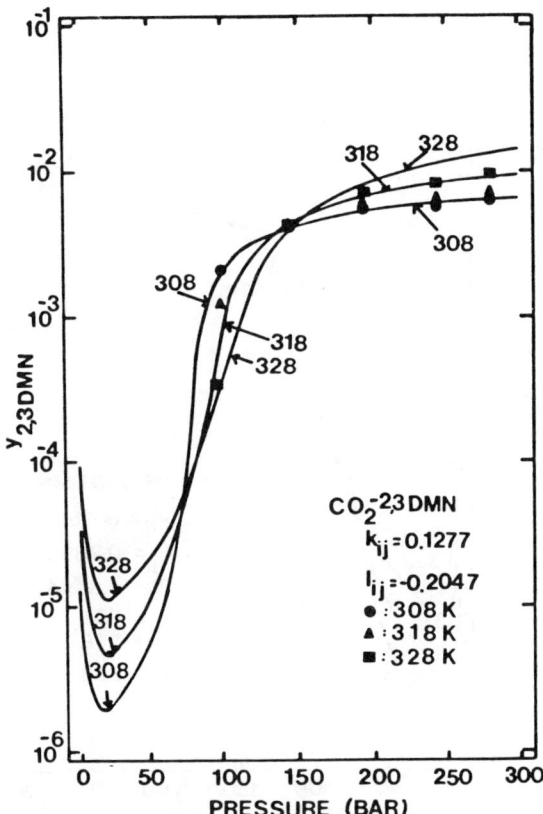

Figure 2. Solubility of 2,3 dimethyl naphthalene (DMN) in supercritical carbon dioxide. ——, Peng-Robinson equation of state with the original mixing rules; ●, ■ and ▲, experimental data (13).

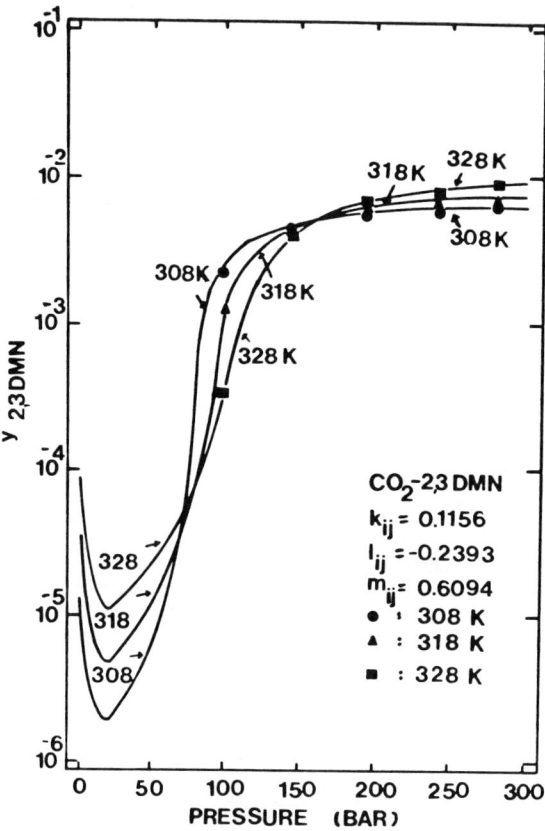

Figure 3. Solubility of 2,3 dimethyl naphtalene (DMN) in supercritical carbon dioxide. ———, Peng-Robinson equation of state with the new mixing rules; •, ■ and ▲, experimental data (13).

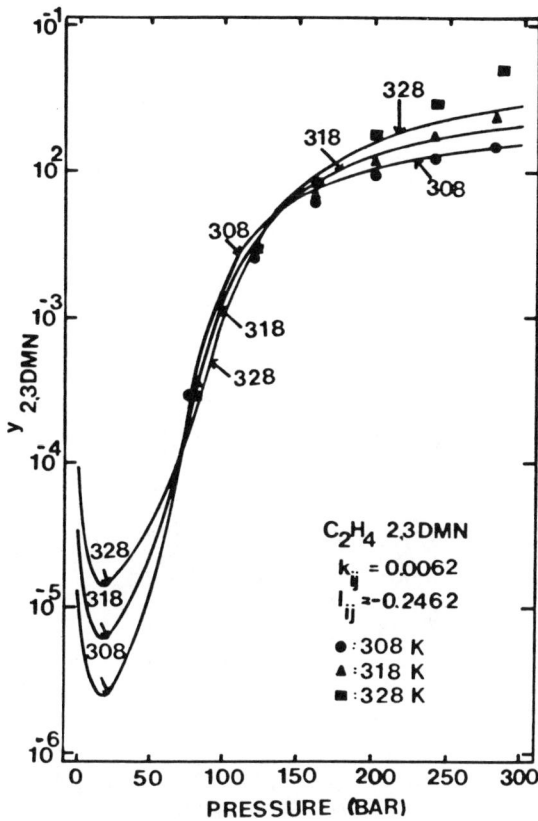

Figure 4. Solubility of 2,3 dimethyl naphtalene (DMN) in supercritical ethylene. ———, Peng-Robinson equation of state with the original mixing rules; •, ■ and ▲, experimental data (13).

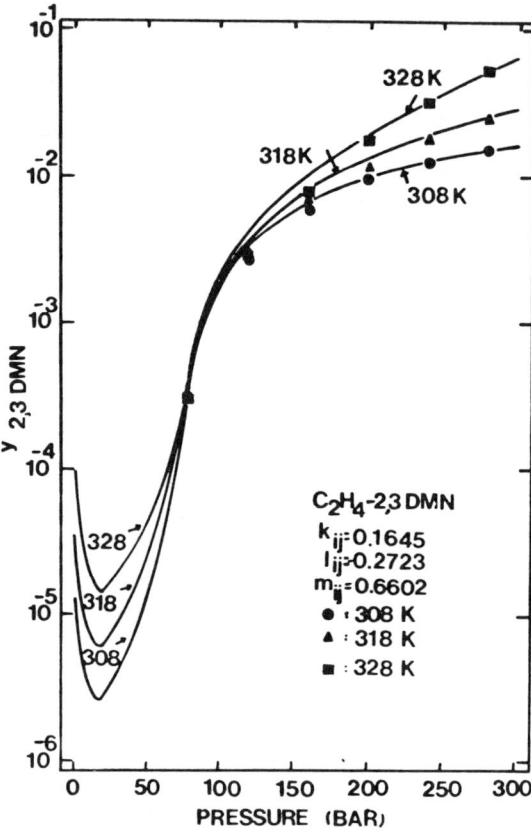

Figure 5. Solubility of 2,3 Dimethyl naphtalene (DMN) in supercritical ethylene. ——, Peng-Robinson equation of state with the new mixing rules; •, ■ and ▲, experimental data (13).

Applications for Vapor-Liquid Equilibrium calculations

When applying an equation of state to both vapor and liquid phases, the vapor-liquid equilibrium predictions depend on the accuracy of the equation of state used and, for multicomponent systems, on the mixing rules. Attention will be given to binary mixtures of hydrocarbons and the technically important nonhydrocarbons such as hydrogen sulfide and carbon dioxide -Figures 6-7.

Thermodynamic Model. In the equilibrium state, the intensive properties -temperature, pressure and chemical potentials of each component- are constant in the overall system. Since the fugacities are functions of temperature, pressure and compositions, the equilibrium condition

$$f_i^V(T,P,\{y\}) = f_i^L(T,P,\{x\}) \quad i=1,2,\ldots,n \quad (31)$$

can be expressed by

$$y_i \phi_i^V = x_i \phi_i^L \quad i=1,2,\ldots,n \quad (32)$$

The expression for the fugacity coefficient ϕ_i depends on the equation of state that is used and is the same for the vapor and liquid phases. In calculating the mixture properties with the Peng-Robinson equation of state we have used the following combining rule:

$$c_{ij} = (1 - k_{ij})[c_{ii} \cdot c_{jj}/b_{ii} \cdot b_{jj}]^{1/2} b_{ij} \quad (33)$$

A three parameter search routine was used to evaluate the binary interaction parameters which minimize the following objective function:

$$OF = \sum_{i=1}^{M} \left[\frac{P(exp) - P(cal)}{P(exp)}\right]_i^2 \quad (34)$$

where M is the number of experimental points considered.

Acknowledgments

This research is supported by the National Science Foundation Grant CPE - 8306808.

Legend of Symbols

a, b, c, d	: equation of state parameters
f	: fugacity
k, l, m	: binary interaction parameters
n	: number of components
N_o	: Avogadro number
OF	: objective function to be minimized
P	: pressure
T	: temperature

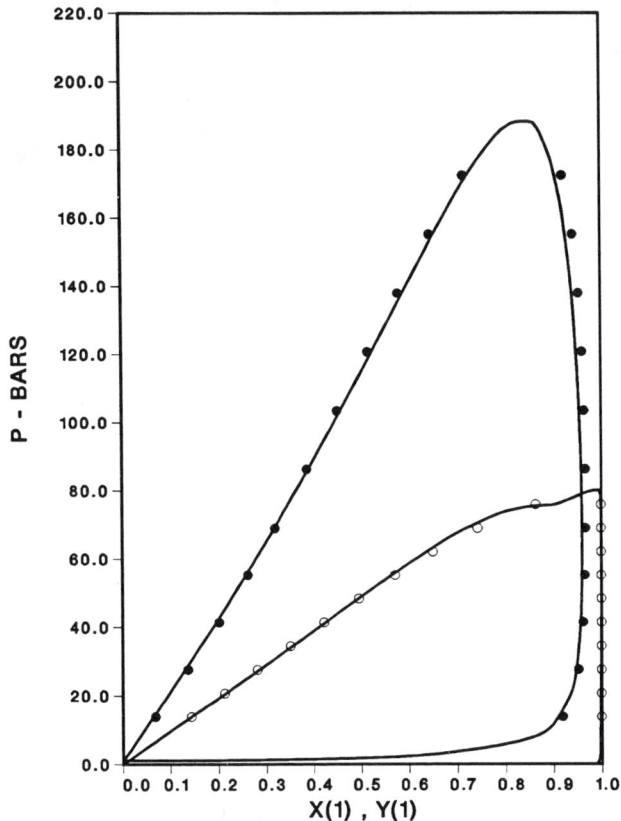

Figure 6. Pressure-composition diagram of carbon dioxide and n-decane. ——, Peng-Robinson equation of state with the new mixing rules and three fitted parameters (k_{ij}=0.4290, l_{ij}=0.3658 and m_{ij}=-0.3217); •(444.26 K), o (310.93 K), experimental data (16).

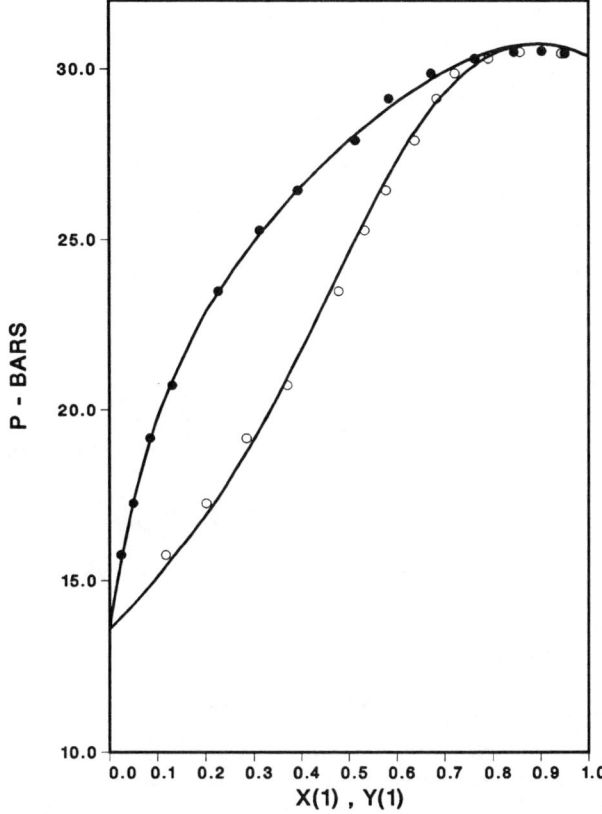

Figure 7. pressure-composition diagram of ethane and hydrogen sulfide at 283.15 K. ——, Peng-Robinson equation of state with the new mixing rules and three fitted parameters (k_{ij}=-0.2060, l_{ij}=-0.6119 and m_{ij}=0.1365); • and o experimental data (17).

u	:intermolecular potential function
v	:molar volume
x	:mole fraction
y	:mole fraction in the vapor phase
Z	:compressibility factor
ϕ	:fugacity coefficient
σ	:intermolecular distance parameter
ϵ	:interaction energy parameter
ω	:acentric factor
κ	:a function of the acentric factor

Subscripts

c	:critical property
i, j	:component identification
2	:solute

Literature Cited

1. Alem, A. H.; Mansoori, G. A. AIChE J. 1984, 30, 468.
2. Nauman, K. H.; Leland, T. W. Fluid Phase Equilibria 1984, 1, 18.
3. Renon, H., Ed.; Fluid Phase Equilibria 1983, 13.
4. Van der Waals, J. D. Ph.D. Thesis, Leiden, 1873.
5. Rowlinson, J. S.; Swinton, F. L. In "Liquid and Liquid Mixtures"; 3rd Ed.; Butterworths: Wolborn, Mass., 1982.
6. Redlich, O.; Kwong, J. N. S. Chem. Rev. 1949, 44, 233.
7. Peng, D. Y.; Robinson, D. B. Ind. Eng. Chem. Fund. 1976, 5, 59.
8. Leland, T. W.; Chappelear, P. S. Ind. Eng. Chem. 1968, 15, 60.
9. Leland, T. W.; Rowlinson, J. S.; Sather, G. A. Trans. Faraday Soc. 1968, 64, 1447.
10. Leland, T. W.; Rowlinson, J. S.; Sather, G. A.; Watson, I. D. Trans. Faraday Soc. 1969, 65, 2034.
11. Mansoori, G. A.; Ely, J. F. J. Chem. Phys. 1985, 82, 406.
12. Prausnitz, J. M. In "Molecular Thermodynamics of Fluid Phase Equilibria"; Prentice-Hall: Englewood Cliffs, NJ, 1969.
13. Kurnik, R. T.; Holla, S. J.; Reid, R. C. J. Chem. Eng. Data 1981, 26, 47.
14. Katz, D. L.; Firoozabadi, A. J. Petrol. Tech. 1978, Nov., 1649.
15. Firoozabadi, A.; Hekim, Y.; Katz, D. L. Canadian J. Chem. Eng. 1978, 56, 610.
16. Reamer, H. H.; Sage, B. H. J. J. Chem. Eng. Data 1963, 8, 508.
17. Kalra, H.; Robinson, D. B.; Krishnan T. R. J. Chem. Eng. Data 1977, 22, 85.

RECEIVED June 24, 1986

Chapter 10

High-Pressure Phase Equilibria in Ternary Fluid Mixtures with a Supercritical Component

A. Z. Panagiotopoulos and R. C. Reid

Department of Chemical Engineering, Massachusetts Institute of Technology, Cambridge, MA 02139

> Experimental results are presented for high pressure phase equilibria in the binary systems carbon dioxide - acetone and carbon dioxide - ethanol and the ternary system carbon dioxide - acetone - water at 313 and 333 K and pressures between 20 and 150 bar. A high pressure optical cell with external recirculation and sampling of all phases was used for the experimental measurements. The ternary system exhibits an extensive three-phase equilibrium region with an upper and lower critical solution pressure at both temperatures. A modified cubic equation of a state with a non-quadratic mixing rule was successfully used to model the experimental data. The phase equilibrium behavior of the system is favorable for extraction of acetone from dilute aqueous solutions using supercritical carbon dioxide.

The potential of supercritical extraction, a separation process in which a gas above its critical temperature is used as a solvent, has been widely recognized in the recent years. The first proposed applications have involved mainly compounds of low volatility, and processes that utilize supercritical fluids for the separation of solids from natural matrices (such as caffeine from coffee beans) are already in industrial operation. The use of supercritical fluids for separation of liquid mixtures, although of wider applicability, has been less well studied as the minimum number of components for any such separation is three (the solvent, and a binary mixture of components to be separated). The experimental study of phase equilibrium in ternary mixtures at high pressures is complicated and theoretical methods to correlate the observed phase behavior are lacking.

One important potential application of supercritical extraction is the recovery of polar organic compounds from aqueous solutions. Such mixtures arise frequently as products of biochemical syntheses. In many cases, the costs of the energy-intensive separations are high and there is substantial incentive for the development of more efficient processes.

Previous work in the area of high-pressure phase equilibrium of aqueous solutions of organic compounds with supercritical fluids includes the pioneering work of Elgin and Weinstock (1) carried out in the 1950's. Francis (9) presented a large number of qualitative results for ternary systems of liquid carbon dioxide at 298 K. In the more recent years, Paulaitis et al. (2,3), Kuk and Montagna (4), McHugh et al. (5) and Radosz (6) have reported measurements in systems with some of the first alcohols, water and supercritical fluids.

Experimental

The experimental setup used is shown in Figure 1. The main element of the equipment is a high pressure optical cell (Jerguson gage model 19-TCH-40) with an internal volume of about 50 cm^3, that serves as a mixing and separating vessel, as well as for the visual observation of the number and quality of the coexisting phases. Connections at the bottom, top and side of the vessel permit withdrawal of the lower, upper and middle phases, any two of which can be recirculated externally with a dual high pressure Milton-Roy Mini Pump. An on-line Mettler-Paar vibrating tube density meter (DMA 60 with a DMA 512 cell) is used to measure the density of one of the recirculating phases. The density meter can be switched to sample any of the recirculating streams.
   The Jerguson gage is immersed in a constant temperature bath with silicon fluid as the heat-transfer medium, that also thermostats the density meter. The bath temperature is controlled to ± .01 K with a Thermomix 1460 temperature regulator and temperature is measured with a calibrated mercury-in-glass thermometer to within .01 K. The lines external to the bath are maintained at the bath temperature with the help of heating tapes and temperature at several points is monitored with thermocouples. Pressure is measured with two calibrated Heise pressure indicators to within ±.16 bar for pressures up to 160 bar and to within ±.4 bar for pressures up to 350 bar.
   Sampling is performed with two high pressure switching valves with internal volume .5 µl (for the upper phase) and .2 µl for the lower or middle phases. The samples are directly depressurized into a He carrier gas stream and analyzed with a Perkin Elmer Sigma 2 Gas Chromatograph, using a Porapak Q column supplied by Supelco. The response factors for the materials used were found to be close to those reported by Dietz (7). Typical reproducibility of the analysis is ±.003 in mole fraction, with somewhat larger deviations for the gas-phase compositions at low pressures.
   After purging and evacuation of the equipment, the cell is charged with a liquid mixture of known composition and the supercritical component up to the desired pressure. The recirculation pumps are started, and approach to equilibrium monitored by the stability of pressure, density and composition measurements with time. For the mixtures studied, a typical equilibration time is 15 min, but at least 30 min are allowed before the final sampling. At least two samples are taken from each phase at equilibrium. A new pressure point can be established immediately by introducing or removing supercritical fluid or liquid, so that the level of the interfaces in the cell at the new desired pressure is appropriate for sampling through the available ports. Entrainment of the phases is not normally a problem, except near a critical point. Selective flashing and adsorption on the switching valves is sometimes a difficulty for the upper(gas) phase sampling loop, but is prevented by heating the valve.

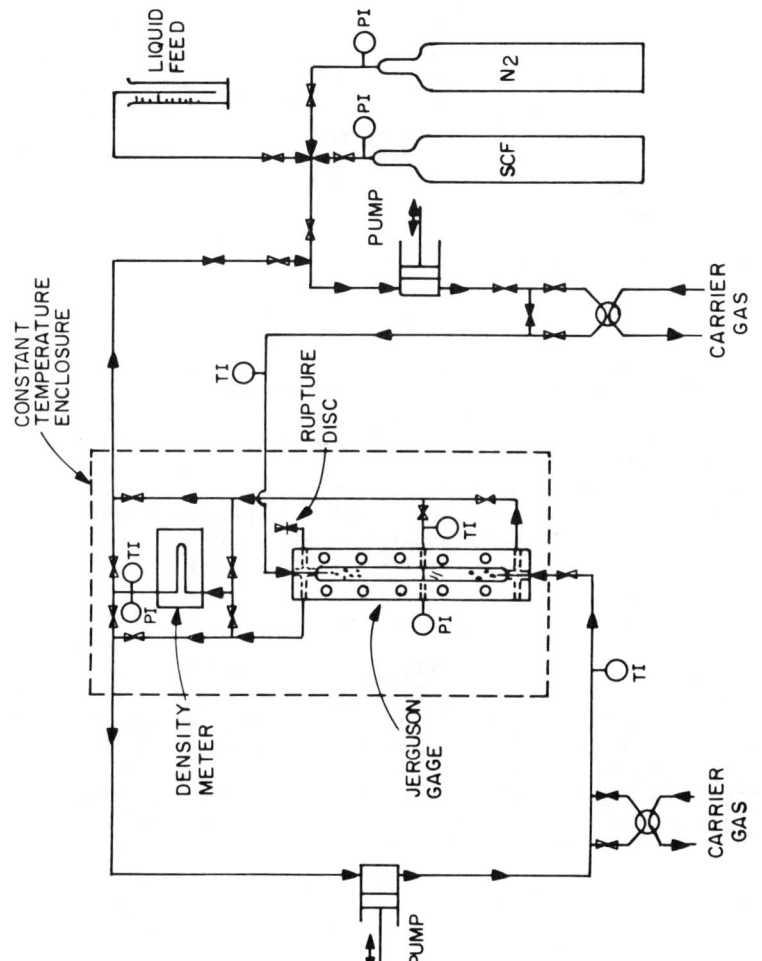

Figure 1. Schematic drawing of the equipment.

Results

Binary Systems. As a test of the validity of the equipment design and experimental procedures, we first investigated a binary system for which results have been previously published. Knowledge of the behavior of the binary systems is in itself important, since the data provide the appropriate limits for the corresponding ternary systems.

In Figure 2, we present experimental results and literature data for the system carbon dioxide - acetone at 313 and 333 K (literature results are only available for the lower temperature). The agreement between the two sets of results is excellent. Also shown on the same figure as continuous lines are the results of fitting the experimental data using a modified Peng-Robinson equation of state described in the Appendix. The agreement between experimental and predicted phase compositions is within the experimental uncertainty of the data, except in the critical region.

Several of the features of the phase behavior shown on Figure 2 are quite general. In particular, it can be seen that the solubility of carbon dioxide in the liquid phase is high even at moderate pressures. The system critical pressure is comparable to the critical pressure of carbon dioxide and increases with temperature at the temperature range studied. Although the solubility of acetone in the gas phase is low in the two-phase region, there is complete miscibility between CO_2 and acetone at relatively moderate pressures. Carbon dioxide at liquid-like densities is an excellent solvent for a wide range of organic compounds, as was first observed by Francis (9) for near critical liquid CO_2.

One additional example of the same general type of behavior is demonstrated by the carbon dioxide - ethanol system. Three measured isotherms for this system are presented in Figure 3. No comparable set of experimental results appears to be available in the open literature. Again, agreement between experimental results and model predictions is good.

Ternary Systems. As one of a series of model systems, we studied the carbon dioxide - acetone - water ternary system at 313 and 333 K. The most interesting feature of the system behavior is an extensive three-phase region at both temperatures. The three-phase region is first observed at a pressure of less than 30 bar at 313 K and approximately 35 bar at 333 K, extending up to approximately the critical pressure of the binary carbon dioxide - acetone system. Table I summarizes our experimental results for the composition of the three phases at equilibrium as a function of pressure and temperature.

The physical picture that underlies this behavior, as pointed out first by Elgin and Weinstock (1), is the 'salting out' effect by a supercritical fluid on an aqueous solution of an organic compound. As pressure is increased, the tendency of the supercritical fluid to solubilize in the organic liquid results in a phase split in the aqueous phase at a lower critical solution pressure (which varies with temperature). As pressure is further increased, the second liquid phase and the supercritical phase become more and more similar to each other and merge at an upper critical solution pressure. Above this pressure only two phases can coexist at equilibrium. This pattern of behavior was also observed by Elgin and Weinstock for the system ethylene - acetone - water at 288 K. In addition, the same type of

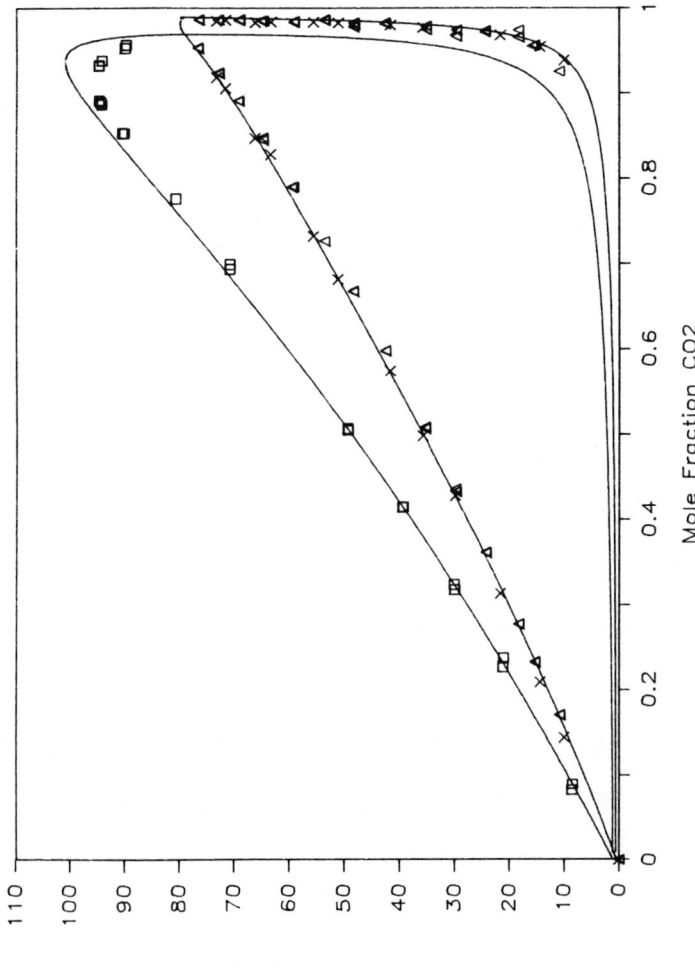

Figure 2. Phase equilibrium behavior for the binary system carbon dioxide - acetone. Experimental, 313 K (△); experimental, 333 K (□); literature, 313 K (×); modified Peng-Robinson equation of state (—). Literature data are from reference (10).

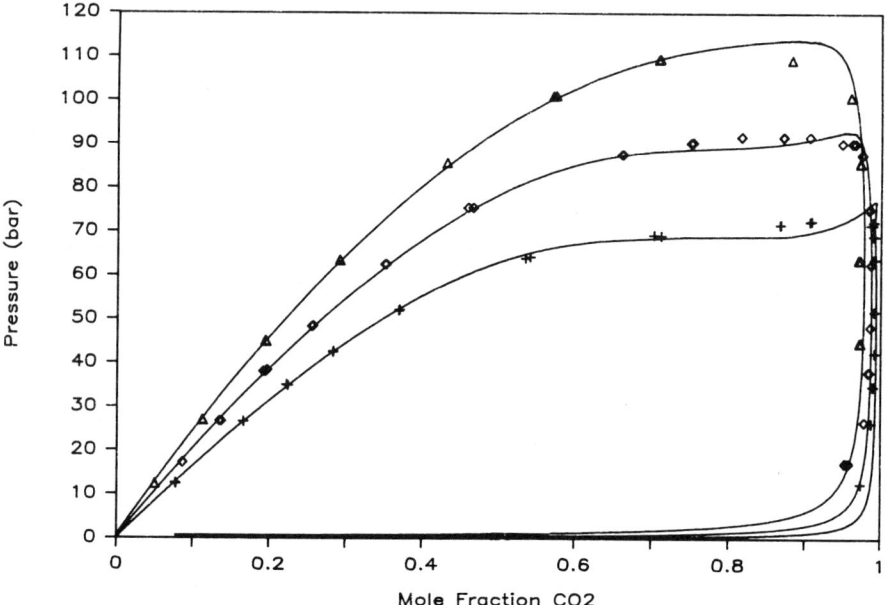

Figure 3. Phase equilibrium behavior for the system carbon dioxide - ethanol. Experimental, 308.1 K (+); 323.1 K (◊); 338.1 K (∆); modified Peng-Robinson equation of state (—).

TABLE I : Three phase equilibrium compositions for the system Water (1) - Acetone (2) - CO_2 (3)

P(bar)	lower phase			upper phase			middle phase		
	x_1	x_2	x_3	y_1	y_2	y_3	z_1	z_2	z_3
T = 313 K									
29.3	0.810	0.153	0.037	0.005	0.025	0.970	0.294	0.454	0.252
35.9	0.864	0.106	0.030	0.006	0.019	0.975	0.197	0.427	0.376
43.2	0.904	0.075	0.021	0.003	0.017	0.980	0.153	0.360	0.487
55.9	0.920	0.052	0.028	0.002	0.015	0.983	0.119	0.230	0.651
61.1	0.942	0.037	0.021	0.004	0.015	0.981	0.114	0.191	0.695
65.8	0.944	0.032	0.024	0.002	0.015	0.983	0.130	0.148	0.722
75.2	0.959	0.017	0.024	0.002	0.014	0.984	0.020	0.069	0.911
79.6	0.966	0.011	0.023	0.002	0.015	0.983	0.049	0.035	0.916
T = 333 K									
39.4	0.795	0.163	0.042	[1]			0.400	0.398	0.202
51.1	0.880	0.091	0.029	[1]			0.182	0.418	0.400
59.4	0.906	0.068	0.026	0.006	0.030	0.964	0.173	0.359	0.468
70.3	0.925	0.049	0.026	0.008	0.035	0.957	0.117	0.300	0.583
79.2	0.939	0.039	0.022	0.006	0.032	0.962	0.089	0.234	0.677
92.6	0.946	0.029	0.025	0.010	0.046	0.944	0.084	0.127	0.789

[1] We were not able to obtain results for these state conditions.

behavior, but at quite different pressures relative to the pure solvent critical pressure was reported by Paulaitis et al.($\underline{3}$) for the system carbon dioxide - isopropanol - water, by McHugh et al. ($\underline{5}$) for the system ethane - ethanol - water and by Paulaitis et al. ($\underline{2}$) for the carbon dioxide - ethanol - water system. This behavior appears then to be quite common.

A more complete picture of the phase equilibrium behavior is given in Figure 4 for the system under study at 333 K. At this temperature, we measured tie lines in the two-phase region in addition to the three-phase compositions. The darker lines and squares on Figure 4 represent these experimental measurements. The light lines are model predictions, using values of interaction parameters determined solely from regression of binary data (see Appendix). The two sets of results are in good agreement with each other. The ability to predict ternary data from binary data only is of great practical importance, since it permits the evaluation of a large range of process alternatives from limited experimental information.

An interesting feature of the phase equilibrium behavior is the relative insensitivity of the phase envelope and positions of the tie-lines to variations in pressure in the two-phase region, as evidenced by comparison of the diagrams at 100 and 150 bar in Figure 4. This would seem to imply that similar separations can be achieved by operation at a range of pressures above the upper critical solution pressure of ~95 bar.

From an engineering point of view, the most important quantities in the evaluation of a separation scheme based on a phase behavior pattern such as the one shown, are the selectivity of the separation with respect to the desired component, as well as the loading of the desired component in the extractant phase. It is well known that those two factors usually increase in different directions. Figures 5 and 6 present experimental data and model predictions for the distribution coefficient of acetone and the selectivity ratio α of acetone over water (defined as the ratio of distribution coefficients of acetone and water in the liquid and fluid phases). The distribution coefficient shows a maximum at both temperatures, with a higher value at the higher temperature, whereas the selectivity decreases smoothly from the limiting value of ~300 at low acetone concentrations to 1 as the plait point is approached. The experimental data show the same trends, although there is some scatter due to the fact that small absolute errors in the low supercritical phase concentrations of acetone and water result in large relative errors for the distribution coefficient and selectivity factors. Selectivities are generally higher for the lower temperature, but loadings are lower.

In Figure 7, we present the w/w concentration of acetone in the supercritical phase on a CO_2-free basis versus the corresponding concentration of acetone in the liquid phase. The curve shows a broad maximum at a range of concentrations of acetone in the aqueous phase between 2% and 15% w/w. The maximum concentration of acetone in the supercritical phase is close to 95% w/w. We can, therefore, obtain almost pure acetone from a dilute aqueous solution using a single-step extraction with supercritical carbon dioxide.

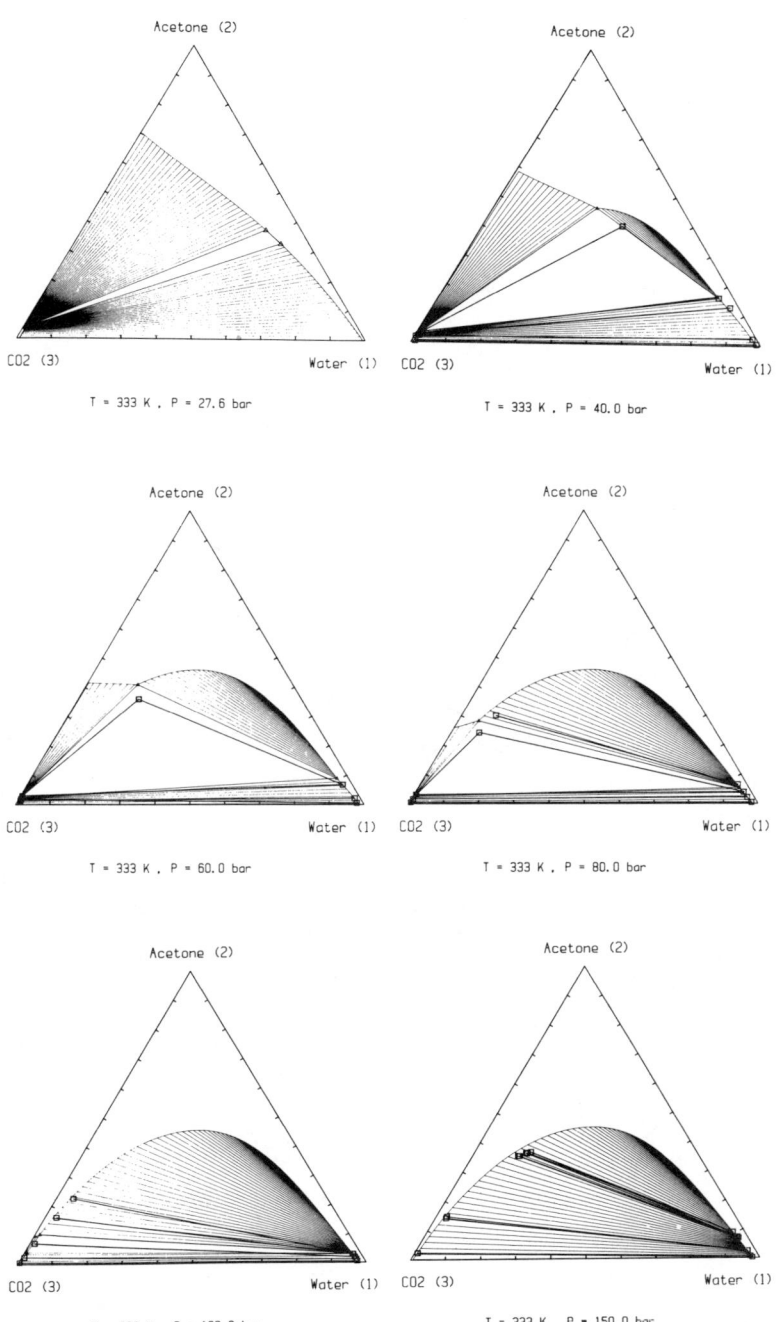

Figure 4. Phase equilibrium behavior for the system water (1)-acetone (2) - carbon dioxide (3) at 333 K : experimental phase compositions (□) and tie-lines (—) ; predicted tie-lines (—) ; predicted three-phase equilibrium compositions (Δ).

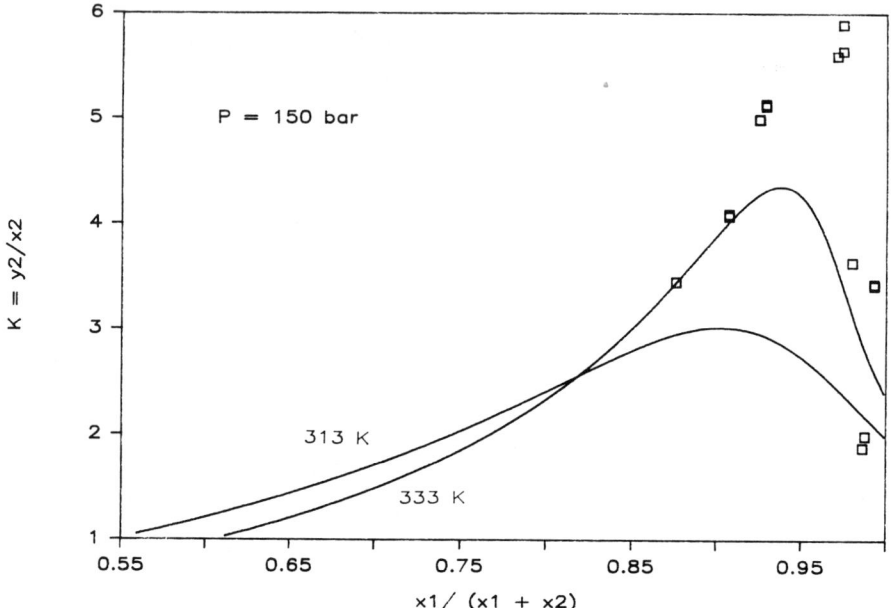

Figure 5. Distribution coefficient of acetone for the system water (1) - acetone (2) - carbon dioxide (3) at 150 bar, as a function of the water concentration in the lower phase. Experimental, 333 K (□); predicted, 333 K and 313 K (—).

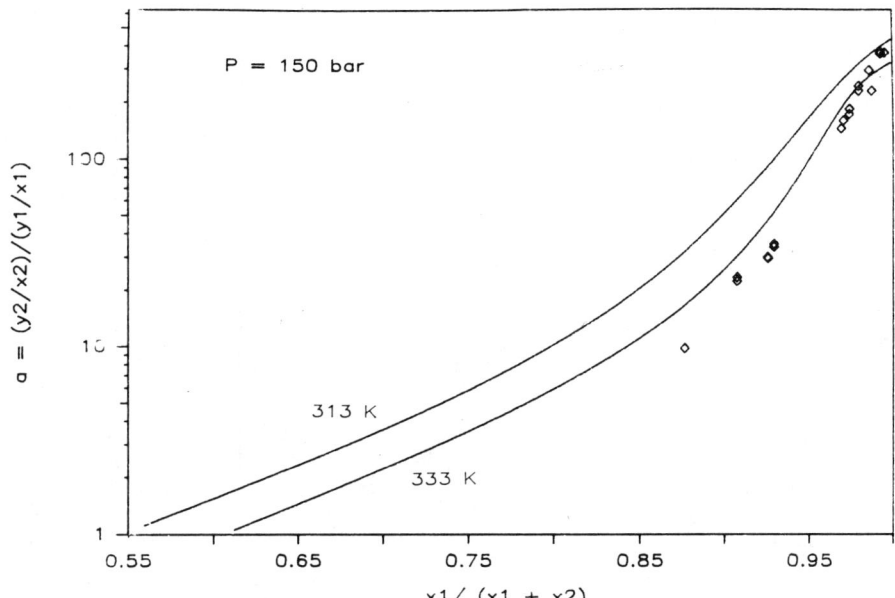

Figure 6. Selectivity factor for acetone over water for the system water (1) - acetone (2) - carbon dioxide (3) system at 150 bar, as a function of the water concentration in the lower phase. Experimental, 333 K (◊); predicted, 333 K and 313 K (—).

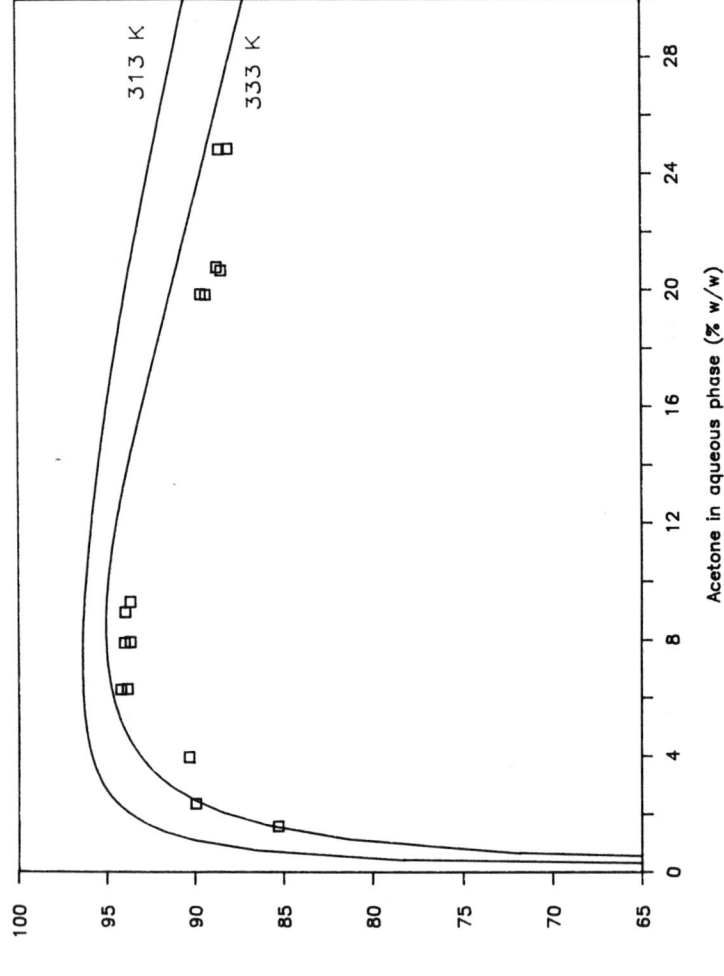

Figure 7. Concentration of acetone in the supercritical fluid phase as a function of the concentration of acetone in the liquid phase (concentration units: % w/w on a CO_2-free basis). Experimental, 333 K (□); predicted, 333 K and 313 K (——).

APPENDIX

Equation of State used

For the correlation of the experimental results, we used a modification of the Peng-Robinson equation of state developed by the authors. The essence of the method, which is presented in detail in (8) is the use of a non-quadratic mixing rule for the attractive parameter a_m. The mixing rule involves two binary interaction parameters to be determined from regression of experimental data. The equations for the pressure, the mixing rule used and the resulting expression for the chemical potential of a component in a mixture are given below for the case of a general cubic equation of state (for the Peng-Robinson equation of state, $u = 2$ and $w = -1$).

$$P = \frac{RT}{V - b_m} - \frac{a_m}{V^2 + uVb_m + wb_m^2} \quad (1)$$

$$a_m = \sum_i \sum_j x_i x_j a_{ij} \quad (2)$$

$$b_m = \sum_i x_i b_i \quad (3)$$

$$a_{ij} = \sqrt{a_i a_j}\,[1 - k_{ij} + (k_{ij} - k_{ji})x_i] \quad (4)$$

$$\ln\phi_k = \ln\frac{\hat{f}_k}{x_k P} = \frac{b_k}{b_m}\left(\frac{PV}{RT} - 1\right) - \ln\frac{P(V - b_m)}{RT} +$$

$$+ \left[\frac{\sum_i x_i(a_{ik} + a_{ki}) - \sum\sum x_i^2 x_j (k_{ij} - k_{ji})\sqrt{a_i a_j} + x_k \sum_i x_i (k_{ki} - k_{ik})\sqrt{a_k a_i}}{a_m} - \frac{b_k}{b_m}\right] \times$$

$$\times \frac{a_m}{\sqrt{u^2 - 4w}\, b_m RT}\,\ln\frac{2V + b_m(u - \sqrt{u^2 - 4w})}{2V + b_m(u + \sqrt{u^2 - 4w})} \quad (5)$$

To estimate the pure component parameters, we used the technique of Panagiotopoulos and Kumar (11). The technique provides parameters that exactly reproduce the vapor pressure and liquid density of a subcritical component. Table II presents the pure component parameters that were used. For the supercritical components, the usual acentric factor correlation was utilized.

It is important to note, that the interaction parameters between the components (two per binary) were estimated solely from binary phase equilibrium data, including low-pressure VLE data for the binary acetone - water; no ternary data were used in the fitting. The values of the interaction parameters obtained are shown in Table III.

TABLE II : Pure component parameters

Component	T(K)	a (J×m³/mol)	10⁶ b (m³/mol)
Acetone	313.15	2.1772	62.35
	333.15	2.1165	62.65
Ethanol	308.15	2.0292	48.23
	323.15	1.9679	48.44
	338.15	1.9069	48.61
Water	313.15	.81614	16.07
	333.15	.79123	15.99

TABLE III : Interaction parameters

System	T (K)	k_{12}	k_{21}
$CO_2(1)$ - acetone(2)	313.15	-0.02	0.00
	333.15	-0.02	0.00
$CO_2(1)$ - ethanol(2)	308.15	0.072	0.069
	323.15	0.093	0.077
	338.15	0.089	0.061
$CO_2(1)$ - water(2)	313.15	-0.205	0.162
	333.15	-0.185	0.160
acetone(1) - water(2)	313.15	-0.310	-0.150
	333.15	-0.293	-0.132

Conclusions

A recirculation apparatus for the determination of high pressure phase equilibrium data for mixtures of water, polar organic liquids and supercritical fluids was constructed and operated for binary and ternary systems with supercritical carbon dioxide.

The system carbon dioxide - acetone - water was investigated at 313 and 333 K. The system demonstrates several of the general characteristics of phase equilibrium behavior for ternary aqueous systems with a supercritical fluid. These include an extensive LLV region that appears at relatively low pressures. Carbon dioxide exhibits a high selectivity for acetone over water and can be used to extract acetone from dilute aqueous solutions.

A model based on a modified mixing rule for the Peng-Robinson equation of state was able to reproduce quantitatively all features of the observed phase equilibrium behavior, with model parameters determined from binary data only. The use of such models may substantially facilitate the task of process design and optimization for separations that utilize supercritical fluids.

Acknowledgments

Financial support for this research was provided by a grant from the National Science Foundation. One of the authors (AZP) was also supported by a Halcon Inc. Fellowship. We gratefully acknowledge both sponsors.

Literature Cited

1. Elgin, J.C.; Weinstock, J.J. J. Chem. Eng. Data 1959, 4(1), 3-12.
2. Paulaitis, M.E.; Gilbert, M.L.; Nash, C.A. "Separation of Ethanol - Water Mixtures with Supercritical Fluids", paper presented at the 2nd World Congress of Chemical Engineering, Montreal, Canada, Oct. 5, 1981.
3. Paulaitis, M.E.; Kander, R.G.; DiAndreth, J.R. Ber. Bunsenges. Phys. Chem. 1984, 88, 869-875.
4. Kuk, M.S.; Montagna, J.C. Ch. 4 in Paulaitis. M.E. et al. (ed), "Chemical Engineering at Supercritical Fluid Conditions", Ann Arbor Science Publishers, Ann Arbor, Mich., 1983 pp. 101-111.
5. McHugh, M.A.; Mallett, M.W.; Kohn, J.P. "High Pressure Fluid Phase Equilibria of Alcohol - Water - Supercritical Solvent Mixtures", paper presented at the 1981 annual AIChE meeting, New Orleans, Louisiana, Nov. 9, 1981.
6. Radosz, M. Ber. Bunsenges. Phys. Chem. 1984, 88, 859-862.
7. Dietz, W.A. J. Gas Chrom. 1967, 5(2) 68-71.
8. Panagiotopoulos, A.Z.; Reid, R.C., in "Equations of State-Theories and Applications", Robinson, R.L; Chao, K.C. (ed.), ACS Symposium Series No. 300, American Chemical Society, Washington, D.C., 1986; pp. 571-582.
9. Francis, A.W. J. Phys. Chem. 1954, 58, 1099-1114
10. Katayama, T.; Oghaki, K.; Maekawa, G.; Goto, M.; Nagano, T. J. Chem. Eng. Jpn. 1975, 8(2), 89-92.
11. Panagiotopoulos, A.Z.; Kumar, S. Fluid Phase Equil. 1985, 22, 77-88.

RECEIVED June 24, 1986

Chapter 11

Solubilities of Five Solid *n*-Alkanes in Supercritical Ethane

Iraj Moradinia and Amyn S. Teja

School of Chemical Engineering, Georgia Institute of Technology, Atlanta, GA 30332-0100

The effect of solid structure on the solubilities of n-alkanes in supercritical ethane has been investigated at a temperature just above the critical point of ethane. Solubilities of n-alkanes containing 28 to 33 carbon atoms in ethane at 308.15K and pressures up to 20 MPa are reported in this work. The enhancement factor is shown to exhibit a regular trend with the number of carbon atoms in the n-alkane, although different trends are exhibited by the odd and even members of the series.

The n-alkanes are an interesting homologous series because they display great regularity in their behavior. Many of their fluid phase properties, for example, can be correlated with the number of carbon atoms in the molecules (1,2). In order to develop general relations for supercritical extraction, therefore, we have studied the solubilities of solid n-alkanes containing 28 to 33 carbon atoms in supercritical ethane.

An additional reason for studying the n-alkane series is that even-numbered n-alkanes exhibit different trends in their solid phase properties (e.g. sublimation pressure, heat of fusion, etc.) than the odd-numbered members of the series (3-6). This is shown for the heat of fusion in figure 1, and is a consequence of the different packing arrangements in the solid phase. It may be possible using supercritical extraction to exploit these differences in order to separate close-boiling members of the series. Any generalization for supercritical extraction behavior must, therefore, take account of these differences in behavior. The solubilities of solid n-Octacosane (n-$C_{28}H_{58}$), n-Nonacosane (n-$C_{29}H_{60}$), n-Triacontane (n-$C_{30}H_{62}$), n-Dotriacontane (n-$C_{32}H_{66}$), and n-Tritriacontane (n-$C_{33}H_{68}$) in supercritical ethane are reported below. The solubilities of the even-numbered members of the series have been obtained from our previous work (7).

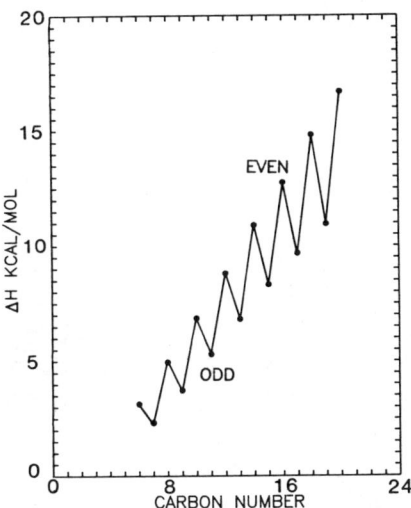

Figure 1. Heat of fusion vs. carbon number for the n-alkanes.

EXPERIMENTAL

The apparatus used in this study was a single-pass flow system shown schematically in figure 2. It is similar in principle to that used by Kurnik et al. (8). The experimental procedure has been outlined in a previous publication (7).

SOURCE AND PURITY OF THE MATERIALS

The solid n-alkanes ($n-C_{28}H_{58}$, $n-C_{30}H_{62}$, and $n-C_{32}H_{66}$) had a stated purity of 99% or better and $n-C_{29}H_{60}$ had a stated purity of 98% or better. The n-alkanes were obtained from Wiley Organics. Solid n-$C_{33}H_{68}$ was obtained from Fluka Chemical Corp. and had a stated purity of 97% or better. The solids were used without further purification. Ethane was furnished by the Matheson Gas Co. with 99+% purity and was also used without further purification.

RESULTS

Our results for the binary systems $C_2H_6 + n-C_{28}H_{58}$, $C_2H_6 + n-C_{29}H_{60}$, $C_2H_6 + n-C_{30}H_{62}$, $C_2H_6 + n-C_{32}H_{66}$, and $C_2H_6 + n-C_{33}H_{68}$ are given in table 1.

Table 1. Experimental Results at 308.15K

P(MPa)	$y_2 \times 10^3$ $C_2H_6(1)+$ $n-C_{28}H_{58}(2)$	$y_2 \times 10^3$ $C_2H_6(1)+$ $nC_{29}H_{60}(2)$	$y_2 \times 10^3$ $C_2H_6(1)+$ $nC_{30}H_{62}(2)$	$y_2 \times 10^3$ $C_2H_6(1)+$ $nC_{32}H_{66}(2)$	$y_2 \times 10^3$ $C_2H_6(1)+$ $nC_{33}H_{68}(2)$
6.57	1.89	2.32	0.549	0.216	.371
10.10	3.38	4.32	1.24	0.713	.963
12.02	6.43	8.29	1.45	0.801	1.14
13.64	7.53	9.91	1.71	0.959	1.36
16.67	10.80	14.20	2.24	1.26	1.64
20.20	15.18	-----	3.20	1.81	2.37

DATA CORRELATION

Experimental data were correlated using the Patel-Teja (9) equation of state. For a pure solid phase in equilibrium with a supercritical gas phase, we may write (10).

$$P_2^{vap} \phi_2^{vap} \text{EXP} \left\{ \int_{P_2^{vap}}^{P} \frac{V_2^s}{RT} dP \right\} = y_2 \phi_2 P \qquad (1)$$

where P_2^{vap} is the sublimation (vapor) pressure of the solute 2, ϕ_2^{vap} is the fugacity coefficient of the solute at its sublimation pressure, V_2^s is the molar volume of the pure solute, y_2 is the composition (solubility) of the solute in the supercritical solvent, ϕ_2 is fugacity coefficient of the solute in the supercritical solvent, P is the pressure and all properties are evaluated at the system temperature T. Since the sublimation pressure is usually very small, we may assume $\phi_2^{vap} \approx 1$ and the integral to be evaluated from zero pressure to the pressure P.

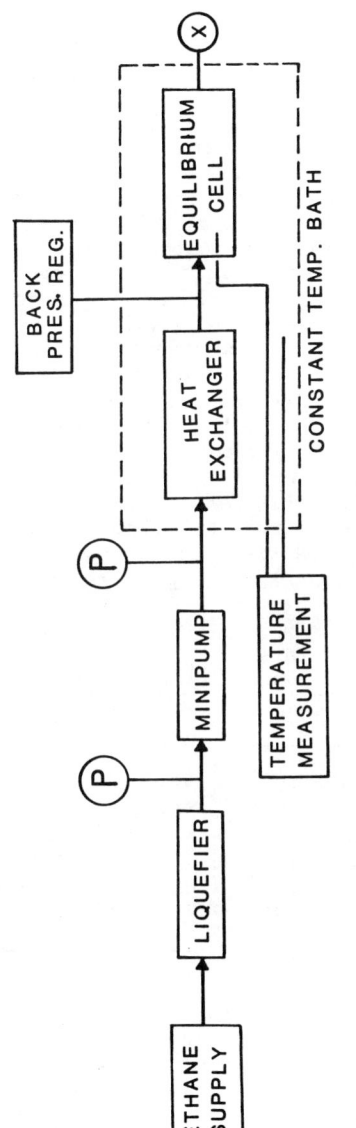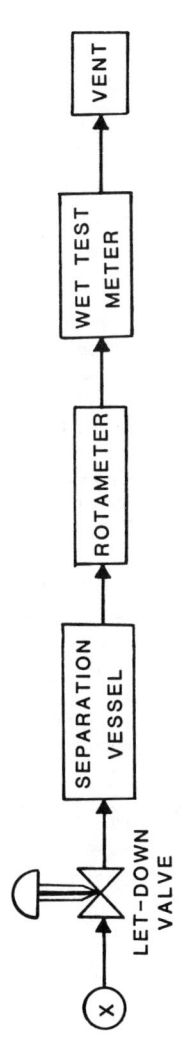

Figure 2. Single-pass supercritical flow apparatus.

Also, since the solid volume is approximately constant with pressure, we can integrate and rewrite eq. (1) as follows:

$$y_2 = \frac{P_2^{vap}}{P} \frac{EXP[PV_2^s/RT]}{\phi_2} \qquad (2)$$

Thus, if the solid phase properties (densities and sublimation pressures) are known, then the solubility of the solute in the supercritical solvent at any pressure and temperature can be calculated provided an equation of state is available for the calculation of ϕ_2. The results for five binary systems using Patel-Teja equation of state shown in figures 3 and 4.

The Patel-Teja equation of state is able to correlate the data for the binary systems reasonably well provided a binary interaction coefficient (k_{ij}) is included in the calculations. It is interesting to note that the binary interaction coefficients obtained from correlation of data for the odd members of the series are an order of magnitude smaller than those obtained for the even members of the series and that they show regular behavior with carbon number. These differences are due to differences in sublimation pressures and lead to the conclusion that different correlations will be required for odd and even numbered members of any homologous series. A plot of enhancement factor (E) also shows this behavior with carbon number (Figure 5).

Enhancement factors are given in Table 2. Solid densities and sublimation pressures used in the solubility calculations (Eq. 2) are given in Table 3. The densities were supplied by the manufacturers whereas the sublimation pressures were extrapolated using data on other n-alkanes [7].

Table 2. Enhancement Factors at T = 308.15K

P(MPa)	$n-C_{28}H_{58}$	$n-C_{29}H_{60}$	$n-C_{30}H_{62}$	$n-C_{32}H_{66}$	$n-C_{33}H_{68}$
6.57	1.947×10^{11}	1.216×10^{11}	9.977×10^{11}	6.926×10^{12}	3.057×10^{12}
10.10	5.354×10^{11}	3.481×10^{11}	3.464×10^{12}	3.514×10^{13}	1.220×10^{13}
12.02	1.212×10^{12}	7.951×10^{11}	4.821×10^{12}	4.699×10^{13}	1.719×10^{13}
13.64	1.611×10^{12}	1.077×10^{12}	6.452×10^{12}	6.364×10^{13}	2.327×10^{13}
16.67	2.824×10^{12}	1.889×10^{12}	1.037×10^{13}	1.025×10^{14}	3.429×10^{13}
20.2	4.809×10^{12}	---------	1.788×10^{13}	1.784×10^{14}	6.004×10^{13}

Table 3. Solid Properties Required in Eqn. (2)

Substance	Lit./Mol. v_2^s	$P_2^{vap} \times 10^{14}$ MPa 308.15K
$n-C_{28}H_{58}$	0.4894	6.37619
$n-C_{29}H_{60}$	0.5058	12.53244
$n-C_{30}H_{60}$	0.5222	0.36151
$n-C_{32}H_{66}$	0.5550	0.02049
$n-C_{33}H_{68}$	0.5714	0.07973

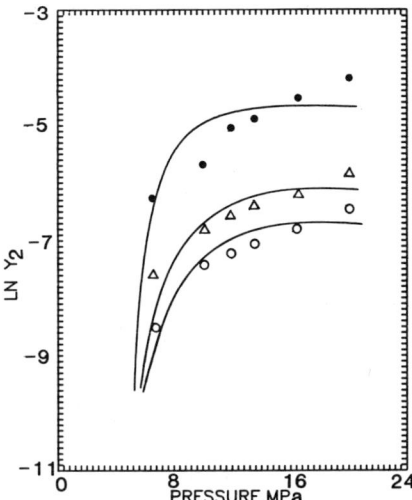

Figure 3. Experimental and calculated solubilities of even-numbered n-alkanes in supercritical ethane at 308.1K. ● experimental data for n-$C_{28}H_{58}$ (k_{ij} = -·0405); △-experimental data for (n-$C_{30}H_{62}$) (k_{ij} = -·0306); ○-experimental data for n-$C_{32}H_{66}$ (k_{ij} = -·0383). Calculations using the Patel-Teja equation of state are shown by the solid lines.

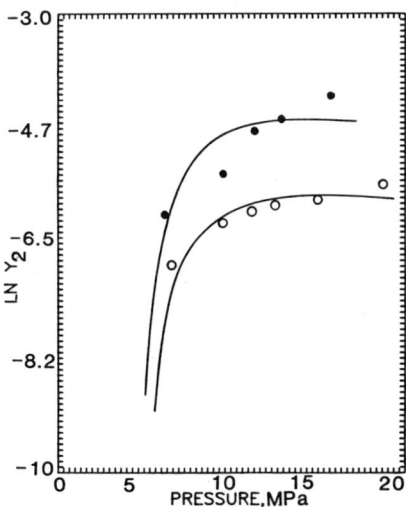

Figure 4. Experimental and calculated solubilities of odd-numbered n-alkanes in supercritical ethane 308.1k. ● experimental data for n-$C_{29}H_{60}$ (k_{ij} = -·0049); ○ experimental data for n-$C_{33}H_{68}$ (k_{ij} = -·0037). Calculations using the Patel-Teja equation of state are shown by the solid lines.

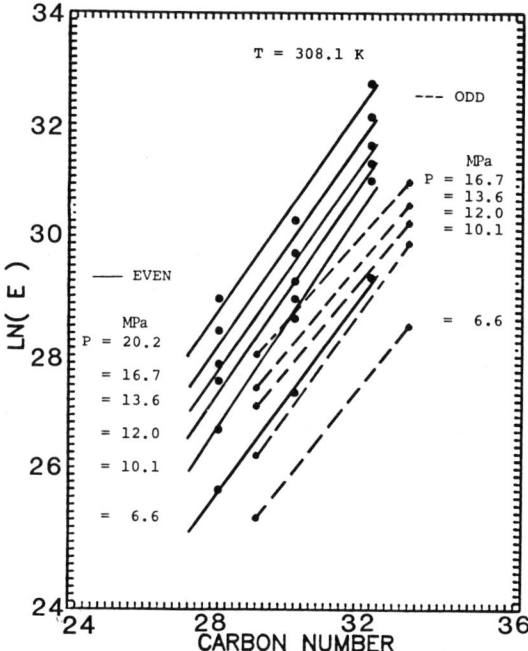

Figure 5. Enhancement factor (E) vs. Carbon number for ethane + n-alkane systems.

Literature Cited

1. Ambrose, D., National Physical Laboratory Report Chem. 57, December 1976.

2. Chase, J. D., Chem. Eng. Prog., 1984, 80, 63-66.

3. Morawetz, E., J. Chem. Thermo., 1972, 4, 139.

4. Bondi, A., J. Chem. Eng. Data, 1963, 8, 371.

5. Broadhurst, M.G., J. Res. Nat. Bur. Stand. A, 1962, 66, 241.

6. Müller, A., Proc. Roy. Soc. Lond. A, 1929, 124, 317.

7. Moradinia, I. and Teja, A. S. Fluid Phase Equil. in press 1986.

8. Kurnik, R. T., Holla, S. J., and Reid, R. C., J. Chem. Eng. Data, 1981, 26, 47-51.

9. Patel, N. C., and Teja, A. S., Chem. Eng. Sci. 1982, 37, 463-468.

10. Prausnitz, J. M., "Molecular Thermodynamics of Fluid Phase Equilibria," Prentice Hall, Englewood-Cliffs N.J., 1969.

RECEIVED August 11, 1986

Chapter 12

Solubility of *meso*-Tetraphenylporphyrin in Two Supercritical Fluid Solvents

T. R. Bergstresser and Michael E. Paulaitis

Department of Chemical Engineering, University of Delaware, Newark, DE 19716

>   Solubilities of meso-tetraphenylporphyrin (normal
>   melting temperature = 444°C) in pentane and in toluene
>   have been measured at elevated temperatures and
>   pressures. Three-phase, solid-liquid-gas equilibrium
>   temperatures and pressures were also measured for these
>   two binary mixtures at conditions near the critical
>   point of the supercritical-fluid solvent. The
>   solubility of the porphyrin in supercritical toluene is
>   three orders of magnitude greater than that in
>   supercritical pentane or in conventional liquid
>   solvents at ambient temperatures and pressures. An
>   analysis of the phase diagram for toluene-porphyrin
>   mixtures shows that supercritical toluene is the
>   preferred solvent for this porphyrin because (1) high
>   solubilities are obtained at moderate pressures, and
>   (2) the porphyrin can be easily recovered from solution
>   by small reductions in pressure.

The enhanced solubility of solids in compressed gases at elevated temperatures and pressures was first noted more than one hundred years ago. Hannay and Hogarth (1) observed that the solubility of salts in compressed ethanol was considerably greater than expected based on the vapor pressure of the salts. Since this early investigation, numerous authors have discussed the phase behavior of solids in dense fluids at elevated temperatures and pressures. Review articles by Paulaitis et al. (2), Luks (3), Streett (4), and Rowlinson and Richardson (5) document many of the investigations that have been made.

The phase behavior of solids in supercritical fluids has practical significance as well as academic interest. Since the mid-1970's, it has been recognized that supercritical fluids can be useful as solvents for commercial-scale extractions. While a variety of applications are documented in the literature (2,6), supercritical-fluid (SCF) extraction has been particularly useful in upgrading petroleum fractions (7), extracting volatile components from coal (8), and deashing oil shale (9) and coal liquids (10). The application of SCF extraction in these processes requires solubilities for the mixtures of interest in SCF solvents. Solubilities can be difficult to predict, however, since these mixtures are highly complex in the sense that numerous constituents are involved which differ significantly in molecular size, shape, structure, and polarity. Experimental solubilities on well-defined, model systems are required to develop better predictive methods.

In this paper, we present the results of an experimental study on the phase behavior of well-characterized binary mixtures which represent the more complex mixtures that arise in SCF extractions of petroleum residua and coal liquids. These binary mixtures consist of pentane and toluene with meso-tetraphenylporphyrin (TPP). Porphyrins occur naturally in crude oils (11,12) and represent an important class of high molecular-weight constituents of these oils, including those which contain heavy metals, such as nickel and vanadium. Pentane and toluene were selected as SCF solvents because these hydrocarbons have critical temperatures convenient to the desired operating temperatures for processing petroleum residua and coal liquids, respectively.

Several different measurements were made as a result of the different phase behavior observed for the two binary mixtures. Gas-solid equilibrium was observed for mixtures of TPP and pentane at conditions near the critical point of pentane. Hence, solid solubilities for TPP in supercritical pentane were measured. However, for mixtures of TPP and toluene, a third (liquid) phase forms in the presence of the gas and the solid, at pressures well below the critical pressure of toluene. At higher pressures, gas-liquid and solid-liquid equilibria were observed, rather than gas-solid equilibrium. Thus, phase compositions for gas-liquid equilibrium were measured for this binary mixture to give TPP solubilities in each of the fluid phases. Pressures and temperatures for three-phase, solid-liquid-gas equilibrium were also measured for both binary mixtures.

Experimental Methods

The following experimental techniques were used to measure the pressures and temperatures for solid-liquid-gas equilibrium, phase compositions (bubble and dew points) for gas-liquid equilibrium, and solid solubilities in supercritical pentane. Experimental procedures and the apparatus are described in detail elsewhere (13).

Solid-liquid-gas (SLG) equilibrium temperatures and pressures were measured in a constant-volume view cell. A known amount of porphyrin was loaded into the cell, which was then attached to the solvent delivery system and heated to the desired operating temperature. Once thermal equilibrium was obtained, solvent was metered into the cell until the pressure within the cell reached the SLG pressure. Equilibrium was obtained when the pressure stablized and all three phases could be observed. This pressure and the corresponding temperature were recorded as one point on the SLG equilibrium line for the binary mixture. Additional points were obtained by setting a new temperature and repeating the procedure.

Bubble points for gas-liquid equilibrium were measured at constant temperature by observing the pressure at which the equilibrium gas phase disappeared upon injection of small amounts of solvent into the view cell. The equilibrium composition of the liquid phase was obtained from the known composition in the cell. Other pressures and corresponding compositions at this temperature were obtained by repeating the procedure for different porphyrin loadings.

Dew points were also measured using the procedure described above, except that the disappearance of the liquid phase was observed as solvent was added to the view cell. Dew point measurements were limited by low porphyrin concentrations in the gas phase, which required loading very small amounts of porphyrin into the cell.

Measurements of solid solubilities in supercritical pentane were, in principle, identical to the bubble point or dew point measurements described above. The equilibrium pressure and corresponding solid solubility at a fixed temperature were determined from the measured pressure and known mixture composition in the cell when the last crystal of solid dissolved. These measurements were limited at low solubilities by the low porphyrin loadings, and at high solubilities by the dark purple color of the fluid phase which obscured observation of the solid phase.

Meso-tetraphenylporphyrin (>97% chlorin free) was purchased from Man-Win Chemical Company and was used without further purification. Pentane and toluene (ACS certified grade) were purchased from Fischer Scientific Company and were degassed before use.

Experimental Results

The accuracy of the pressure and temperature measurements was verified by measuring the vapor pressure curves and critical points for pentane and for toluene. Vapor pressures were measured by observing the formation of a liquid phase as pentane or toluene was injected into the constant-volume view cell under isothermal conditions. The observation of critical opalescence was used to determine the critical point. The measured vapor pressures and critical points are given in Table I. Vapor pressures deviate from

Table I. Measured Vapor Pressures for Pentane and Toluene

Pentane		Toluene	
Temperature (°C)	Vapor Pressure (atm)	Temperature (°C)	Vapor Pressure (atm)
84.8	4.6	120.9	1.4
104.5	6.4	159.9	3.4
125.7	10.2	191.9	6.3
134.7	12.2	196.9	7.0
158.3	18.5	224.7	11.1
184.9	28.4	248.9	16.1
196.1	33.9 CP	274.8	23.1
		287.6	27.6
		299.6	32.1
		318.5	40.3 CP

CP = Critical Point

literature values (14,15) by an average of 0.32 atm for pentane and 0.10 atm for toluene. Critical temperatures and pressures are in close agreement with literature values of 197.5°C and 33.6 atm for pentane (14), and 318.57°C and 40.6 atm for toluene (15).

Three-phase, SLG equilibrium temperatures and pressures for binary mixtures of pentane and toluene with TPP are given in Tables II and III, respectively. A lower critical endpoint (LCEP) was observed for pentane-TPP mixtures, and is also denoted in Table II.

Table II. Measured Three-Phase SLG Temperatures and Pressures for Binary Mixtures of Pentane

Temperature (°C)	Pressure (atm)
173.4	23.0
180.4	25.6
186.9	28.3
192.3	30.8
195.0	32.1
197.7	33.4
198.3	33.6 LCEP

Table III. Measured Three-Phase SLG Temperatures and Pressures for Binary Mixtures of Toluene and TPP

Temperature (°C)	Pressure (atm)
286.2	22.2
292.9	24.2
298.8	24.9
299.4	25.1
304.1	26.2
310.6	27.3
315.1	27.7
319.9	28.5
325.0	29.6
325.6	29.5
337.8	30.2
345.2	31.0
350.0	31.1

This LCEP is a gas-liquid critical point in the presence of the solid phase. A LCEP was not observed for toluene-TPP mixtures at temperatures below 350°C. Measurements were not made at higher temperatures because of thermal degradation of the porphyrin. The results in Tables I-III are also shown on PT projections in Figures 1 and 2. The mixture critical points in Figure 2 are obtained from Figure 4.

Solid solubilities for TPP in pentane at temperatures of 207°C and 250°C are given in Table IV and presented in Figure 3. The maximum experimental uncertainties are 3×10^{-5} for porphyrin mole fraction, 1 atm for pressure, and 1 degree C for temperature.

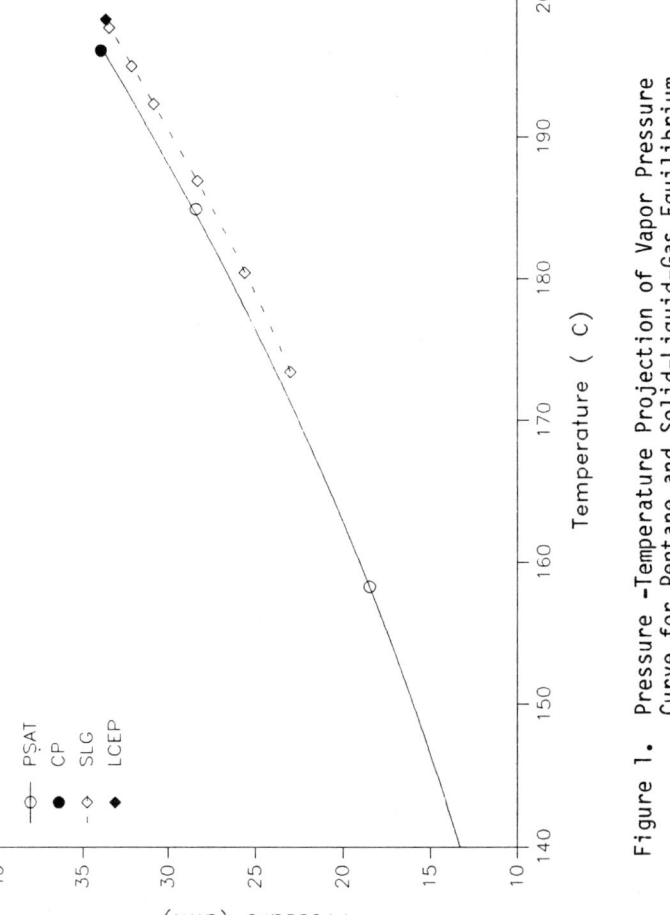

Figure 1. Pressure-Temperature Projection of Vapor Pressure Curve for Pentane and Solid-Liquid-Gas Equilibrium Curve for Pentane-TPP Mixtures.

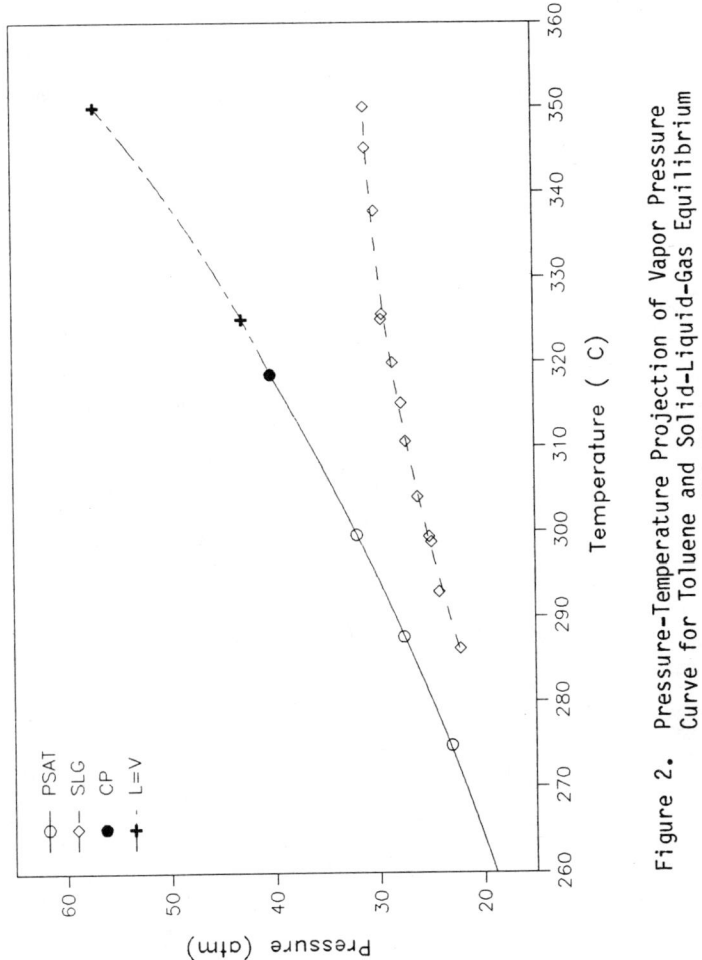

Figure 2. Pressure-Temperature Projection of Vapor Pressure Curve for Toluene and Solid-Liquid-Gas Equilibrium Curve for Toluene-TPP Mixtures.

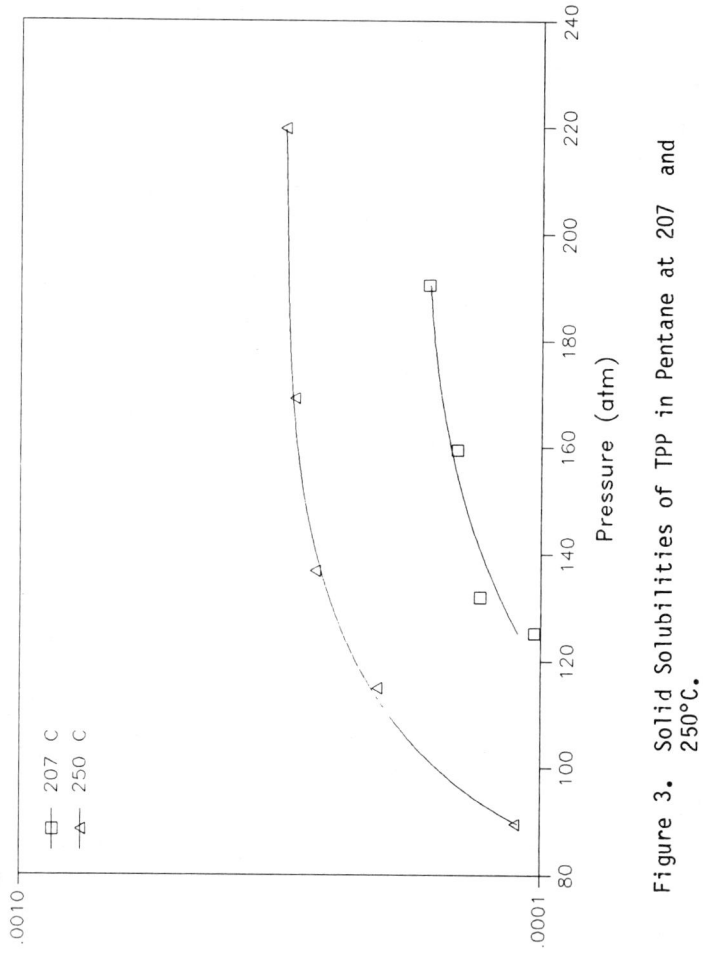

Figure 3. Solid Solubilities of TPP in Pentane at 207 and 250°C.

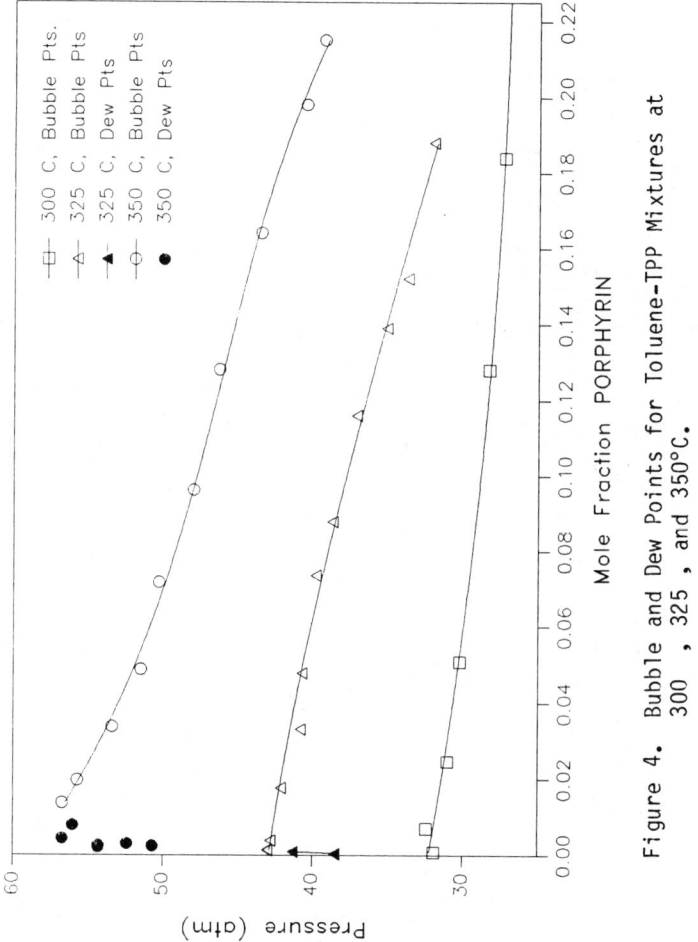

Figure 4. Bubble and Dew Points for Toluene-TPP Mixtures at 300, 325, and 350°C.

Table IV. Measured Solid Solubilities of TPP in Compressed Pentane

Pressure (atm)	Mole Fraction TPP x 10^4
T=207°C	
125	1.0
132	1.3
159	1.4
190	1.6
T=250°C	
89	1.1
115	2.1
137	2.7
169	3.0
220	3.1

The normal melting point of TPP was measured using a differential scanning calorimeter, and was found to be 444°C. The experimental results in Figure 2 indicate, however, that TPP will melt at temperatures of 300-350°C in the presence of compressed toluene at pressures of 20-30 atm (well below the critical pressure of toluene). At higher pressures, gas-liquid and solid-liquid equilibria, rather than gas-solid equilibrium, will be obtained for toluene-TPP mixtures. Bubble points and dew points for this binary mixture were measured at temperatures of 300°, 325°, and 350°, and pressures above the corresponding SLG equilibrium pressures. The results are given in Table V and shown in Figure 4. The maximum

Table V. Measured Bubble and Dew Points for Binary Mixtures of Toluene and TPP

T=300°C		T=325°C		T=350°C	
Pressure (atm)	Mole Fraction TPPx10^2	Pressure (atm)	Mole Fraction TPPx10^2	Pressure (atm)	Mole Fraction TPPx10^2
		Bubble Points		Bubble Points	
27	18.40	32	18.80	39	21.50
28	12.80	34	15.20	40	19.80
29	8.26	35	13.90	43	16.40
30	5.08	37	11.60	46	12.80
31	2.46	39	8.78	48	9.62
32	.70	40	7.36	50	7.18
32	.08	41	4.78	51	4.88
		41	3.31	53	3.38
		42	1.77	56	1.97
		43	.38	57	1.37
		43	.15	Dew Points	
		43	.12	56	.79
		Dew Points		57	.44
		41	.09	52	.30
		38	.04	54	.26
		38	.02	51	.24
				54	.21

experimental uncertainties for bubble and dew point mole fractions are 10% of the reported values. Maximum experimental uncertainties are 1 atm for pressure and 2 degrees for temperature at 350°C.

Discussion

The melting behavior for TPP in the presence of compressed pentane (Figure 1) is characterized by an interrupted three-phase, SLG equilibrium line which terminates at a LCEP. This behavior is characteristic of a gas and a solid with low mutual solubility, and is expected when the triple-point temperature of the solid is much greater than the critical temperature of the gas (3). At temperatures just above the LCEP temperature, TPP does not melt in the presence of compressed pentane, and gas-solid equilibrium is observed at pressures up to two hundred atmospheres (Figure 3). The solubility of TPP in supercritical pentane at these elevated pressures is also quite low. At 250°C, the maximum solubility is approximately 2.7×10^{-4} mole fraction units.

The melting behavior for TPP in the presence of compressed toluene is significantly different (Figure 2). The three-phase, SLG equilibrim curve does not show a LCEP, and probably continues without interruption to the triple point of TPP. This behavior is characteristic of binary mixtures with relatively high mutual solubilities. For SLG equilibrium, the liquid phase would have high TPP concentrations; whereas the gas phase would be mostly toluene and the solid phase would be essentially pure TPP. The pressure-composition diagram depicting gas-liquid equilibrium for toluene-TPP mixtures (Figure 4) indicates that liquid-phase concentrations of TPP can be on the order of 20 mol % at 350°C and 40 atm.

A comparison of solid solubilities for TPP in supercritical pentane, supercritical toluene, and various liquid solvents is given in Table VI. The solid solubility of TPP in toluene corresponds to the liquid-phase concentration for SLG equilibrium, estimated from the results in Figures 2 and 4. Several important conclusions can be

Table VI. Comparison of Solid Solubilities of TPP in Various Solvents

Solvent	Mole Fraction TPP x 10^4	Conditions
Toluene	2210.00	T=325°C,P=30 atm
Pentane	2.71	T=250°C,P=137 atm
Pyridine (16)	6.30	ambient
Benzene (16)	5.20	ambient
Acetone (17)	1.40	ambient
Ethanol (17)	.01	ambient

drawn from this comparison. First, TPP solubility in supercritical toluene is three orders of magnitude greater than that in supercritical pentane and in the liquid solvents at ambient temperatures and pressures. Obviously, the higher temperature associated with supercritical toluene must be a significant factor. However, the chemical nature of the solvent is also important, as demonstrated by the fact that the enhancement factor (2) calculated for TPP in toluene is about one order of magnitude greater than that calculated for TPP in supercritical pentane. Second, the high solubility in toluene is obtained at a moderate pressure, 30 atm, well below the critical pressure of toluene. This pressure corresponds to the SLG equilibrium pressure at 325°C. Third, the SLG pressure of 30 atm represents the minimum pressure for high TPP solubilities in toluene at this temperature -- i.e., the minimum pressure for the existence of the liquid phase. At lower pressures, only the gas and solid phases will form, corresponding to essentially complete separation of toluene and the porphyrin. In summary, the comparison in Table VI indicates that supercritical toluene would be the preferred solvent for meso-tetraphenylporphyrin based upon the high TPP solubilities, moderate operating pressures, and the ease with which the porphyrin can be recovered from solution by reducing the pressure below the SLG equilibrium pressure.

Acknowledgments

Financial support from the National Science Foundation (CPE 8000276) is gratefully acknowledged.

Literature Cited

1. Hannay, J. B.; Hogarth, J. Proc. Roy. Soc. (London) 1879, 29, 324.
2. Paulaitis, M. E.; Krukonis, V. J.; Kurnik, R. T.; Reid, R. C. Rev. in Chem. Eng. 1983, 1, 179.
3. Luks, K. D. Proc. 2nd Intern. Conf. on Phase Equilibria and Fluid Properties in the Chemical Industry, Berlin, 1980.
4. Streett, W. B. Can. J. Chem. Eng. 1974, 52, 92.
5. Rowlinson, J. S.; Richardson, M. J. "Advances in Chemical Physics. Vol. II", Interscience Publishers, Inc.: New York, 1959.
6. Williams, D. F. Chem. Eng. Sci. 1981, 36, 1769.
7. Gearhart, J. A.; Garwin, L. Hydrocarb. Proc. 1976, 55, 125.
8. Maddocks, R. R.; Gibson, J. Chem. Eng. Prog. 1977, 73, 59.
9. Clapper, T. W. U. S. Patent 4 162 965, 1979
10. Baldwin, R. A.; Davis, R. E.; Wing, H. F. paper presented at the 4th Intern. Conf. on Coal Gasification, Liquefaction, and Conversion to Electricity, Pittsburgh, 1977.
11. Baker, E. W.; Palmer, S. E. "Geochemistry of Porphyrins" Part A; Dolphin, D.; Ed.; Academic Press: New York, 1978.
12. Didyk, B.; Alturki, Y. I. A.; Pillinger, C. T.; Eglinton, G. Chem. Geology 1975, 15, 193.
13. Bergstresser, T. R.; Ph.D. Thesis, University of Delaware, 1986.
14. Sage, B. H.; Lacey, W. N. Ind. Eng. Chem. 1942, 34, 730.
15. Ambrose, D.; Broderick, B. E.; Townsend, R. J. Chem. Soc. A 1967, 633.
16. Koifman, O. I.; Berezin, B. D.; Zelov, V. V.; Nikitina, G. E. Russ. J. Phys. Chem. 1978, 52, 1032.
17. Berezin, B. D.; Koifman, O. I.; Zelov, V. V.; Nikitina, G. E. Russ. J. Phys. Chem. 1978, 52, 1281.

RECEIVED August 27, 1986

CHROMATOGRAPHY

Chapter 13

Supercritical Fluid Adsorption at the Gas-Solid Interface

Jerry W. King[1]

CPC International, Moffett Technical Center, Argo, IL 60501

Adsorption of a supercritical fluid upon the surface of an adsorbent significantly alters the gas-solid interface thereby permitting the removal and migration of bound adsorbates. Such a process can contribute substantially to the regeneration of adsorbents and the chromatographic separation of solutes as reported in the scientific and patent literature. In this paper, we have characterized the adsorption of dense gases at solid interfaces in terms of their reported adsorption isotherms. Particular emphasis has been placed on the pressure at which maximum differential adsorption of the gas occurs and its importance in assuring rapid breakthrough of sorbates from packed columns. An experimental approach is presented for determining adsorbate breakthrough volumes as a function of gas compression using elution pulse chromatographic techniques. The derived retention volume data define distinct regions of solute breakthrough characteristics which are determined by the relative uptake of the compressed fluid on the sorbent surface. The results obtained in this study permit an estimation of the physical conditions required to effectively remove aliphatic and aromatic hydrocarbons from such sorbents as alumina and porous organic resins with minimal compression of the supercritical fluid phase.

The utilization of supercritical fluids in conjunction with adsorbents and active solids is well documented in the technical literature. The most frequently cited applications involve the use of dense gases for the regeneration of adsorbents ([1]) and as mobile phases in supercritical fluid chromatography ([2]). Numerous

[1]Current address: Northern Regional Research Center, Agricultural Research Service, U.S. Department of Agriculture, 1815 North University Street, Peoria, IL 61604

0097-6156/87/0329-0150$06.50/0
© 1987 American Chemical Society

patents also contain examples of the removal of specific solutes, such as caffeine or nicotine, from supercritical fluid media by activated carbon (3, 4) and ion exchange resins (5, 6). Fractionation of solutes from dense gas streams has also been demonstrated by Barton and Hajnik (7) utilizing molecular sieves as a sorbent. Less well known, but of historical importance, is the sorption of methane by coal (8), an equilibria of importance in preventing mine explosions. Recently, Findenegg (9) has suggested that radioactive gases may be stored to advantage on zeolites at elevated pressures.

Regeneration of adsorbents was initially demonstrated in the mid-1970's in both Japan (10) and the United States (11, 12). In research sponsored by the Environmental Protection Agency (13-15), Critical Fluid Systems, Inc. (Arthur D. Little) demonstrated the feasibility of regenerating activated carbon and organic resins using supercritical carbon dioxide. Typical conditions in the above studies utilizing CO_2 for the removal of selected sorbates from activated carbon are shown on a plot of reduced state in Figure 1. For the extractions cited, a wide range of reduced temperature and pressure have been employed depending upon the chemical nature of the compounds being desorbed. Presumably, these regeneration conditions were selected empirically or are based on independent measurements of sorbate (solute) solubility in the supercritical fluid.

More recently, Kander and Paulaitis (16) have studied the adsorption of phenol onto activated carbon and measured its sorption equilibria from dense CO_2. These researchers found that temperature controlled the adsorption equilibria and that phenol uptake was negligibly effected by changes in the gas phase density. Such a result indicates that factors other then a solute's solubility in a dense gas play a key role in defining the adsorption equilibrium which accompany such processes.

There is a synergism between adsorption and chromatographic processes which is clearly demonstrated in the supercritical fluid literature. Research in supercritical fluid chromatography can usually be divided into analytical applications and the measurement of physicochemical data. Early analytical separations methodology performed at pressures close to ambient conditions showed that the elution order of injected solutes could be altered by changing the nature of the carrier gas (17-20). Later, adsorbents and crosslinked polymeric packings were utilized in analytical studies to circumvent the problem of stationary phase volatility in the highly compressed carrier gas. For example, Sie and Rijnders (21, 22) demonstrated the advantages of using such adsorbents as alumina and microporous polymers for fractionating a wide variety of mixtures including polycyclic aromatics hydrocarbons, alkaloids, and epoxy resin oligomers. Results obtained in the above study indicated that the solute distribution coefficients could be varied as a function of carrier gas pressure in a rather dramatic fashion and that elution peak shape was substantially modified by the interaction of the supercritical fluid with the sorbent surface. Kobayashi and co-workers (23-28) using tracer perturbation chromatography have substantiated the above findings and determined equilibrium K-values, isotherms, and thermodynamic functions of adsorption for light hydrocarbons on several adsorbents.

In this paper, we have utilized literature data in tandem with pulse chromatographic measurements to define the role of supercritical fluid adsorption at the gas-solid interface. It should be appreciated that the adsorption of the supercritical fluid on the adsorbent surface is appreciable at higher pressures leading to the formation of a condensed phase (29) in the microporous adsorbents commonly used in regeneration experiments. The formation of such a "liquid-like" film at the interface can have a profound effect in determining the uptake of another adsorbate at the solid surface or in the conditions required for regeneration of the adsorbent surface. The measurements reported in this study extend out to 1600 atmospheres, a compression regime previously unexplored by gas-solid chromatography. Such experiments have permitted us to correlate chromatographic retention volume measurements with adsorbate breakthrough characteristics as a function of pressure for several adsorbent/adsorbate systems. The role of the supercritical fluid at the gas-solid interface has been invoked to explain these breakthrough characteristics, therefore a brief discussion of the fundamentals governing the adsorption of high pressure gases at the adsorbent interface will be given.

Fundamentals of Supercritical Fluid Adsorption

Fundamental studies on the adsorption of supercritical fluids at the gas-solid interface are rarely cited in the supercritical fluid extraction literature. This is most unfortunate since equilibrium shifts induced by gas phase non-ideality in multiphase systems can rarely be totally attributed to solute solubility in the supercritical fluid phase. The partitioning of an adsorbed specie between the interface and gaseous phase can be governed by a complex array of molecular interactions which depend on the relative intensity of the adsorbate-adsorbent interactions, adsorbate-adsorbate association, the sorption of the supercritical fluid at the solid interface, and the solubility of the sorbate in the critical fluid. As we shall demonstrate, competitive adsorption between the sorbate and the supercritical fluid at the gas-solid interface is a significant mechanism which should be considered in the proper design of adsorption/desorption methods which incorporate dense gases as one of the active phases.

In general, adsorption of a high pressure gas follows a Type II BET isotherm. The shape of the isotherm is partly dependent on the units chosen for the isotherm coordinates, however there is a major difference in the character of the isotherm depending on whether absolute or Gibb's adsorption is chosen for the abscissa. Whereas absolute adsorption is a measure of the total mass or volume of the gas adsorbed onto the surface, Gibb's adsorption (or the surface excess) measures the amount of gas adsorbed on the solid surface in excess of that corresponding to the density of gas in the fluid phase at the same temperature and pressure. This latter definition provides some insight as to the molecular interactions taking place at the interface as a function of gas density. A typical adsorption isotherm illustrating both the absolute and differential (Gibbs) adsorption is shown in Figure 2.

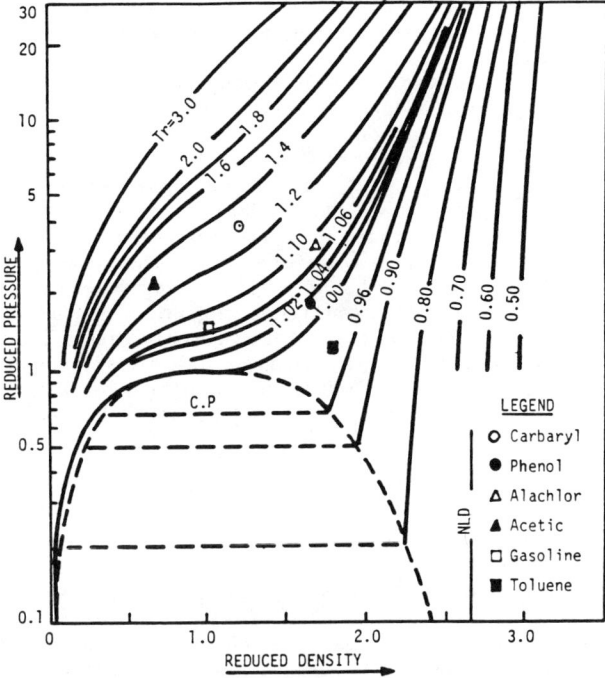

Figure 1. Reported conditions for the supercritical fluid desorption of selected adsorbates from activated carbon.

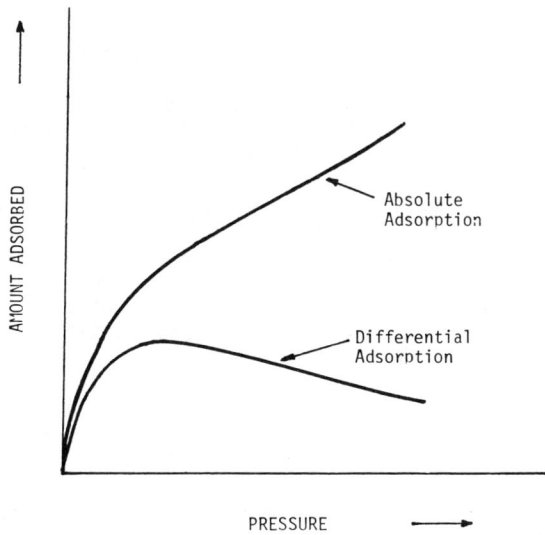

Figure 2. Typical adsorption isotherm for a supercritical gas in which absolute and differential adsorption are express as a function of pressure.

Menon has noted in his comprehensive review (30) that an adsorption maxima occurs as a specific gas pressure, p_{max}, when the Gibbs adsorption is plottted versus the gas pressure. These adsorption maxima are quite pronounced when the supercritical fluid is adsorbed at reduced temperatures approaching unity. Further compression of the critical fluid beyond p_{max} results in a density increase of the adsorbed film and a concomitant decrease in the surface excess as the gas phase approaches a density equal to or exceeding that of the adsorbed phase. As we shall show later, attainment of this specific pressure value is critical for saturating the adsorbent surface with the compressed fluid and thereby leading to conditions which favor the desorption of the bound solute from the surface region. Menon (31) has correlated the occurence of p_{max} for various gases in terms of their respective critical properties and has proposed the following empirical equation which is applicable for adsorption on macroporous adsorbents (pore diameters greater than 20 Å):

$$p_{max} = P_c T_r^2 \qquad (1)$$

where P_c = critical pressure of the gas
T_r^c = reduced temperature at which adsorption occurs

The above relationship predicts a monotonic increase in p_{max} with increasing temperature as shown in Figure 3 for three different gases. Hence, the maximum amount of gas (by the Gibbs definition) adsorbed on the surface of the adsorbent can be attained at a lower pressure by operating close to the critical temperature of the adsorbed gas. Application of even higher pressures then p_{max} will result in a large increase in the cohesional energy of the adsorbed gas leading to the formation of a liquid-like layer on the surface of the porous adsorbent (27).
Additional studies by Menon (32) have indicated the p_{max} can occur at lower pressures then those predicted by Equation 1 depending on the pore structure associated with the adsorbent. Empirically, adsorbents possessing microporosity exhibit a p_{max} that is 0.6-0.8 of the value predicted by Equation 1. This observation is attributed to the overlapping of potential fields in the adsorbent pores, thereby enhancing sorption of the gas at lower pressures. Experimental studies by Ozawa (33) have verified this trend as shown in Figure 4 for the CO_2/activated carbon system. Here the adsorption maxima for the gas occurs at a lower pressure than the critical pressure of carbon dioxide. It should also be noted that the amount of gas adsorbed is decreased at higher reduced temperatures and that additional compression is required to reach a defined adsorption maxima (i.e., at very high values of T_r it is sometimes difficult to discern a well-defined adsorption maximum). The above trend has also been found for other adsorbent/adsorbate systems, such as silica gel/CO_2.
The dependence of gas adsorption on the pore structure of the sorbent has been extensively studied by Ozawa, Kusumi, and Ogino (34) and theoretically correlated by application of the Pickett equation. Data obtained from this study has been plotted

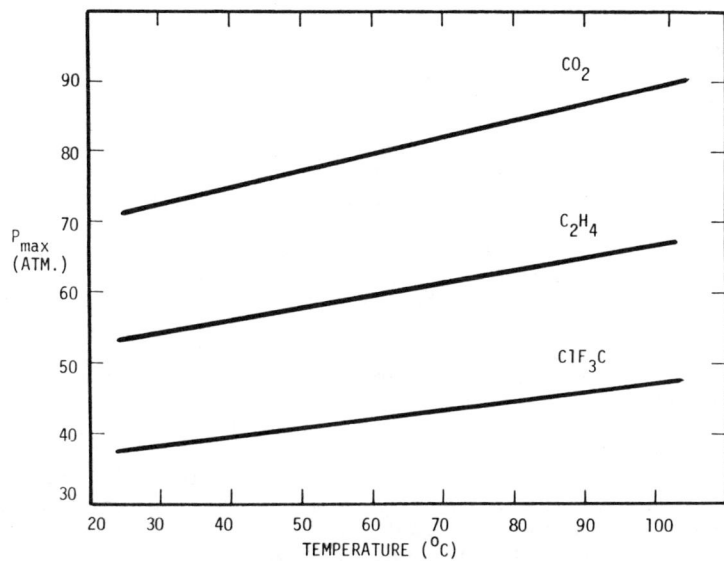

Figure 3. Dependence of p_{max} on temperature for various gases according to Menon (30) on macroporous adsorbent.

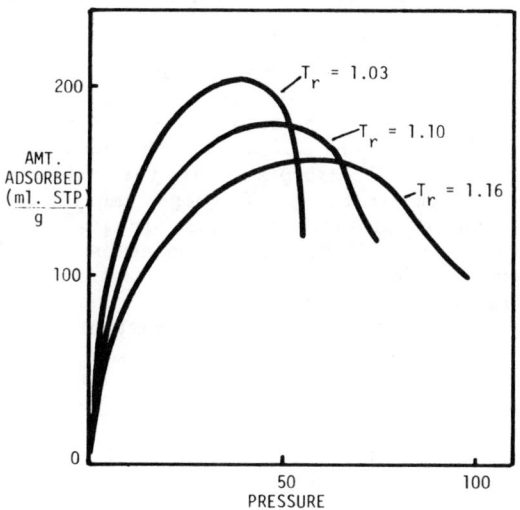

Figure 4. Adsorption isotherms for the carbon dioxide/activated carbon system. Reproduced with permission from Ref. 33. Copyright 1972, Chemical Society of Japan.

in Figure 5 where the p_{max} value versus the mean pore diameter of the adsorbent (activated carbon) is shown for CO_2 sorption. The adsorption maxima in this case, increases with increasing pore diameter, asympotically approaching a constant value at large pore diameters. The value predicted by Menon's relationship (Equation 1) has also been included for the purpose of comparison. Although it would seem prudent to utilize an adsorbent of low porosity to avoid additional gas compression, mitigating factors such as adsorption capacity and stereo-selectivity also play a seminal role in the final selection of a suitable adsorbent. The pore size dependence of p_{max} at various temperatures has also been verified for above carbons and shows a similar dependence to that exhibited in Figure 3.

Further discussion on the theoretical aspects of supercritical fluid adsorption on solids can be found by consulting the papers of Findenegg (35-37). The importance of the high pressure adsorption maximum as discussed above becomes apparent if we compare typical p_{max} values on a plot of reduced state variables with conditions that are commonly employed for the regeneration of adsorbents (Figure 1). As shown in Figure 6, p_{max} values for CO_2 on typical carbonaceous adsorbents occur at considerably lower reduced pressures (p_r = 0.5-1.0) than those utilized for the desorption of common organic solutes (14) or pesticides (1). This result implies that potentially much lower pressures and volumes of supercritical CO_2 could be used to displace adsorbed solutes from the adsorbent surface by maximizing the competitive adsorption of the dense CO_2 on the solid surface.

To provide some experimental verification of the above hypothesis, we have measured via pulse elution chromatography, breakthrough volumes (the peak maximum retention volume in linear elution chromatography corresponding to the 50% breakthrough volume in frontal analysis) of model adsorbates on two different adsorbents over an extended pressure range using CO_2 as the critical fluid. Our experimental results suggest that there are distinct regions in which the surface of the adsorbent is undergoing modification due to the adsorption of the supercritical fluid carrier gas. The implications of these results have permitted the identification of distinct pressure ranges which can be used for the fractionation of dissolved moieties in the critical fluid, a minimum pressure which should be attained for commencement of adsorbent regeneration, and a upper pressure limit at which sorbate breakthrough volume becomes constant with increasing pressure.

Experimental Measurements

The experimental apparatus utilized for these studies is of similar design to that reported by Giddings, Myers, and King (38) as is depicted in Figure 7. Compression of the mobile phase was accomplished using an Aminco air-actuated diaphragm compressor which was kept in a heated chamber above the critical temperature of the carrier gas. Pump pulsations and flow irregularities were minimized by storing the compressed fluid in a high pressure ballast chamber which served as a source of mobile phase. Pressures up to 3000 psig were adjusted by using

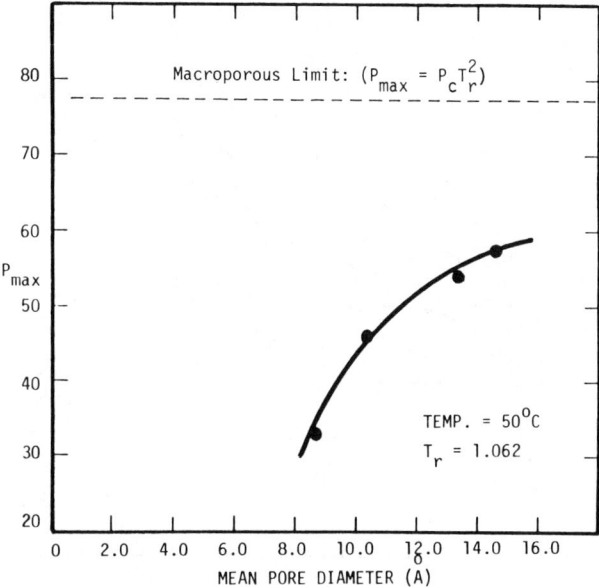

Figure 5. Relationship between the pressure at which maximum differential adsorption of carbon dioxide occurs and the mean pore diameter of activated carbons.

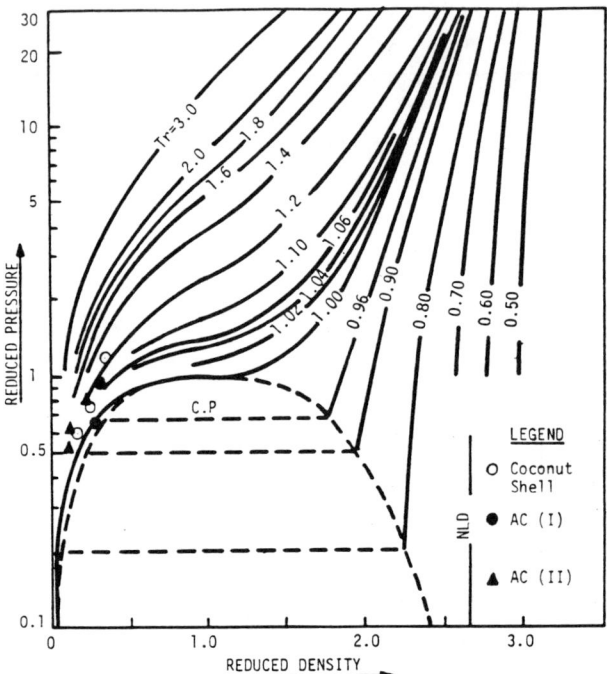

Figure 6. Reported adsorption maxima for carbon dioxide/ adsorbent system on a plot of reduced state. (I) and (II) represent different types of activated carbon (AC).

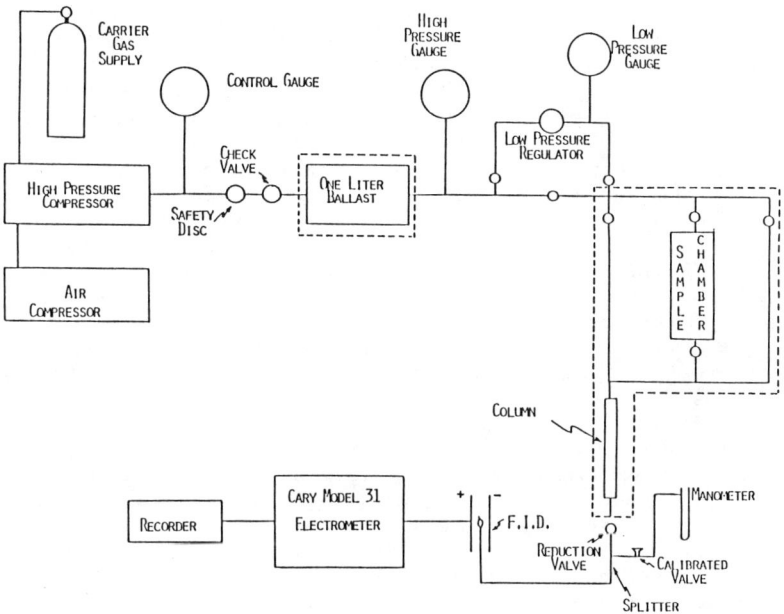

Figure 7. Experimental high pressure gas chromatographic apparatus for retention volume measurements.

an Apco backpressure regulator in conjunction with a Bourdon tube gauge. Regulation of pressures beyond 3000 psig was achieved by isolating the low pressure gauge from the main flow system and subsequently regulating the entire chromatographic system with a dual contact electrical control gauge (modified Aminco Model 47-18330). The sampling chamber/valve and column were placed in a heated, thermostatted oven held at 40°C (dashed line in Figure 7).

Column design and preparation incorporated previously described methods reported in the literature (39). Two different adsorbents were employed: a 100/120 mesh crosslinked styrene/divinylbenzene resin (Polypak P-Waters Associates) and a Woelm aniontropic activity grade alumina. These adsorbents were packed in 300 and 94 cm. stainless steel columns having a 1 mm. internal diameter. Pressure drop across the adsorbent bed was kept to a minimum (<0.02 atm.) by using a heated pressure reduction valve at the end of the column. Typical linear flow velocities through the columns were in the range of 0.27-2.17 cm/sec.

Pneumatic transport of the solute after decompression at the end of the column to the detector orifice was facilitated by application of a heating tape to a small length of narrow bore tubing. Detection of the solute elution profile was accomplished using a fabricated, non-commercial flame ionization detector. Flow rates of the carrier gas were adjusted under ambient conditions by calibrating a U-tube manometer in terms of gas flow rate. This required opening the carrier gas flow splitter so as to divert the entire flow through the detector orifice, subsequently measuring the flow rate and differential pressure on the manometer, and then returning the calibrated metering valve to its initial setting to reestablish the split (1:6-1:10) of the carrier gas flow. Calibration with and without the detector gases (air & H_2) revealed the absence of any backpressure effect due to the detector combustion gases.

Experimental assessment of the column void volume proved to be critical since the solute retention volume approaches the void volume as pressure is increased. Following the recommendations of Kobayashi (24), we used an unretained solute, methane, for this measurement. Values for the void volume determined over an extended pressure range were 1.8 and 0.5 ml. for the crosslinked resin and alumina columns, respectively. These figures were in excellent agreement with void volume approximations of 1.4 and 0.45 ml. based upon the geometric volume of the column assuming a porosity of 0.6 for the packed beds.

The adsorbates and adsorbents in this study were chosen to reflect a range of different types of molecular interactions as well as to observe whether the retention volume data trends could be generalized. In addition, solutes were picked which would rapidly equilibrate with the chosen adsorbents (no hysteresis) and whose distribution coefficients could be measured conveniently over as wide a pressure range as possible. As shown in Table I, the adsorbents corresponded to two distinctly different chemical types as classified by the criterion of Kiselev (40). The alumina represented an adsorbent capable of specific interactions with sorbates having localized peripheral electron density, since it possesses a localized positive charge on its surface. The crosslinked organic polymer is a Type III adsorbent since its negative charge on the surface is due to the presence of an aromatic pi-electron system within its molecular structure. The adsorbates employed represented a homologous series of n-alkanes (Group A sorbates) and two Group B adsorbates, benzene and naphthalene, which have peripheral pi-electron density.

Table I. Adsorbent/Adsorbate Classes Utilized in the Chromatographic Experiments

Adsorbent Type	Adsorbate Type	
Al_2O_3 (Type I)	n-Pentane n-Hexane n-Heptane	(Group A)
Styrene/Divinylbenzene Resin (Type 2)	Benzene Naphthalene	(Group B)

Data Reduction

There is a fundamental relationship described in chromatographic theory between the retention volume of a elution peak and the mid-point of a breakthrough curve achieved by operating the column under frontal analysis conditions (41). In the Henry's Law region of the adsorption isotherm, the net retention volume and its measurement can be used to describe the variation of sorbate breakthrough volume as illustrated in Figure 8. Utilizing the experimental apparatus described in the last section, retention volumes were measured as a function of pressure at 40°C (T_r = 1.03) over as wide a range of pressure as possible to permit elucidation of the breakthrough characteristics for the selected sorbates. Examination of the pulse profiles indicated a high degree of symmetry which was indicative of the absence of non-linear sorption isotherm for the systems under study. Therefore, the reported retention volume measurements can be equated with the breakthrough and/or desorption pattern of the adsorbate during its passage through the column bed.

Retention volume data were computed using Equation 2:

$$V_r = (T_{col}/T_a) \; (Z_a/Z_{col}) \; (p_a - p_{water}/p_{col}) \; F_a \; (t_r - t_o) \quad (2)$$

where V_r = retention volume of the adsorbate

T_{col} = column temperature

T_a = ambient temperature

Z_{col} = compressibility factor at p_c

Z_a = compressibility factor at ambient pressure

P_a = ambient pressure

P_{col} = column pressure

P_{water} = vapor pressure of water

F_a = ambient flow rate

t_r = retention time (peak maxima)

t_o = retention time of unretained adsorbate pulse

The retention data described by Equation 2 are corrected from ambient conditions to the prevailing temperature and pressure conditions within the sorbent column. Use of the expansion valve at the end of the chromatographic column resulted in a negligible pressure drop across the sorbent bed, thereby simplifying the pressure correction factor used in the retention volume computations (23, 39).

The retention volume computed by Equation 2, V_r, can be related to the adsorption coefficient and the surface area value of the sorbent bed by:

$$V_r = K_A A_s \qquad (3)$$

where K_A = equilibrium adsorption coefficient in the Henry's Law region of the adsorption isotherm

A_s = total surface area of the sorbent bed

Unfortunately K_A values cannot be directly calculated from retention volume measurements by Equation 3 since the interfacial surface area is changing as a function of pressure due to the carbon dioxide sorption. However, the relative magnitude of the equilibrium shift of the sorbate from the solid adsorbent into the gaseous phase can be estimated by calculating the capacity factor, k', according to Equation 4 as given below:

$$k' = \frac{V_r - V_o}{V_o} \qquad (4)$$

where V_o is the column void volume.

Results and Discussion

The application of pressure causes a considerable decrease in the retention or breakthrough volume for an adsorbate transversing down a sorbent column. This trend is amply illustrated in Figure 9 where the retention volume for benzene in CO_2 has been plottted as a function of pressure for the crosslinked styrene/ divinylbenzene resin at 40°C. In this figure, there is a considerable decrease initially in V_r over a small pressure interval and the breakthrough volume appears to become constant

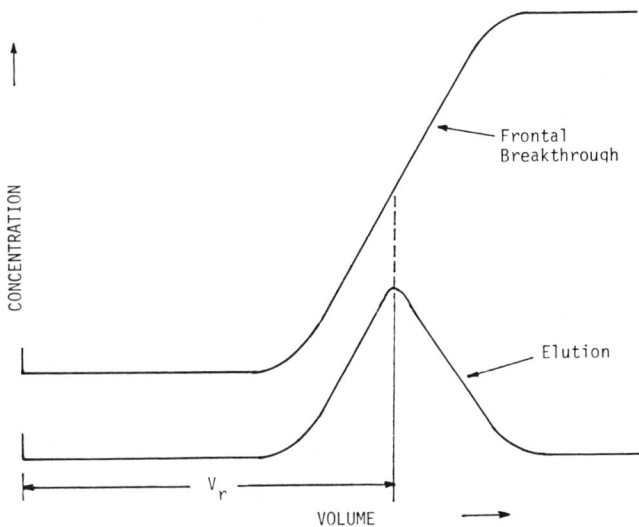

Figure 8. Relationship between the mid-point of the frontal breakthrough profile and the elution peak maxima.

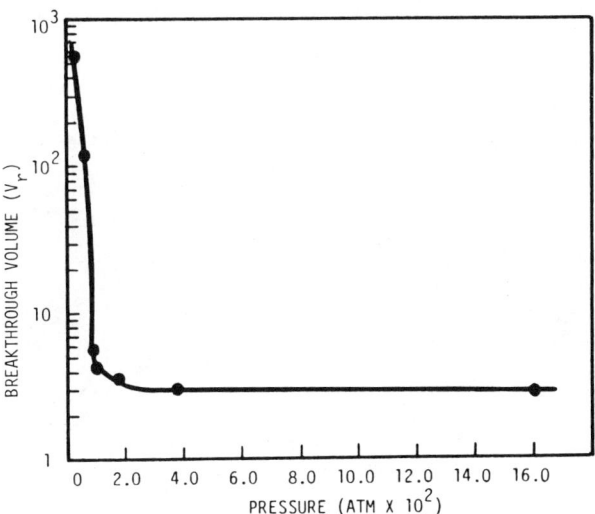

Figure 9. Breakthrough volume as a function of carbon dioxide pressure for benzene on the crosslinked polymeric resin.

beyond 200 atmospheres. The observed trend implies that further application of pressure affords no advantage in removing benzene from the adsorbent column. In general, all of sorbates studied in our experiments exhibited a similar trend upon injection into the dense CO_2 carrier gas.

Close examination of Figure 9 reveals that the rate of change for the retention volume with pressure undergoes an inflexion in the low pressure region of the graph. Therefore, additional measurements were taken to confirm this trend at pressures below 200 atmospheres. Figure 10 shows the results of this extended study for benzene on the resin adsorbent. The data show that the rate of decrease in the breakthrough volume is linear up to a pressure of approximately 70 atmospheres, however beyond this degree of gas compression, the sorbate retention volume decreases even more rapidly with increasing pressure. At a pressure below 100 atmospheres, the breakthrough volume of benzene approaches a constant value equivalent to the void volume of the column bed. Similar conclusions were reached when V_r was plotted as a function of of gas density.

To confirm the above observations, an additional set of experiments were performed using a different sorbate probe, n-heptane. Once again, as with benzene, the retention volume was found to decrease in a similar manner with increasing gas pressure on the crosslinked resin sorbent. As shown in both Figures 10 and 11, the abrupt decrease in breakthrough volume occurs at a gas pressure close to the p_c or p_{max} value as calculated by Equation 1 for carbon dioxide. A similar inflection has also been noted by Sie, Van Beersum, and Rijnders (42) when the logarithm of the partition coefficient is plotted as a function of pressure in high pressure gas-liquid chromatographic studies. The above result suggests that the uptake of supercritical gas by the adsorbent produces a significant change in the breakthrough volume of the sorbate pulse and that observed migration enhancement is in part due to modification of the gas-solid interface by the dense fluid.

Similar experiments were also conducted on the styrene/divinylbenzene adsorbent in which the capacity factor and retention volumes were determined for a series of n-alkanes ($n-C_5-C_7$) at 40°C. It was found that the capacity factors for the homologues decreased linearly by an order of two magnitudes and in a parallel fashion up to a pressure of 1250 psig (85 atm.). At this pressure the solute capacity factors began to change rapidly with pressure, similar to the trend observed in the breakthrough volume for benzene and n-heptane reported above. It was also noted in this transition region, that the capacity factors for the individual alkanes tended to merge together, and that differential migration down the column was not possible in this higher pressure regime. Such a result suggests that fractionation of the above homologues can only be accomplished successfully below the p_{max} value associated with the adsorption of the critical fluid at the gas-solid interface. Apparently, the formation of a multi molecular layer of adsorbed gas with increasing pressure not only assists in desorbing the injected sorbate but significantly nullifies the potential field of the adsorbent to such an extent that differential

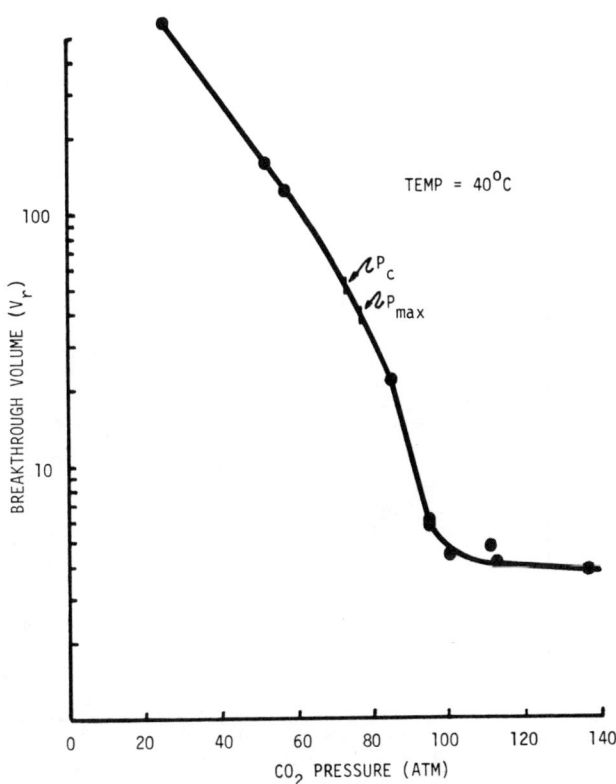

Figure 10. Retention volume for benzene versus carbon dioxide pressure on the styrene/divinylbenzene resin.

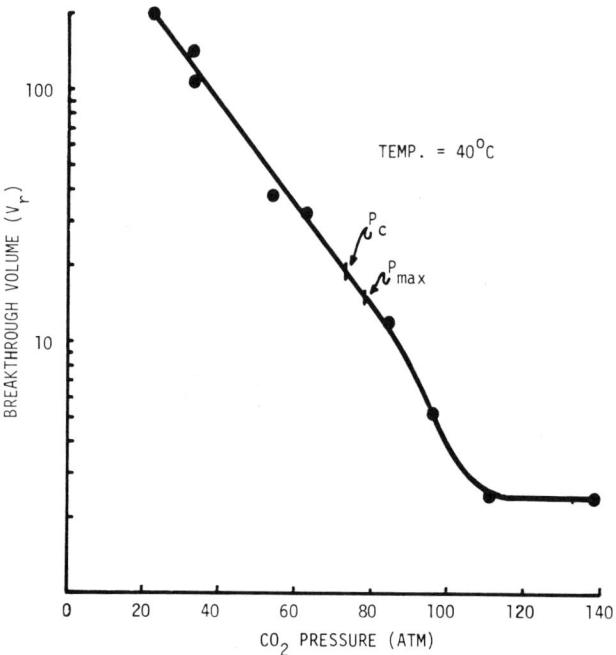

Figure 11. Retention volume for n-heptane as a function of carbon dioxide pressure on the styrene/divinylbenzene resin.

adsorption of various adsorbates is no longer possible. Similar conclusions have been reached by other investigators (43) using elution chromatographic experiments.

Additional evidence for the role of the supercritical fluid in modifying the interface is provide in Figure 12. Here the capacity factor for n-pentane on alumina at 40°C is shown as a function of pressure up to 700 atmospheres. The considerable reduction in k' below 100 atmospheres parallels the solute breakthrough behavior described earlier on the organic resin. Also plotted in Figure 12 is gravimetric adsorption data for three gases adsorbing on alumina; carbon dioxide, nitrogen, and carbon monoxide, as taken from the literature. The isotherm depicted for nitrogen and carbon monoxide were abstracted from the study of Menon (44) and represent the differential adsorption of these gases at 50°C. The maximum sorption of CO_2 on a alumina is designated with an asterisk in Figure 12 for clarity, however the entire isotherm can be found in the paper of Ozawa and Ogino (33). The actual isotherm for CO_2 on alumina at 40°C is quite sharp when compared to the nitrogen and carbon monoxide isotherms, a feature in keeping with the low reduced temperature at which the CO_2 is adsorbed (T_r = 1.03). The adsorption recorded for nitrogen and carbon monoxide was taken at reduced temperatures of 2.48 and 2.35, respectively, hence additional gas compression is required to achieve maximum uptake of the supercritical fluid. Despite the different aluminas used in the cited experimental studies, the gravimetric adsorption data in Figure 12 suggest that the breakthrough volume is controlled by competitive adsorption of the supercritical fluid at the gas-solid interface. The close proximity of the carbon dioxide adsorption maximum to the region of maximum change in V_r supports the above concept. Likewise, the higher pressures required to reach the adsorption maxima in the CO and N_2 cases, would argue that a more gradual change in the breakthrough volume with pressure should occur for similar sorbates which are desorbed at 50°C using these supercritical fluids. Maximum desorption due to competitive adsorption of the supercritical fluid at the gas-solid interface is achieved by operating close to the critical temperature or p_{max} of the displacing gas.

To test the above concept, a rather radical experiment was performed employing helium as the carrier gas at 40°C (T_r = 59.1) with subsequent measurement of the retention volumes for several light hydrocarbons ($n-C_3-C_5$) on the alumina column. Figure 13 shows the reduction in the capacity factor with pressure for n-butane as a sorbate utilizing helium and carbon dioxide as carrier gases. The trend observed in the capacity factor for n-butane as well as the other alkanes in pressurized carbon dioxide is identical to that observed with alkanes of higher carbon number on alumina and the crosslinked organic resin. However, when helium is used as the carrier fluid, there appeared to be a small initial drop in the capacity factor followed by a more gradual decrease in k' with increasing pressure (up to 10,000 psig). Examination of Figure 13 suggests that much higher pressures must be employed to effect a further reduction in the breakthrough volume for n-butane in helium on the oxide sorbent, at least comparable to that observed when using carbon

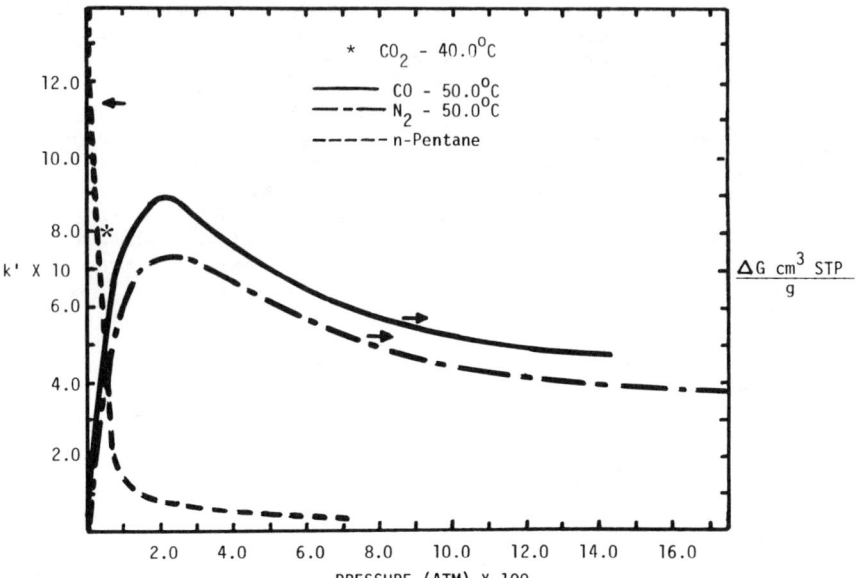

Figure 12. Capacity factor for n-pentane on alumina column at 40°C and the differential adsorption of three gases as a function of gas compression.

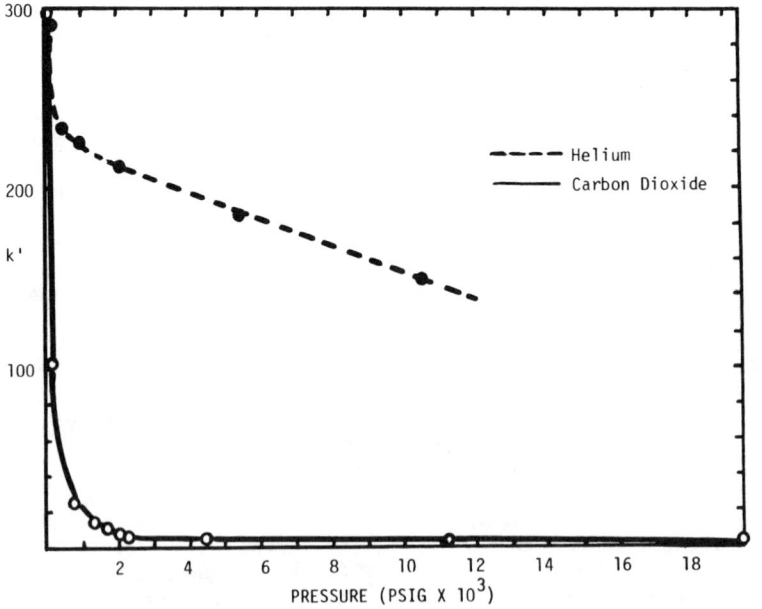

Figure 13. Capacity factor for n-butane in carbon dioxide and helium as a function of column pressure. Temp. = 40°C.

dioxide as the supercritical fluid phase. We believe that the observed difference in sorbate behavior in the two candidate gases is a reflection of the degree of surface modification caused by the different amounts of adsorption exhibited by the two supercritical gases. Sorption of helium at such a high reduced temperature is very low, but not entirely negligible at the pressures employed in this study. This later factor combined with the weak solvent power of helium gas [the solubility parameter for helium at 14,000 psig is 1.6 Hildebrands (45)] most likely accounts for the recorded gradual breakthrough of the n-butane pulse from the alumina column.

Conclusions

In this study, we have examined the role of a supercritical fluid adsorbing at the gas-solid interface and the possible implications of such a phenomena on adsorption/desorption processes. The above results strongly support the hypothesis that displacement of a bound adsorbate via competitive adsorption of the supercritical fluid at the interface is a major factor in regenerating adsorbents and in controlling the partition of the solute (sorbate) into the fluid phase. Additional support for this concept can be found in the regeneration studies of Eppig and co-workers (14) who demonstrated the recovery of ethanol, methyl ethyl ketone, and toluene from activated carbon beds using modest CO_2 pressures (less than 100 atms.). Adsorption/desorption studies of light hydrocarbon/CO_2 mixtures on molecular sieve (46) also indicate that CO_2 is preferentially adsorbed over the alkane moieties, supporting a similar concept advanced in this study.

The retention volume data presented in this study imply that there are distinct regions of sorbate breakthrough behavior that are in part defined by the extent of compressed gas adsorption at the solid interface. The region for most profitable fractionation of adsorbates would appear to lie below the p_{max} value as defined by the adsorption isotherm of the supercritical fluid. Plots of retention volume versus pressure also are of great aid in defining a maximum pressure beyond which no further reduction in breakthrough volume is possible through increased gas compression. Such data is of value in specifying the most efficient conditions for the rapid and economical supercritical fluid regeneration of adsorbent beds. It would appear that the p_{max} value discussed above, is a minimal pressure which investigators should endeavor to attain when attempting to desorb compounds with supercritical fluid media. However, extra gas compression may be required above the p_{max} value to solvate and strip strongly adsorbed components from adsorbent matrix. Additional experimental studies should be undertaken in the future to better define and verify the concepts presented in this study, particularly utilizing other supercritical fluid displacing agents and adsorbates of lower volatility and a polar nature.

Literature Cited

1. deFilippi, R. P.; Krukonis, V. J.; Robey, R. J.; Modell, M. "Supercritical Fluid Regeneration of Activate Carbon for Adsorption of Pesticides," EPA Report 600/2-80-054, 1980.
2. Klesper, E. In "Extraction with Supercritical Gases"; Scheneider, G. M.; Stahl, E.; Wilke, G., Eds.; Verlag Chemie: Deerfield Beach, Florida, 1980; pp. 115-139.
3. Zosel, K. U.S. Patent 4 156 688, 1979.
4. Hubert, P.; Vitzthum, O. U.S. Patent 4 411 923, 1983.
5. Sirtl, W. U.S. Patent 4 344 974, 1982.
6. Stegan, G.; DeWilt, H. British Patent 2 008 921, 1979.
7. Hajnik, D. F. M.S. Thesis, Pennsylvania State University, State College, Pennsylvania, 1980.
8. Yang, R. T.; Saunders, J. T. Fuel 1985, 64, 616.
9. Findenegg, G. H.; Korner, B.; Fischer, J.; Bohn, M. Ger. Chem. Eng. 1983, 6, 80-84.
10. Shimokobe, I. Japanese Patent 31 679, 1976.
11. Modell, M. U.S. Patent 4 061 566, 1977.
12. Modell, M. U.S. Patent 4 124 528, 1978.
13. Modell, M.; deFilippi, R. P.; Krukonis, V. J. In "Activated Carbon Adsorption of Organics from the Aqueous Phase"; Stuffet, I. H.; McGuire, M. J., Eds.; Ann Arbor Science Publishers, Ann Arbor, Michigan, 1980; Vol. 1, pp. 447-461.
14. Eppig, C. P.; deFilippi, R. P.; Murphy, R. A. "Supercritical Fluid Regeneration of Activated Carbon Used for Volatile-Organic-Compound Vapor Adsorption," EPA Report 600/2-82-067, 1982.
15. deFilippi, R. P.; Robey, R. J. "Supercritical Fluid Regeneration of Adsorbents," EPA Report 600/2-83-038, 1983.
16. Kander, R. G.; Paulaitis, M. E. In "Chemical Engineering at Supercritical Fluid Conditions"; Paulaitis, M. E.; Penninger, J. M. L.; Gray, R. D.; Davidson, P., Eds.; Ann Arbor Science Publishers: Ann Arbor, Michigan, 1983; pp. 461-476.
17. Greene, S. A.; Roy, H. E. Anal. Chem. 1957, 29, 569.
18. Janak, J. Ann. N.Y. Acad. Sci. 1959, 72, 606.
19. Bachmann, L.; Bechtold, E.; Cremer, E. J. Catalysis 1962, 1, 113.
20. Rabbini, G. S. M.; Rusek, M.; Janak, J. J. Gas Chromatog. 1968, 6, 399.
21. Sie, S. T.; Rijnders, G. W. A. Separation Sci. 1967, 2, 757.
22. Sie, S. T.; Bleumer, J. P. A.; Rijnders, G. W. A. In "Gas Chromatography-1968"; Harbourn, C. L. A., Ed.; Elsevier, Amsterdam, 1969, pp. 235-251.
23. Gilmer, H. B.; Kobayashi, R. AIChE J. 1964, 10, 797.
24. Gilmer, H. B.; Kobayashi, R. AIChE J. 1965, 11, 702.
25. Masukawa, S.; Kobayashi, R. J. Chem. Eng. Data 1968, 13, 197.
26. Masukawa, S.; Kobayashi, R. AIChE J. 1969, 15, 190.
27. Rolniak, P. D., Kobayashi, R. AIChE J. 1980, 26, 616.
28. Kobayashi, R.; Kragas, T. J. Chromatog. Sci. 1985, 23, 11.

29. Brady, B. O.; Kao, C. C.; Gambrell, R. P.; Dooley, K. M.; Knopf, F. C. Ind. Eng. Chem. Process Des. Dev. To be published.
30. Menon, P. G. Chem. Rev. 1968, 68, 277.
31. Menon, P. G. J. Phys. Chem. 1968, 72, 2695.
32. Menon, P. G. In "Advances in High Pressure Research"; Bradley, R. S., Ed.; Academic Press, New York, 1979; Vol. 3, pp. 313-365.
33. Ozawa, S.; Ogino, Y. Nippon Kagaku Kaishi 1972, 1.
34. Ozawa, S.; Kusumi, S.; Ogino, Y. Proc. 4th International Conference on High Pressure - Kyoto - 1974, 1975, pp. 580-587.
35. Specovius, J.; Findenegg, G. H. Ber. Bunsenges. Phys. Chem. 1978, 82, 174.
36. Specovius, J.; Findenegg, G. H. Ber. Bunsenges. Phys. Chem. 1980, 84, 690.
37. Blumel, S.; Koster, F.; Findenegg, G. H. J. Chem. Soc. Faraday Trans. 2 1982, 78, 1753.
38. Giddings, J. C.; Myers, M. N.; King, J. W. J. Chromatog. Sci. 1969, 7, 276.
39. Myers, M. N.; Giddings, J. C. Separation Sci. 1966, 1, 761.
40. Kiselev, A. V.; Yashin, Y. I. "Gas-Adsorption Chromatography"; Plenum Press, New York; 1969; pp. 11-16.
41. Conder, J. R.; Young, C. L. "Physicochemical Measurement by Gas Chromatography"; John Wiley & Sons: New York, 1979; p. 389.
42. Sie, S. T.; Van Beersum, W.; Rijnders, G. W. A. Separation Sci. 1966, 1, 459.
43. Sie, S. T.; Rijnders, G. W. A. Anal. Chim. Acta. 1967, 38, 31.
44. Menon, P. G. J. Am. Chem. Soc. 1965, 87, 3057.
45. Giddings, J. C.; Myers, M. N.; McLaren, L.; Keller, R. A. Science 1968, 162, 67.
46. Basmadian, D.; Wright, D. Chem. Eng. Sci. 1981, 36, 937.

RECEIVED July 1, 1986

Chapter 14

Mechanism of Solute Retention in Supercritical Fluid Chromatography

C. R. Yonker, R. W. Wright, S. L. Frye, and R. D. Smith

Chemical Methods and Separations Group, Chemical Technology Department, Pacific Northwest Laboratory, Battelle Memorial Institute, Richland, WA 99352

The complicated dependence of retention in supercritical fluid chromatography as a function of temperature and pressure is examined. Simple thermodynamic relationships are derived and discussed which allow the calculation of the slope of solute retention as a function of both temperature and pressure. These models are compared to experimental data obtained over a wide range of temperatures and pressures.

Solute retention as a function of pressure at constant temperature is dependent on the partial molar volume of the solute in the stationary phase ($\overline{V}_i^{\infty \text{-stat}}$), the solubility of the solute in the fluid and the isothermal compressibility of the fluid. The model was compared to the experimental retention data of naphthalene and biphenyl and fit the data quite well.

Solute retention as a function of temperature at constant pressure is seen to be dependent on the partial molar enthalpy of solute transfer ($\Delta \overline{H}_{i(T_0)}^{\infty}$) between the mobile and stationary phases, the heat capacity of the supercritical fluid mobile phase and the volume expansivity of the fluid. The model was compared to chromatographic retention data for solutes in n-pentane and CO_2 as the fluid mobile phase and was seen to fit the data well.

The effect of temperature, pressure and density on solute retention (k') in supercritical fluid chromatography (SFC) has been well studied.([1-6](#)) Retention in SFC depends upon both solute solubility in the fluid and solute interaction with the stationary phase. The functional relationship between retention and pressure at constant temperature has been described by Van Wasen and Schneider.([1](#)) The trend in retention is seen to depend on the partial molar volume of

0097–6156/87/0329–0172$06.00/0
© 1987 American Chemical Society

the solute in the mobile and stationary phases coupled with the isothermal compressibility of the fluid mobile phase.

The effect of temperature on retention has been described experimentally,(4-8) but the functional dependence of k' with temperature has only recently been described.(9) A thermodynamic model was outlined relating retention as a function of temperature at constant pressure to the volume expansivity of the fluid, the enthalpy of solute transfer between the mobile phase and the stationary phase and the change in the heat capacity of the fluid as a function of temperature.(9) The solubility of a solid solute in a supercritical fluid has been discussed by Gitterman and Procaccia (10) over a large range of pressures. The combination of solute solubility in a fluid with the equation for retention as a function of pressure derived by Van Wasen and Schneider allows one to examine the effect of solubility on solute retention.

In this work we derive simple relationships between temperature, solute solubility and retention. The simple thermodynamic models developed predict the trend in retention as a function of pressure, given the solubility of the solute in the fluid mobile phase at constant temperature and the trend in k' as a function of temperature at constant pressure. Our aim is to examine the complicated dependence of retention on the thermodynamic and physical properties of the solute and the fluid, providing a basis for consideration of more subtle effects in SFC.

Theory

In SFC the basic assumption of infinitely dilute solutions of the solute in the mobile and stationary phases is valid. The concentration of the solute in these phases respectively is $C_i = X_i/V_m$, where X_i is the mole fraction of solute (i) and V_m is the molar volume of the pure mobile or stationary phase.(1) Solute retention is determined from a dimensionless retention factor, k', where,

$$k' = (C_i^{stat}/C_i^{mob}) \cdot (V^{stat}/V^{mob}) \quad (1)$$

C_i^{stat} and C_i^{mob} are the concentration of solute (i) in the stationary and mobile phases respectively, V^{stat} and V^{mob} are the volumes of the stationary and mobile phase. Substituting for concentration into eq. 1,

$$k' = (X_i^{stat}/X_i^{mob}) \cdot (V_m^{mob}/V_m^{stat}) \cdot (V^{stat}/V^{mob}) \quad (2)$$

Taking the natural logarithm of both sides of eq. 2 one obtains,

$$\ln k' = \ln (X_i^{stat}/X_i^{mob}) + \ln (V_m^{mob} V^{stat}/V_m^{stat} V^{mob}) \quad (3)$$

At equilibrium; the solute chemical potential in the respective phases are equal: $\mu_i^{stat} = \mu_i^{mob}$.(11) Therefore,

$$\mu_i^{stat} = \mu_i^{mob} = \mu_i^{stat\,\infty} + RT \ln X_i^{stat} = \mu_i^{mob\,\infty} + RT \ln X_i^{mob} \quad (4)$$

where μ_i^∞ is the chosen standard state of unit molar concentration of solute (i) in the two phases. It should be pointed out that in the case of the solute in the stationary phase, this choice of standard state is a hypothetical one which exhibits the properties of an infinite dilute solution.(12) Rearranging eq. 4,

$$\ln (X_i^{stat}/X_i^{mob}) = (\mu_i^{mob\,\infty} - \mu_i^{stat\,\infty})/RT \quad (5)$$

Substituting eq. 5 into 3,

$$\ln k' = (\mu_i^{mob\,\infty} - \mu_i^{stat\,\infty})/RT + \ln (V_m^{mob} V^{stat}/V_m^{stat} V^{mob}) \quad (6)$$

An assumption can be made that the second term on the right-hand side (RHS) of eq. 6 is independent of pressure except for V_m^{mob}, the molar volume of the fluid mobile phase. Therefore differentiation of eq. 6 with respect to pressure at constant temperature yields,

$$(\partial \ln k'/\partial P)_T = 1/RT \; [(\partial \mu_i^{mob\,\infty}/\partial P)_T - (\partial \mu_i^{stat\,\infty}/\partial P)_T] + (\partial \ln V_m^{mob}/\partial P)_T \quad (7)$$

The partial molar volume of the solute is defined as $(\partial \mu_i/\partial P)_T$.(11) The second term on the RHS of eq. 7 is the isothermal compressibility (K) of the fluid mobile phase.(13) Thus, eq. 7 reduces to,

$$(\partial \ln k'/\partial P)_T = 1/RT \; [\bar{V}_i^{mob\,\infty} - \bar{V}_i^{stat\,\infty}] - K \quad (8)$$

where $\bar{V}_i^{mob\,\infty}$ and $\bar{V}_i^{stat\,\infty}$ are the partial molar volumes of the solute (i) in the mobile and stationary phases at infinite dilution, respectively. Equation 8 is the same as obtained by Van Wasen and Schneider (1) in their derivation of the trend in retention as a function of pressure for SFC.

The solubility of a solid in a supercritical fluid has been described by Gitterman and Procaccia.(10) The region of interest chromatographically will be for infinitely dilute solutions whose concentration is far removed from the lower critical end point (LCEP) of the solution. Therefore the solubility of the solute in a supercritical fluid at infinite dilution far from criticality can be approximated as,

$$(\partial \ln X_i^{mob}/\partial P)_T = 1/RT \; [V^s - \bar{V}_i^{mob\,\infty}] \quad (9)$$

where V^s is the molar volume of the pure solid solute.(10) Equation 9 is based on the assumption that the solute's partial molar volume at infinite dilution roughly approximates that of a saturated solution at low mole fraction values of the solute in the fluid phase (see Figures 1 and 5). Solving eq. 9 for $\bar{V}_i^{mob\,\infty}$,

$$\bar{V}_i^{mob\,\infty} = -RT\,(\partial \ln X_i^{mob}/\partial P)_T + V^s \quad (10)$$

Substituting eq. 10 into eq. 8 and rearranging,

$$(\partial \ln k'/\partial P)_T = (V^s - \bar{V}_i^{stat\,\infty})/RT - (\partial \ln X_i^{mob}/\partial P)_T - K \quad (11)$$

Equation 11 should be the relationship between retention-solubility and pressure at constant temperature for infinitely dilute solutions. The RHS of eq. 11 consists of three terms, the first term will be a constant whose value depends on the partial molar volume of the solute in the stationary phase. The second term is the solubility of the solute in the supercritical fluid mobile phase. The last term, the fluid's isothermal compressibility, can be reasonably predicted from a two-parameter, cubic equation of state (EOS) such as the Redlich-Kwong EOS or the Peng-Robinson EOS.(14,15)

The fluid's isothermal compressibility was determined using the Redlich-Kwong EOS to evaluate the derivative $(\partial V_m^{mob}/\partial P)_T$ in eq. 12 due to the ease of finding an analytical solution,

$$(\partial \ln V_m^{mob}/\partial P)_T = (1/V_m^{mob}) \cdot (\partial V_m^{mob}/\partial P)_T = -K \quad (12)$$

From the Redlich-Kwong EOS,

$$(\partial V_m^{mob}/\partial P)_T = \frac{V_m^{mob}\,T^{0.5}\,(b^2 - (V_m^{mob})^2)}{P\,T^{0.5}(3\,(V_m^{mob})^2 - b^2) - RT^{3/2}\,(2V_m^{mob} + b) + a} \quad (13)$$

where R is the gas constant, P is pressure, and T is temperature in K. The constants a and b of the Redlich-Kwong EOS are,

$$a = 0.4278\,R^2\,T_c^{2.5}/P_c \quad (14a)$$

$$b = 0.0867\,RT_c/P_c \quad (14b)$$

where P_c and T_c are the critical pressure and temperature of the fluid. The molar volume of the fluid was determined by an EOS, thus allowing one to solve for the isothermal compressibility of the fluid. Therefore, using eq. 11, we can calculate the trend in retention as a function of pressure at constant temperature given the solubility of the solute in the supercritical fluid, the isothermal compressibility of the fluid and the partial molar volume of the solute in the stationary phase at infinite dilution.

The derivation of the dependence of the trend in solute retention with temperature at constant pressure is similar to that described above for the trend in solute retention as a function of solute solubility and pressure at constant temperature. Once again making the same assumptions that led to eq. 6 and assuming temperature has a negligible effect on V^{stat}, V_m^{stat}, and V^{mob}, differentiation of eq. 6 with respect to temperature at constant pressure yields

$$\left(\frac{\partial \ln k'}{\partial T}\right)_P = \frac{-1}{R}\left(\frac{\partial(\Delta\mu_i^\infty/T)}{\partial T}\right)_P + \left(\frac{\partial \ln V_m^{mob}}{\partial T}\right)_P \tag{15}$$

The first term on the right hand side of eq. 15 can be evaluated as (where $\Delta\mu_i^\infty = \mu_i^{mob\,\infty} - \mu_i^{stat\,\infty}$),

$$\frac{\Delta\mu_i^\infty}{T} = \frac{\overline{\Delta H_i^\infty}}{T} - \overline{\Delta S_i^\infty} \tag{16}$$

where $\overline{H_i^\infty}$ and $\overline{\Delta S_i^\infty}$ are the partial molar enthalpy and entropy of transferring the solute molecule from the stationary phase to the mobile phase at infinite dilution. Taking the derivative of eq. 16 with respect to temperature at constant pressure and substituting into eq. 15 one obtains,

$$\left(\frac{\partial \ln k'}{\partial T}\right)_P = \frac{-1}{R}\left[\frac{-\overline{\Delta H_i^\infty}}{T^2} + \frac{1}{T}\left(\frac{\partial \overline{\Delta H_i^\infty}}{\partial T}\right)_P - \left(\frac{\partial \overline{\Delta S_i^\infty}}{\partial T}\right)_P\right] + \left(\frac{\partial \ln V_m^{mob}}{\partial T}\right)_P \tag{17}$$

and upon rearrangement;

$$\left(\frac{\partial \ln k'}{\partial T}\right)_P = \frac{1}{RT^2}\left[\overline{\Delta H_i^\infty} - T\left(\frac{\partial \overline{\Delta H_i^\infty}}{\partial T}\right)_P + T^2\left(\frac{\partial \overline{\Delta S_i^\infty}}{\partial T}\right)_P\right] + \frac{1}{V_m^{mob}}\left(\frac{\partial V_m^{mob}}{\partial T}\right)_P \tag{18}$$

The second and third terms in the bracket on the RHS of eq. 18 are equal and opposite because $(\partial \overline{\Delta H_i^\infty}/\partial T)_P = \overline{\Delta Cp_i^\infty}$ and $(\partial \overline{\Delta S_i^\infty}/\partial T)_P = \overline{\Delta Cp_i^\infty}/T$. Thus eq. 18 becomes on substitution,

$$\left(\frac{\partial \ln k'}{\partial T}\right)_P = \frac{\overline{\Delta H_i^\infty}}{RT^2} + \frac{1}{V_m^{mob}}\left(\frac{\partial V_m^{mob}}{\partial T}\right)_P \tag{19}$$

Equation 19 is valid over a limited temperature range far from the critical point of the solution mixture. Therefore, for the case of typical gas and/or liquid chromatographic conditions, eq. 19 becomes the limiting case where $\overline{\Delta H_i^\infty}$ is constant and independent of temperature. Further, assuming that the second term on the RHS of eq. 19 is small or negligible under normal operating conditions for liquid and gas chromatography, eq. 19 on rearrangement reduces to the familiar case described by Van't Hoff (16)

$$\left(\frac{\partial \ln k'}{\partial(1/T)}\right)_P = \frac{-\overline{\Delta H_i^\infty}}{R} \tag{20}$$

14. YONKER ET AL.   *Mechanism of Solute Retention*   177

Assuming $\overline{\Delta H_i^\infty}$ is a function of temperature, (17)

$$\overline{\Delta H_i^\infty} = \overline{\Delta H_{i(To)}^\infty} + \int_{To}^{T_f} \overline{\Delta C_{Pi}^\infty} dT \qquad (21)$$

where $\overline{\Delta C_{Pi}^\infty}$ is the difference in heat capacity of the solute in the mobile phase as compared to the solute in the stationary phase. Equation 21 is valid over the entire temperature range of interest for chromatographic separations, i.e., liquid, supercritical fluid and gas chromatography. Substituting eq. 21 into eq. 19 one obtains,

$$\left(\frac{\partial \ln k'}{\partial T}\right)_P = \frac{1}{RT^2}\left[\overline{\Delta H_{i(To)}^\infty} + \int_{To}^{T_f} \overline{\Delta C_{Pi}^\infty} dT\right] + \frac{1}{V_m^{mob}}\left(\frac{\partial V_m^{mob}}{\partial T}\right)_P \qquad (22)$$

Equation 22 describes the relationship between temperature and solute retention at constant pressure over any range of temperatures.

A simplifying assumption is that at infinite dilution of the solute in the mobile phase, $\overline{\Delta C_{Pi}^\infty}$ in eq. 22 can be approximated by the heat capacity of the mobile phase. The dependence of heat capacity on temperature and pressure near the critical point of a fluid is complicated.(18) In this work we formulate an empirical approximation to the integral in eq. 22 for the heat capacity as a function of temperature near the critical point of the fluid mobile phase.(9)

The second term on the RHS of eq. 22 is the volume expansivity which can be calculated from a two-parameter, cubic equation of state such as the Redlich-Kwong EOS.(14) We chose the Redlich-Kwong EOS because the "a" term is independent of temperature, which simplifies the procedure of solving for the analytical solution. Using the method of implicit differentiation with the Redlich-Kwong equation of state,

$$\left(\frac{\partial V_m^{mob}}{\partial T}\right)_P = \frac{(V_m^{mob})^2 R + V_m^{mob} Rb + 0.5 V_m^{mob} aT^{-1.5} - 0.5 abT^{-1.5}}{3(V_m^{mob})^2 P - 2V_m^{mob} RT - Pb^2 - RTb + aT^{-0.5}} \qquad (23)$$

where R is the gas constant, P is pressure, and T is temperature in K. The constants a and b of the Redlich-Kwong equation have already been defined in eqs. 14a and 14b. The determination of the derivative in eq. 23 was accomplished by the use of an EOS to solve for the compressibility of the fluid. The molar volume of the fluid mobile phase can then be calculated from its compressibility.

The enthalpy of transfer $\left(\overline{\Delta H_{i(To)}^\infty}\right)$ of the solute between phases can be determined from experiment by a Van't Hoff plot of lnk' versus 1/T at constant density.(19) Therefore, the RHS of eq. 22 can be evaluated for a particular fluid mobile phase-solute combination and the slope of lnk' at constant pressure as temperature is varied can be predicted.

Experimental

The experimental apparatus and technique has been described in detail elsewhere.(20,21) The retention factors of naphthalene and biphenyl under isothermal conditions at various pressures were obtained using capillary columns coated with a cross-linked phenyl polymethylphenylsiloxane stationary phase with carbon dioxide as the fluid mobile phase. A Varian 8500 syringe pump was operated under computer control providing accurate, pulsefree control of the fluid pressure. The retention times of the solute as a function of pressure were determined by a reporting integrator with an accuracy of a tenth of a second. Solubility data for the solutes in CO_2 was obtained from the literature.(22)

The effect of temperature on retention was studies using n-hexadecane on a 20 m, 50μ I.D. fused silica capillary column coated with an OV-17 phase using FID detection. The OV-17 was cross-linked in-situ to decrease its solubility in the supercritical fluid. The stationary phase film thickness was calculated to be ~0.25 μm. The enthalpy of transfer ($\Delta H_{i(T_0)}^{\infty}$) for the n-alkane (n-heptadecane, Alltech Associates) was obtained on a column containing a cross-linked OV-17 stationary phase over a range of temperatures at constant density (0.38 g/cm^3). The retention times of the solutes for the Van't Hoff plots and solute retention as a function of temperature were determined by a reporting integrator.

Results and Discussion

Solute retention as a function of pressure has been determined experimentally for a wide number of solutes over a range of temperatures and pressures.(1,23-25) The trend in retention of a solute with pressure can be predicted within the limitations of our assumptions using eq. 11. The calculations are then compared with experimental data for the retention of naphthalene and biphenyl at various temperatures and pressures were obtained with supercritical CO_2.

For naphthalene the temperatures investigated were 35.0°C, 60.4°C and 64.9°C with pressures ranging from 75 atm to 120 atm. The solubility of naphthalene at these temperatures in supercritical CO_2 has been reported by McHugh and Paulaitis.(22) This data allows the trend in retention to be predicted from the calculation of the slope of $(\partial \ln k'/\partial P)_T$ for naphthalene at these temperatures. A plot of the data ($\ln x_i^{mob}$ versus pressure) from McHugh and Paulaitis (22) is shown in Figure 1. The slope $(\partial \ln x_i^{mob}/\partial P)_T$ at the above-mentioned temperatures was obtained through interpolation between the data points. Where necessary, the data range was extended to lower pressures than given in reference 22 as shown in Figure 2 by fitting the data using the method described by Reid et al.(26) This technique allows the modeling of the trend in retention as a function of pressure at constant temperatures for a wider range in pressures. The isothermal compressibility of supercritical CO_2 was calculated for the temperature and pressure range of interest using eq. 13 and a two-parameter cubic equation of state to predict the molar volume of the fluid. The solute partial molar volume in the stationary phase was assumed constant and independent of pressure.(27) The molar

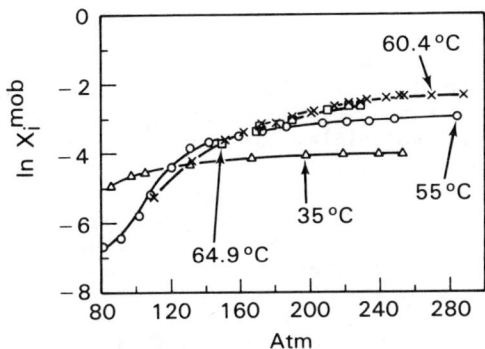

Figure 1 The solubility of naphthalene in CO_2 as a function of pressure. Data from McHugh and Paulaitis.([22](#))

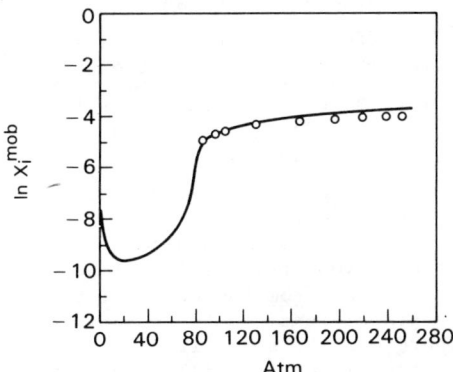

Figure 2 The solubility of naphthalene in CO_2 at 35.0°C as a function of pressure using the technique of Reid et al. ([26](#)) to extend data to lower pressures.

volume of naphthalene was determined to be $\sim$130.8 cm^3/mole. Therefore, with the above simplifying assumptions, experimental solubility data and the prediction of the isothermal compressibility of the fluid as a function of pressure at constant temperature, the trend in solute retention could be calculated based on an appropriate choice of $\bar{V}_i^{\infty,stat}$ which gives the best fit to the experimental data. The experimental data for naphthalene at 35.0°C and 64.9°C are given in Figures 3 and 4, where the experimentally obtained lnk' versus pressure is plotted against the predicted slope of the retention of naphthalene from eq. 11 (solid line). The simple thermodynamic model was found to fit the data satisfactorily for this solute.

The experimental solubility for biphenyl in supercritical CO_2 at different temperatures as a function of pressure are shown in Figure 5.(22) The slopes of these plots were used to determine $(\partial \ln X_i^{mob}/\partial P)_T$ for biphenyl in supercritical CO_2, and these values were supplemented with solubilities predicted for lower pressures from theory as needed.(26) The experimental results of biphenyl at the temperatures 35.8°C, 45.5°C, and 55.2°C are shown in Figures 6 and 7. The solid line in these figures is the theoretical prediction of the slope of the retention of the solute based on eq. 11 and the $\bar{V}_i^{\infty,stat}$ value which gives the best fit to the experimental data. Once again the agreement between theory and experiment is seen to be quite good for biphenyl, as it was for naphthalene.

The effect of the partial molar volume of the solute in the stationary phase at infinite dilution can be seen in Figure 8. This figure shows how $\bar{V}_i^{\infty,stat}$ affects the theoretical slope of the solute retention as a function of pressure. For the particular case of biphenyl at 55.2°C, increasing $\bar{V}_i^{\infty,stat}$ increases the slope predicted from eq. 11, due to $(V^s - \bar{V}_i^{\infty,stat})$ becoming more dominant as $\bar{V}_i^{\infty,stat}$ increases. Thus, the combination of solubility and chromatographic data may provide a simple approach to study the partial molar volume of solutes in condensed phases.

On examination of eq. 11, one can deduce that as the isothermal compressibility of the solvent becomes less important (temperature and pressure further removed from the critical temperature and pressure), $(\partial \ln k'/\partial P)_T$ is proportional to the solubility of the solute in the fluid phase. Therefore, if solubility is found to be a linear function of density, then retention will mirror this behavior and also be a linear function of density. Further, the farther one is from the critical pressure and temperature of the solvent, the more likely one obtains a constant slope [$(\partial \ln k'/\partial P)_T$ = constant].

The effect of temperature on solute retention in SFC has been investigated experimentally (5,8) and described theoretically.(9) The effect of temperature as discussed in the theory section, based on the assumptions outlined, resulted in eq. 22, where the trend in retention is dependent on the volume expansivity of the fluid, the enthalpy of transfer of the solute from the mobile phase to the stationary phase and the change in the heat capacity of the fluid mobile phase.

The volume expansivity leads to the retention maxima seen in SFC near the critical point of the supercritical fluid mobile phase. The

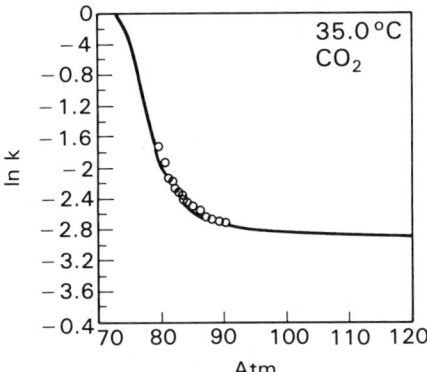

Figure 3 Retention as a function of pressure for naphthalene at 35.0°C. Solid line was predicted from eq. 11.

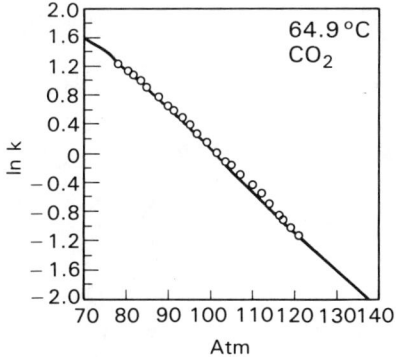

Figure 4 Retention as a function of pressure for naphthalene at 64.9°C. Solid line was predicted from eq. 11.

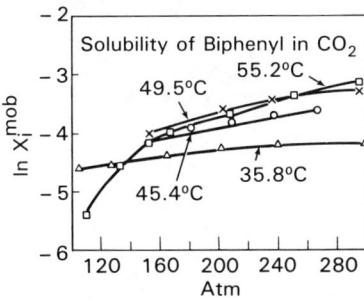

Figure 5 The solubility of biphenyl in CO_2 as a function of pressure. Data from McHugh and Paulaitis.(22)

182   SUPERCRITICAL FLUIDS

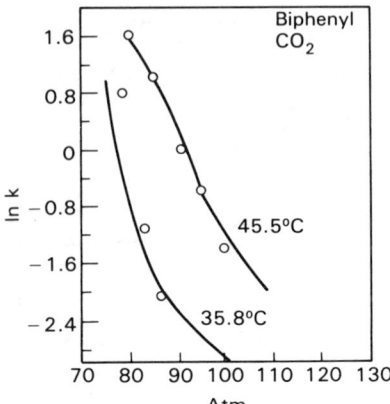

Figure 6. Retention as a function of pressure for biphenyl at 35.8 and 45.5 °C. Solid lines were predicted from eq. 11.

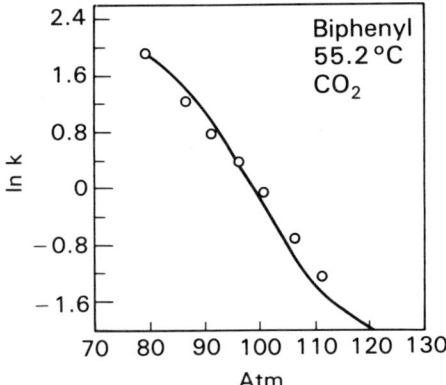

Figure 7. Retention as a function of pressure for biphenyl at 55.2 °C. Solid line was predicted from eq. 11.

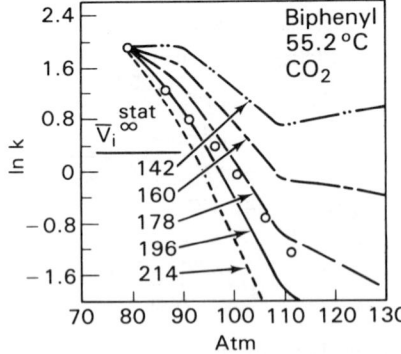

Figure 8. Retention as a function of pressure for biphenyl at 55.2 °C. Solid linew were predicted from eq. 11 using different $\bar{V}_i^{stat\,\infty}$ values.

effect of temperature on the partial derivative in the volume expansivity of the fluid is shown in Figure 9. The partial differential of the fluid molar volume with respect to temperature is plotted for n-pentane at 35.5 atm (Case 1) and 69.1 atm (Case 2). One can see that near the critical point for n-pentane (33.3 atm, 196.4°C) the differential term becomes dominant. At pressures further removed from the fluid's critical pressure the differential becomes negligible, as seen in Figure 9.

Figure 10 depicts experimental data from reference 5 plotted as lnk' versus temperature. The solid line is the slope in retention predicted from eq. 22. The enthalpy of transfer [$\overline{\Delta H}_{i(T_0)}^{\infty}$] was assumed independent of temperature and equal to -5.0 Kcal/mole. The thermodynamic relationship outlined in the theory section predicts the major features for the temperature dependence of retention. However, further discussion is needed regarding the apparent correlation between the retention maximum and molecular weight evident in the data of Klesper and co-workers.(5,8) The assumption that $\overline{\Delta H}_i^{\infty}$ is invariant with temperature is poor near the critical point of the solvent. Equation 22 incorporates a correction term ($\int_{T_0}^{T_f} \overline{\Delta C}_{p_i}^{\infty} dT$) to $\overline{\Delta H}_i^{\infty}$ which dominates near the critical point of the fluid. Therefore, two contributions compete near the solvent's critical point; (a) the volume expansivity of the solvent and (b) $\overline{\Delta C}_p^{\infty}$. Density effects both of these terms through its effect on the solvent molar volume and the number of intermolecular interactions occurring in the critical region. As stated earlier, with conditions of solute infinite dilution which are approximated by SFC, $\overline{\Delta C}_p^{\infty}$ can be assumed to be equal to the heat capacity of the mobile phase. For carbon dioxide at its critical point, the heat capacity follows a sharply increasing gaussian function.(18) Since no simple mathematical relationship presently exists for $\overline{\Delta C}_p^{\infty}$ at the pressures relevant to SFC, this term can only be estimated crudely.(9) Overall, this approach predicts the trend in retention quite well. The addition of the correction term to the enthalpy of solute transfer making it dependent on temperature, allows simulation of the effect of the change in the number of intermolecular interactions between the solute and the solvent molecules as the density decreases rapidly near the critical point of the solvent.

Figure 11 contains the experimental data for chrysene (5) replotted as lnk' versus temperature with the solid line being the theoretical prediction of the slope from eq. 22. The data in Figures 10 and 11 are both with n-pentane as the fluid mobile phase; the difference between the two sets of data is the experimental pressure. The model predicts the trend in retention for both cases quite well. In Figure 11 one can see the decrease in importance of the volume expansivity and heat capacity of the fluid on retention as the experimental conditions are further removed from the critical temperature and pressure of the fluid.

Additional evidence supporting the thermodynamic model was obtained using a capillary column with supercritical CO_2 and the solute hexadecane; modeling was undertaken using a $\overline{\Delta H}_{i(T_0)}^{\infty}$ value of

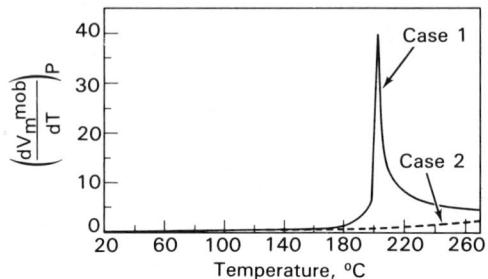

Figure 9 Plot of $(\partial V_m^{mob}/\partial T)_P$ versus temperature for n-pentane at 35.5 atm (Case 1) and 69.1 atm (Case 2).

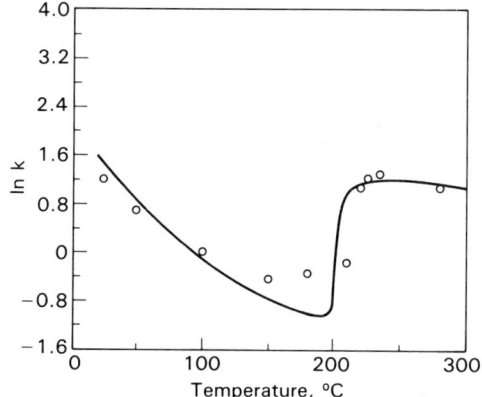

Figure 10 Experimental data (Ref. 5) and theoretical model (solid line) for chrysene with n-pentane, 35.5 atm, $\Delta \bar{H}_i^{-\infty}(T_o)$ = 5.0 Kcal/mole.

−10.6 Kcal/mole obtained from a "Van't Hoff" (lnk' vs. 1/T) plot of heptadecane on the same column. This data is shown in Figure 12, with the solid line being the theoretical prediction from eq. 22. Once again the fit of the experimental data by the model is quite good.

Conclusion

The thermodynamic relationship described in this article have been shown to describe the features of solute retention as a function of temperature at constant pressure and as a function of pressure at constant temperature (given the solubility of the solute in the fluid mobile phase). The dependence of retention upon temperature can apparently be ascribed to a combination of two effects. These effects include the rapid change in the number of intermolecular interactions of the solute-solvent molecules as one progresses through the critical point for the solvent ($\overline{\Delta C_p^\infty}$) and on the volume expansivity of the solvent. Recent work in this laboratory using solvatochromic probes has directly explored these solvent-solute interactions and clarified the nature of the fluid phase solvation phenomena.(28)

The $\overline{RHS}$ of eq. 22 reduces to the limiting case for both liquid and gas chromatography, where $\overline{\Delta H_i^\infty}$ is a constant over a narrow temperature range ($\overline{\Delta C_{p_i}^\infty} = 0$) and the effect of the volume expansivity of the mobile phase is negligible. For this case, solute retention is dominated by the partial molar enthalpy of solute transfer at infinite dilution ($\overline{\Delta H_i^\infty}$). The volume expansivity and ΔC_p^∞ become dominant near the critical temperature of the solvent, contributing to the physical and chemical phenomenon controlling retention in the SFC regime. In deriving this relationship we have made the simplifying assumption that $\Delta H_{i(T_0)}^\infty$ is the same in both the high temperature (gas chromatography) and low temperature (liquid chromatography) regions. While this assumption is not strictly correct, (29) it provides a basis for treatment of the intermediate supercritical fluid regime.

Retention as a function of pressure at constant temperature points out the importance of the relationship between solute solubility in the fluid mobile phase to the trend in solute retention. The trend in solute retention is demonstrated to be dependent on isothermal compressibility, solubility in the fluid and the partial molar volume of the solute in the stationary phase.

This work points to the possibility of determining partial molar volumes of different solute molecules in a wide range of different stationary phases at infinite dilution, determining solute solubility in the infinite dilution regime and the functional dependence of the heat capacity of the fluid mobile phase using relatively simple chromatographic experiments. In future work, a more universal approach in fluids with an EOS is being considered, which will extend the prediction of the trend in retention to a wider range of solutes.

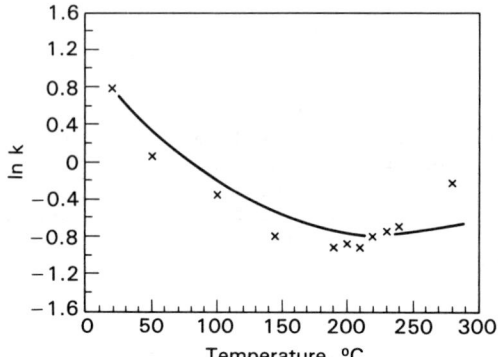

Figure 11 Experimental data (Ref. 5) and theoretical model (solid line) for chrysene in n-pentane at 69.1 atm, $\Delta H_{i(T_o)}^{-\infty}$ = -3.0 Kcal/mole.

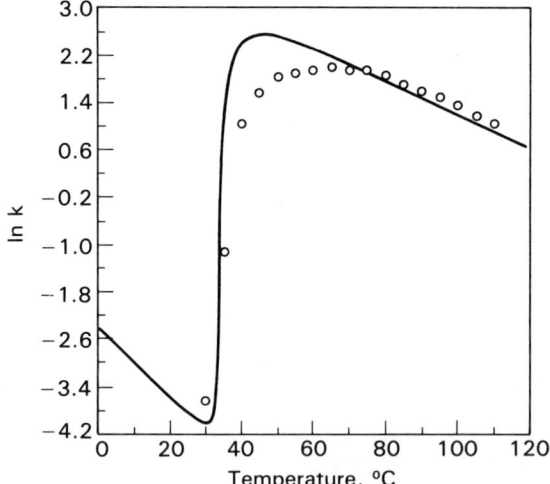

Figure 12 Experimental data for hexadecane on OV-17 with CO_2, 76.5 atm and theoretical model (solid line), $\Delta H_{i(T_o)}^{-\infty}$ = -10.6 Kcal/mole.

Acknowledgments

The authors acknowledge the technical assistance of R. W. Gale and the helpful discussions of R. C. Petersen and the support of the U.S. Department of Energy, Office of Basic Energy Science, under Contract DE-AC06-76RLO 1830.

References

1. Van Wasen, U.; Schneider, G. M., Chromatographia, (1975), 8, 274.
2. Geve, D. R.; Board, R.; McManigill, D., Anal. Chem., (1982), 54, 736.
3. Van Wasen, U.; Swaid, I.; Schneider, G. M., Agnew. Chem. Int. Ed. Engl., (1980), 19, 575.
4. Schmitz, F. P.; Hilger, H.; Leyendecker, D.; Lorenschat, B.; Setzer, U.; Klesper, E., J. of High Res. Chrom. & Chrom. Comm.; (1984), 7, 590.
5. Schmitz, F. P.; Leyendecker, D.; Klesper, E., Ber. Bunsenges. Phys. Chem., (1984), 88, 912.
6. Smith, R. D.; Chapman, E. G.; Wright, B. W., Anal. Chem., (1985), 57.
7. Novotny, M.; Bertsch, W.; Zlatkis, A., J. Chromatogr., (1971), 61, 17.
8. Schmitz, F. P.; Leyendecker, D.; Klesper, E., J. Chromatogr., (1984), 315, 19.
9. Yonker, C. R.; Wright, B. W.; Petersen, R. C.; Smith, R. D., accepted for publication in J. Phys. Chem.
10. Gitterman, M.; Procaccia, I., J. Chem. Phys., (1983), 78, 2648.
11. Klotz, I. M.; Rosenberg, R. M., "Chemical Thermodynamics," 3rd Ed. W. A. Benjamin, Inc. (1972).
12. Meyer, E. F., J. Chem. Ed., (1973), 50, 191.
13. Van Ness, H. C.; Abbott, M. M., "Classical Thermodynamics of Non Electrolyte Solutions with Applications to Phase Equilibria," McGraw-Hill Book Company (1982).
14. Redlich, O.; Kwong, J. N. S., Chem Rev., (1949), 44, 233.
15. Peng, D. Y.; Robinson, D. B., Ind. Eng. Chem. Fundam., (1976), 15, 59.
16. Horvath, C.; Melander, W.; Molnar, I., J. Chromatogr., (1976), 125, 129.

17. Moore, W. J., "Physical Chemistry", 4th Ed., Prentice-Hall, Inc., (1972).

18. Gas Encyclopedia, Elsevier Scientific Publishing Company, (1972).

19. Lauer, H. H.; McManigill, D.; Board, R. D., Anal. Chem., (1983), 55, 1370.

20. Wright, B. W.; Smith, R. D., Chromatographia, (1984), 18, 542.

21. Smith, R. D., Kalinoski, H. T., Udseth, H. R.; Wright, B. W., Anal. Chem., (1984), 56, 2476.

22. McHugh, M.; Paulaitis, M. E., J. Chem. Eng. Data, (1980), 25, 326.

23. Sie, S. T.; Rijnders, G. W. A., Sep. Sci., (1967), 2, 729.

24. Semonian, B. P.; Rogers, L. B., J. Chromatogr. Sci., (1978), 16, 49.

25. Van Wasen, U.; Schneider, G. M., J. Phys. Chem., (1980), 84, 229.

26. Reid, R. C., Kurnik, R. T.; Holla, S. J., J. Chem. Eng. Data, (1981), 26, 47.

27. Wicar, S.; Novak, J., J. Chromatogr., (1974), 95, 1.

28. Yonker, C. R.; Frye, S. L.; Kalkwarf, D. R.; Smith, R. D., to be published in J. Phys. Chem.

29. Yonker, C. R.; Smith, R. D., in press, J. Chromatogr.

RECEIVED July 8, 1986

Chapter 15

Supercritical Fluid Extraction and Chromatography of Nonpolar Nonvolatile Coal-Derived Products

J. W. Jordan, R. J. Skelton, and L. T. Taylor

Department of Chemistry, Virginia Polytechnic Institute and State University, Blacksburg, VA 24061

In recent years supercritical fluid chromatography has gained attention as an alternative technique to high performance liquid chromatography for the separation of nonvolatile or thermally labile compounds, whose analysis with gas chromatography is impossible. The work presented here demonstrates a system that allows supercritical fluid extraction of a high boiling coal-derived material with subsequent direct introduction of a fraction of each extract onto a packed column for analysis via supercritical fluid chromatography. Detection of the eluates included variable wavelength UV and FTIR spectrometry. The coal derived products studied were taken from a bench scale coal liquefaction reactor, in which the same catalyst was used for twenty-five consecutive days. Chemical changes that occur as the catalyst decays were of prime interest. Most significant changes were noted in the aromatic fraction analysis, where a trend towards molecules with higher numbers of condensed rings was observed as the catalyst decayed.

It was suggested in the late 50's that supercritical fluids could be used as mobile phases.([1](#)) Supercritical Fluid Chromatography (SFC) was introduced in 1961 when Klesper([2](#)) demonstrated the use of a supercritical fluid as a mobile phase. These "dense gases" or supercritical fluids have the ability to solubilize nonvolatile compounds, and thus cause them to migrate down a column in partition chromatography.([3](#))

This form of chromatography has several advantages that stem from the physical properties that are exhibited by the mobile phase. In general, diffusivities and viscosities are intermediate between those of a liquid and a gas, allowing more rapid analysis when compared to the analogous methods using a conventional liquid mobile phase.

Many of the same properties that make supercritical fluids advantageous in chromatography also enhance their ability to extract compounds from within a sample matrix.(4) Also, since the solubility of most compounds is dependent on the density of the supercritical fluid, selective extraction is possible.(5) These properties are well known, and have been exploited in some cases where the extraction was formerly done with a liquid. In many cases, the quality of extract is higher, and extractions are of higher efficiency than with liquids. Supercritical fluids, especially CO_2, are also often less expensive, less toxic and less flammable than their organic liquid phase counterparts.

With the demonstration of supercritical fluid extraction, an obvious extension would be to extract or dissolve the compounds of interest into the supercritical fluid before analysis with SFC.(6) This would be analogous to the case in HPLC, where the mobile phase solvent is commonly used for dissolving the sample. The work described here will employ a system capable of extracting materials with a supercritical fluid and introducing a known volume of this extract onto the column for analysis via SFC. Detection of the separated materials will be by on-line UV spectroscopy and infrared spectrometry. The optimized SFE/SFC system has been used to study selected nonvolatile coal-derived products. The work reported here involved the aliphatic and aromatic hydrocarbon fractions from this residuum material. Residua at several times were taken from the reactor and examined which provided some insight into the effects of catalyst decay on the products produced in a pilot plant operation.

Experimental

The samples for study were obtained from Conoco Research Division in Library, PA, as solid materials capped under nitrogen gas. These materials which were subsequently stored at 5°C originated from a Hydrocarbon Research Institute bench scale run (#227-20) of the Catalytic Two-Stage Liquefaction process.(7) The samples, which are designated "pressure-filter liquids," are the solids free portion of the major second stage products that have been "topped" in the atmospheric still, then filtered. The process utilized a donor solvent derived from previous runs and new catalyst material. Illinois 6, Burning Star coal was used. The process was run for 25 consecutive days, with operating conditions held constant, except for the initial start-up period (i.e. days 1-3). Since conditions were constant, any change in the products formed, should be a function of catalyst age. The filtered products were distilled by Conoco into 2 fractions, those boiling under 850°F, and those above. (Actual fractionation was at 320°C Pot/270°C Column/5 Torr (850°F)). Further details of the coal liquefaction process are available elsewhere.(7,8)

The materials boiling above 850°F, were subsequently separated into four fractions by a preparative scale chemical class separation in our laboratory. This was done as follows: 2 gms of the residual material were ground and dissolved in tetrahydrofuran (HPLC grade, Fisher Products, Fairlawn NJ). Ten grams of 30-40µm silica gel ("Sepralyte", Analytichem Int., Harbor City, CA), which was previously washed with methanol and dried, was added to the THF solution. This slurried material was then rotary evaporated to dryness. The dried silica gel with adsorbed coal-derived product was placed at the top of a glass column which was previously slurry-packed with 40 gms of the same silica gel using hexane. A filter paper was used to separate the layers. The glass column had a diameter which allowed the height (4.5") of the packed portion of the column to equal 3 times its diameter (1.5"). The materials were eluted with three column volumes each of hexane, toluene, chloroform and tetrahydrofuran (ca. 300 mL each). The hexane and toluene fractions were taken to be studied here. Three separations of this type were done, representing the 1st, 14th and 25th day of the run. These are samples labeled 44, 57 and 68, respectively.

The fractions were filtered with a 5 μm filter apparatus (Millipore, Bedford, MA) to remove any silica originating from the preparative scale column. The fractions were then rotary evaporated to near dryness in tared round bottom flasks. These were subsequently dried overnight in a vacuum oven at 60°C and weighed. The hexane and toluene fractions were diluted to 5 and 20 mL, respectively, with CCl_4. Both fractions were totally soluble in this solvent as well as in THF.

Analysis of these two fractions was performed by supercritical fluid extraction/chromatography, utilizing the sampling procedures previously described.([6](#)) This was done by evaporating the solvent from the fraction in the presence of dry, washed, ignited sand which was then used to fill the extraction apparatus. Direct injection of some samples (THF solution) was also done for comparison. Experiments were carried out using a Hewlett Packard 1082B Liquid Chromatograph modified for use with supercritical fluids. Columns utilized include a Hewlett Packard 15cm x 4.6mm i.d. C_{18} column with 5μm packing and a Dupont 25cm x 4.6mm i.d. NH_2 column, also with 5μm packing. Pure CO_2 was used as the mobile phase (Scott Specialty Gases, Plumsteadsville, PA). All on-line extraction/chromatograms utilized a 100μL loop size. Traditional sampling utilized a 10μL syringe and 20μL loop.

All experiments utilized a variable wavelength UV detector. Analysis of the hexane fraction was carried out using both the UV detector and an on-line Nicolet 6000 FTIR. The FTIR flow cell interface has been described in detail elsewhere.([9](#)) This same FTIR was also used to gather static spectra on the various fractions, utilizing a liquid cell with KBr windows.

Results and Discussion

Table I shows the results of the class separation of residuum with absolute and percent recovery for each of the four fractions for day 1, 14 and 25. In general, the concentration of the less polar (hexane and toluene) fractions decreased or remained constant with time; while, the more polar ($CHCl_3$ and THF) fractions increased in percentage. This is best noted from data for days 14 and 25. Data from day 1 are somewhat variant from this trend. This is probably due to the inconsistency in operating conditions that occurred during the start-up period. Such assumptions should be viewed cautiously since the percent weight recovery from day 14 and 25 is not constant.

The hexane fractions from runs 44 and 68 have identical spectral in CCl_4. These show aliphatic C-H (CH_3 sym str = 2872 cm^{-1}, asym = 2962 cm^{-1}; CH_2 sym str = 2853 cm^{-1}, asym = 2926 cm^{-1}) and bending modes (1470 cm^{-1} - 1350 cm^{-1}) as well as a band at 1600 cm^{-1} which may be an aromatic ring vibration. No other functionalities

Table I. Results of Residuum Class Separation*

Fraction	Resid 44 (Day 1)	Resid 57 (Day 14)	Resid 68 (Day 25)
Hexane	0.075 gms. (3.8%)*	0.185 gms. (9.4%)	0.037 gms. (2.1%)
Toluene	1.071 gms. (54.2%)	0.939 gms. (48.0%)	0.857 gms. (48.4%)
$CHCl_3$	0.085 gms. (4.3%)	0.082 gms. (4.0%)	0.103 gms. (5.8%)
THF	0.743 gms. (37.6%)	0.755 gms. (38.5%)	0.776 gms. (43.8%)
Total Weight Recovered	1.974 gms.	1.958 gms.	1.772 gms.
% Sample Recovered	(98.7%)	(97.8%)	(88.5%)

*Number in parenthesis equals percent of recovered material.

are present which can be detected. In both cases, the amount of CH_2 stretching is much greater than CH_3 stretching. This indicates that the aliphatic units are either long in length or are in the form of saturated rings.

The toluene fractions from different days are also very similar to each other. Again, by far the most intense modes of absorbance are aliphatic C-H stretching and bending, however a great deal of aromatic C-H stretching between 3000 cm^{-1} and 3100 cm^{-1} also exists. There is also a very small amount of absorbance in the N-H stretching (3740 cm^{-1}) and O-H stretching (3610 cm^{-1} and 3550 cm^{-1}) regions. In general, compounds with these functionalities should elute in the last two fractions. However, very hindered functionalities often elute in the aromatic fraction, and it is probable that absorbances in these regions are due to these molecules. Again, there is more CH_2 than CH_3 absorbance. There is also some increase in aromatic C-H in sample 68 (day 25) when compared to sample 44 (day 1). Further separation of these materials is needed for a better understanding of their composition. SFC of Hexane Fractions. The hexane fractions were analyzed with an amino-propyl derivatized silica column under moderate density CO_2 conditions. Other columns investigated included cyano, phenyl, and C_{18} derivatized silica. No modifiers were used in any of these separations. Figure 1 shows the chromatograms achieved for sample 57 (Day 14) wherein the top chromatogram was produced by monitoring 254 nm and the bottom chromatogram by monitoring 280 nm. Detection at 210 nm was also employed, but no absorbances were seen. Later eluting peaks are more intense in all cases than the early peaks at 280 nm detection (early peaks decrease in intensity while later peaks absorb more intensely for the same amount injected). This would indicate that the later eluting materials probably have more extensive conjugation, and represent molecules like the PAHs. There is, however, very little aromatic C-H stretching in the static infrared spectra of the whole fraction, although there is the band centered at approximately 1600 cm^{-1} which is associated with aromatic ring vibration. The absence of C-H aromatic stretching could be explained if: (a) there is a considerable number of aliphatic substituents on the aromatic nuclei or nucleus or (b) many saturated rings are fused to an aromatic nucleus. An extensive number of aliphatic substituents may have caused increased solubility in hexane, which enabled these materials to elute from the silica.

To gain further information concerning the hexane fractions, on-line supercritical fluid extraction/SFC was coupled with on-line FTIR detection. FTIR should have an advantage in that it is more sensitive to aliphatic hydrocarbons than the UV detector. Basically, a flow cell is placed in the FTIR instrument and infrared spectra are gathered in the form of interferograms ca. every 0.5 second. The data obtained can be used to create a chromatogram, utilizing absorbance over the entire infrared spectrum, rather than a single wavelength. This analysis was done on samples 44 and 68 (1st and 25th day). Figure 2 shows one of these "chromatograms" (day 25), known as Gram-Schmidt Reconstructions (GSR). One major peak is seen, which, when compared with the UV chromatographic trace obtained simultaneously, shows maximum IR absorbance well before any absorbance in the UV. A spectrum consisting of 8 coadded files taken near the peak maximum reveals that the on-line spectrum is almost exactly the same as the static spectrum of the whole sample. The exception is the disappearance of the band at 1600 cm^{-1} in the file spectrum. This observation along with the lack of UV response in this elution

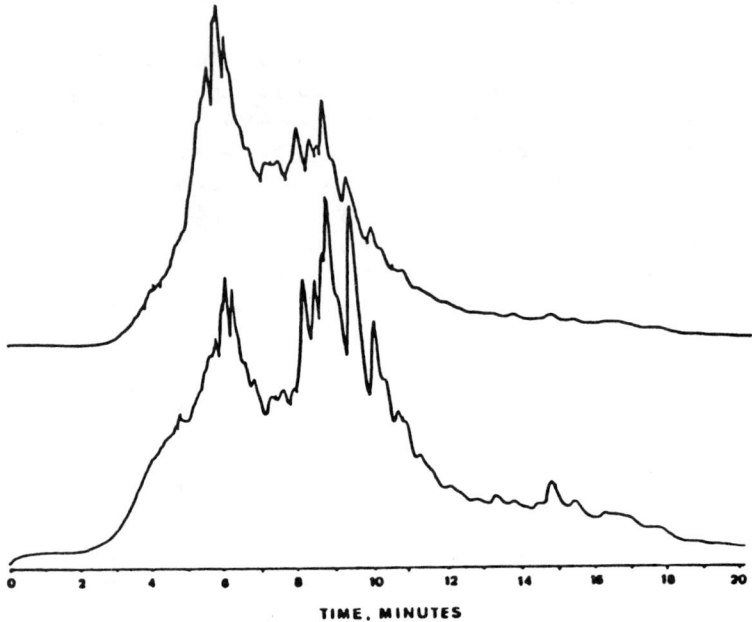

Figure 1. SFC separation (3 mL/min) of hexane fraction Day 14 (Sample 57) on an aminopropyl derivatized 5 μm silica column (25 cm x 4.6 mm, i.d.) at 40° C. Mobile phase is CO_2 at an average pressure of 2800 psi. Top trace corresponds to detection at 254 nm; bottom trace = 280 nm.

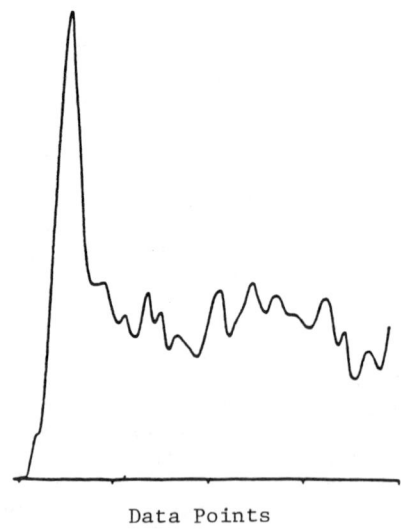

Figure 2. Gram-Schmidt Reconstruction obtained from SFE/SFC/FTIR of hexane fraction of Day 25 sample.

volume indicates that these components are totally aliphatic in nature; whereas, the later eluting materials are aromatic and are of insufficient concentration to be detected by on-line FTIR detection.

Although only one peak is indicated by the GSR, a comparison of spectra obtained from the front of the peak with those which elute in the later half of the peak can be made. When this technique was applied to these data, it was found that certain changes in the spectra were observed. Figure 3 shows the first 6 spectra in the C-H stretching region obtained over the major peak in the GSR. Each of these spectra represent 8 coadded files, resulting in a time resolution of about 4 seconds per file. Infrared spectra taken from the front of the peak show the lowest methyl to methylene ratio. As each subsequent file is ratioed to keep the methylene peak constant, the absorbances due to methyls increase over the width of the peak. Subsequent files show a leveling of this ratio.

Unfortunately, analysis of the 1st and 25th day samples did not reveal any significant trends in the products. However, it is possible that the relative amount of each component produced is equivalent, but that the total production suffers as the catalyst becomes less effective.

SFC of Toluene Fractions

To aid in understanding the components in the toluene fraction, a limited SFC study of model condensed-ring aromatic systems was performed. Because few functionalities are IR detectable in this fraction other than C-H units, the toluene eluted fraction of the residuum probably consists primarily of aromatic rings with short aliphatic substituents and/or fused saturated rings. Because of the distillation parameters employed, compounds with 4 or fewer fused rings should have been removed. The toluene fraction should, therefore, represent compounds with at least 5 rings. Figure 4 shows a separation of several PAH compounds on a 15cm ODS column with a CO_2 mobile phase. All compounds with three or greater rings are separated to baseline resolution. Relatively high density CO_2 and a high flow rate were used to elute large molecules faster. These compounds elute logarithmically with increasing ring numbers which is expected for a homologous series with isocratic elution.

The toluene fractions of the residuum samples were subjected to the same analysis conditions as the PAH model compounds. The chromatogram corresponding to Day 14 is shown in Figure 5. Traditional off-line sampling (top) and an on-line CO_2 extraction (bottom) were employed. Detection at 300 nm wavelength was used to emphasize the larger PAHs which were expected to be present. This wavelength was chosen for the chromatographic presentation here; however, during development, other wavelengths were used. At 254 nm, for example, the early eluting peaks absorb intensely, indicating that nonconjugated systems are also present. Several interesting features are seen in these chromatograms. The later eluting peaks become more intense at later run-days. This is most readily noted in the second grouping of peaks in each chromatogram. The third grouping of peaks, which elute where coronene elutes, are

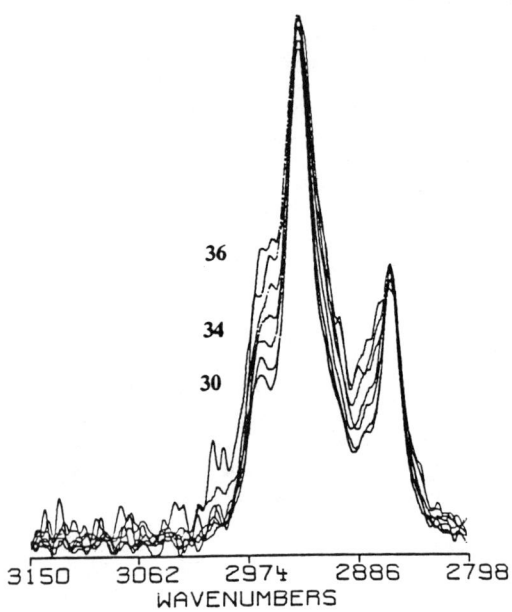

Figure 3. Stack plot of C-H stretch region for the 7 files (Nos. 30-36) taken across the major GSR peak in Figure 2.

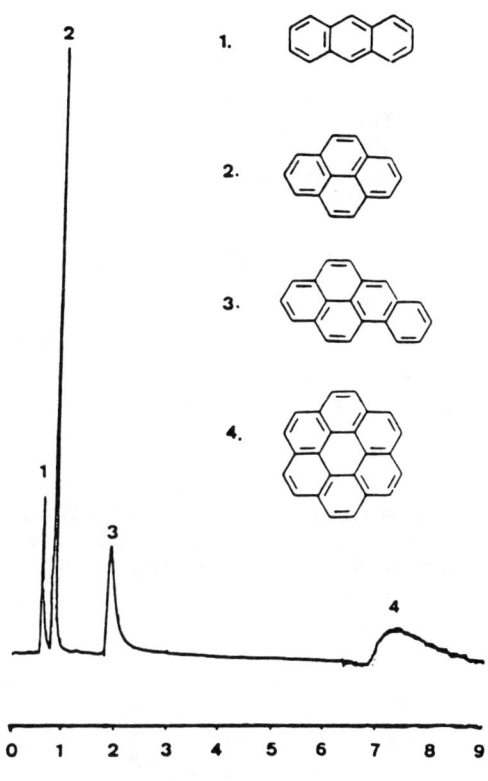

Figure 4. Separation of some PAH compounds: ODS column (5 μm), pressure: 4200 inlet, 3700 outlet. Mobile phase: CO_2 at 3.7 mL/min. Detection: 300 nm, 500 nm reference. Column: 15cm x 4.6 mm ODS.

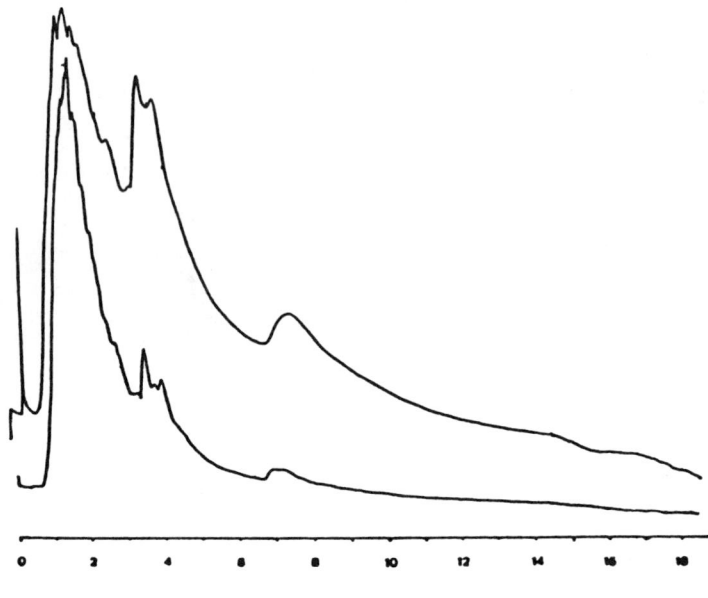

Figure 5. SFC separation (2 mL/min) of toluene fraction of residuum Day 14 comparing on-line (bottom) and traditional (top) sampling methods. ODS (5 μm) column (15 cm x 4.6 mm, i.d.), 40°C, CO_2 (3800 psi), 300 nm.

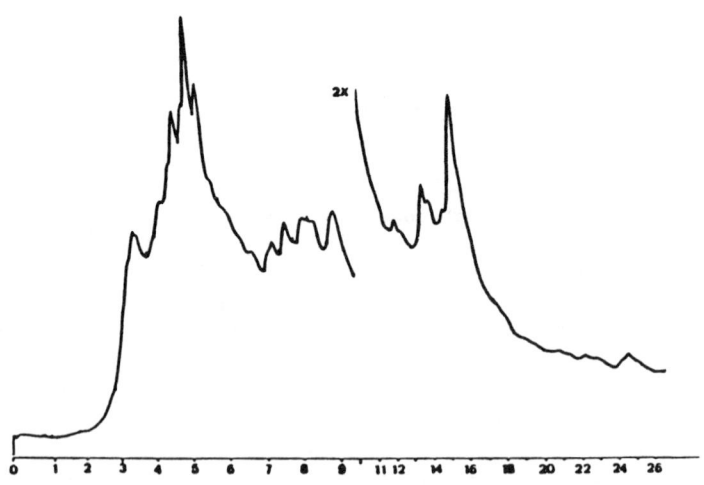

Figure 6. SFC separation after SFE of toluene fraction of residuum Day 25 (Sample 68): NH_2 column (5 μm, 25 cm x 4.6 mm, i.d.), 2 mL/min, 40°C, CO_2 (3200 psi), 300 nm.

probably six-ring compounds. A broad late eluting peak (ca. 15 minutes) is probably a seven-ring compound. Retention time for this component fits to the previously alluded logarithmic plot very well. The second group of peaks, which elutes before the six-ring compounds are possibly five-ring compounds whose rings are oriented in a "stretched" orientation, or the eluted material may be components which contain five aromatic rings and one partially hydrogenated ring. Other detectors, such as diode array UV and fluorescence detectors may yield more selective information about these structures.

The increase in apparent concentration of the larger aromatics as a function of time is probably a direct result of catalyst fouling. The new catalyst demonstrated the greatest ability to hydrogenate the fragments formed by thermolysis; however, this process probably becomes less efficient with time, allowing for reformation of C-C bonds which results in higher aromaticities and larger PAH compounds.

The chromatograms generated with the C_{18} column were specifically used to determine ring size. These columns do not show a high degree of resolution. An aminopropyl column was used at a somewhat lower density to spread the peaks of the toluene fraction out over a larger period of time. Other workers have shown this material to separate compounds in the order of ring number also. Figure 6 shows the chromatogram achieved with the fraction from Day 25. This was obtained with UV 254nm detection, rather than 300nm, as large aromatics will not elute in a reasonable amount of time under these conditions. There are still many unresolved peaks, which alludes to the extreme complexity of the sample. Large aromatic molecules have many isomers, and excellent resolution of these with a packed column may be impossible.

The on-line extraction technique does not perform as well as traditional methods for the toluene fraction in contrast to the hexane fraction. The later eluting peaks are much more intense for the traditional sampling in all cases. It seems that CO_2 favors the lower molecular weight molecules, even at high densities. On line FTIR detection did not yield any new information for the aromatic fractions. Highly aromatic material does not absorb infrared radiation greatly. In contrast, UV detectors are quite sensitive and yield significant information for aromatic compounds.

Acknowledgment

The financial assistance of the Commonwealth of Virginia and Department of Energy Grant DE-FG22-84PC70799 is greatly appreciated.

Literature Cited

1. N. M. Karayannis, Rev. Anal. Chem., 1971, 1, 43.
2. E. Klesper, A. H. Corwin, D. A. Turner, J. Org. Chem., 1962, 27, 700.
3. D. R. Gere, Science, 1983, 222, 253.
4. C. Grimmet, Chem. Ind., 1981, 6, 359.
5. E. Stahl, K. W. Quirin, A. Glatz, D. Gerard, G. Rau, Ber. Bunsenges Phys. Chem., 1984, 8, 900.
6. R. J. Skelton, L. T. Taylor, Chromatographia, 1986, 21, 3.
7. A. G. Conolli, J. B. MacArthur, J. B. McLean, "HRI's Two Stage Catalytic Coal Liquefaction Program - A Status Report", presented at the 1984 DOE Direct Liquefaction Contractors' Review Meeting, Albuquerque, NM, October 1984.
8. F. P. Burke, R. A. Winschel, "Recycle Slurry Oil Characterization - Second Annual Report", DOE Contract No. DE-AC22-80PC30027, August, 1983.
9. J. W. Jordan, C. C. Johnson and L. T. Taylor, Chromatographia, 1985, 20, 717.

RECEIVED July 1, 1986

FRACTIONATION AND SEPARATION

Chapter 16

Supercritical Carbon Dioxide Extraction of Lemon Oil

Steven J. Coppella[1] and Paul Barton

Department of Chemical Engineering, Pennsylvania State University, University Park, PA 16802

```
Vapor-liquid-equilibrium was measured for lemon oil
and carbon dioxide from 303 to 313 K and from 4 to 9
MPa using a nonvisual, constant-volume static cell.
Two-phase behavior was verified visually in separate
experiments. Phase samples were depressurized into
cold traps; carbon dioxide was metered through a wet
test meter, and liquids were analyzed by gas
chromatography. The Peng-Robinson equation of state
reproduced the P-x diagram by modeling the system as
a carbon dioxide-limonene binary with a
temperature-dependent interaction parameter.
Solubilities of oil in the vapor phase were in the
range of 1 to 3 wt%. Relative volatility of
limonene to geranial decreased from 2.2 to 1.3 over
this solubility range; relative volatility for
geranial to β-caryophyllene decreased from 1.5 to
1.1.
```

In 1983-1984, Arizona and California produced annually over 10^6 kg of cold-pressed lemon oil. After being concentrated by distillation or liquid extraction, lemon oil is used as a flavoring and/or fragrance agent in beverages and cosmetics. However, distillation thermally degrades lemon oil, and extraction with organic solvents only partially reduces thermal degradation (since the solvents must be recovered by distillation) and introduces solvent contamination. Extraction with supercritical carbon dioxide near its critical point (304.3 K, 7.38 MPa, 0.467 g/cm^3) offers a cheap, nontoxic solvent that does not impart flavors nor thermally degrade the product. Supercritical extraction is a hybrid unit operation in the domain between extractive distillation and liquid extraction. Solvent recovery is accomplished by depressurization at ambient temperature.

[1]Current address: Department of Chemical Engineering, University of Delaware, Newark, DE 19716

Processes for supercritical extraction of oils have been described in numerous literature references, including Paulaitis et al. (1), Ely and Baker (2), Gerard (3), Stahl et al. (4), and Robey and Sunder (5). The literature lacks detailed phase equilibrium data on multicomponent essential oils with supercritical solvents in the proximity of the solvent critical temperature.

The purposes of our research were to evaluate the feasibility of supercritical carbon dioxide extraction of lemon oil near ambient temperature, generate equilibrium data for carbon dioxide with multicomponent essential oil constituents, and evaluate the ability of the Peng-Robinson equation of state (6) to model this multicomponent supercritical system.

Concentrating Lemon Oil

Staroscik and Wilson (7) have analyzed various lemon oils and identified 38 compounds. These compounds can be subdivided into three major classifications: terpenes (C_{10} hydrocarbons), oxy fraction (oxygenated C_8-C_{12} hydrocarbons), and sesquiterpenes (C_{15} hydrocarbons). Lemon oil is usually concentrated to remove the terpenes and sesquiterpenes from the desired C_8-C_{12} oxy fraction (8).

When concentrating the lemon oil by multistage fractional distillation under vacuum, column performance coupled with condenser pressure, column pressure drop, reboiler design, and mode of operation (batch or continuous) dictates the degree of thermal exposure. Overhead temperatures range from 320 to 340 K, reboiler temperatures range from 330 to 370 K, and exposure times of 15 to 20 hours may be expected during batch distillation. In continuous distillation, exposure times are lower but reboiler temperatures may be higher.

In extractive distillation with a supercritical solvent, extraction rather than vacuum is used to volatilize the components into the vapor phase. Operating temperatures can be lower than in conventional distillation, and the desired C_8-C_{12} oxy fraction will then have been subjected to less thermal degradation. Carbon dioxide has a critical temperature that meets this goal.

In order to make data correlation practicable in supercritical carbon dioxide extraction, it is convenient to represent each major chemical classification by a single compound. Each select compound should have available good vapor pressure data and should be a predominant constituent in its group with regard to structure and concentration. For correlation purposes, we selected limonene, geranial, and β-caryophyllene. Their structures are shown in Figure 1.

Stahl et al. (4) presented solubility data for limonene and caryophyllene with carbon dioxide; Gerard (3) included carvone. Temperatures in the range of 279 to 377 K, and pressures in the range of 1.5 to 11 MPa were covered. Robey and Sunder (5) provided solubility and relative volatility data for carbon dioxide with concentrated lemon oil and with limonene and citral (geranial/neral) from 323 to 353 K and 9.4 to 10.6 MPa.

Gerard (3) described a continuous multistage column process for carbon dioxide extraction (distillation) of essential oils at ambient temperature and 8 MPa, with solvent recovery at 273 K and 3 MPa. Robey and Sunder (5) proposed a lemon oil fractionator which would operate at 333 K and 10 MPa with an efficiency equivalent to 12 stages; solvent recovery would be at 293 K and 5.5 MPa.

Experimental Section

Cold-pressed oil from Arizona early desert lemons was supplied by A. M. Todd Company, Kalamazoo, MI. Degassed lemon oil and dry carbon dioxide were charged into the one-liter isothermal constant volume cell shown in Figure 2. After the operating temperature was reached, the system was stirred for one hour, then allowed to settle for 15 minutes before sampling. A couple of experiments were made with longer mixing and settling times to confirm that equilibrium had been reached. Parallel phase visualization experiments were conducted in a sight gauge to insure that operation was in the two-phase region and that settling time was adequate. Experiments were performed from 303 to 313 K and from 4 to 9 MPa.

A sample of the equilibrated liquid phase was removed (after purging) by depressurization through a valve and hypodermic tubing into a two-stage trap cooled by dry ice-acetone. A wet test meter measured the carbon dioxide off-gas. A sample of the vapor phase was then similarly removed. Purge and sample sizes were kept small to minimize disturbances of equilibrium. Pressure changes in the cell were 0-0.1 MPa during sampling of liquid phase and 0-1.2 MPa during sampling of vapor phase. Special high-pressure sample valves with microliter-sized traps were tried in an effort to reduce pressure disturbance, but reliability was inadequate.

Estimated relative errors are 0.2% for temperature, 5% for pressure, 4% for carbon dioxide mole fraction in the liquid phase, and 10% for lemon oil mole fraction in the vapor phase. Relative error is defined as experimental error divided by sample average value.

The recovered lemon oil samples were analyzed by gas chromatography. A 0.5 mm i.d. x 30 m thin film (0.1 μm) SE-30 glass capillary column (Supelco, Inc., Bellefonte, PA) was used with a flame ionization detector. The temperature was programmed from 348 to 473 K. Peak identification was based on information of Supelco, Inc., A. M. Todd Company, and Staroscik and Wilson (7). Staroscik (9) provided us with the response values used in his work and we assumed that our detector would give proportionate responses. Staroscik found in his work that relative standard deviations of the response values were generally less than 3%.

Results

Lemon oil-carbon dioxide equilibrium was measured at 303, 308, and 313 K and in the pressure range of 4 to 9 MPa. Below 6 MPa, there was insufficient lemon oil in the vapor phase to obtain good samples for analysis. Above 9.0 MPa at 313 K, above 7.8 MPa at 308 K, and above 7.4 MPa at 303 K, the system exhibited a single phase. Nine experiments provided two-phase, vapor-liquid equilibrium data

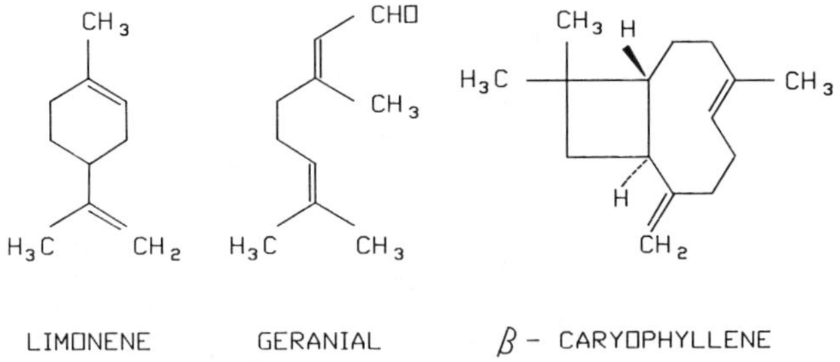

16. COPPELLA AND BARTON *Carbon Dioxide Extraction of Lemon Oil* 205

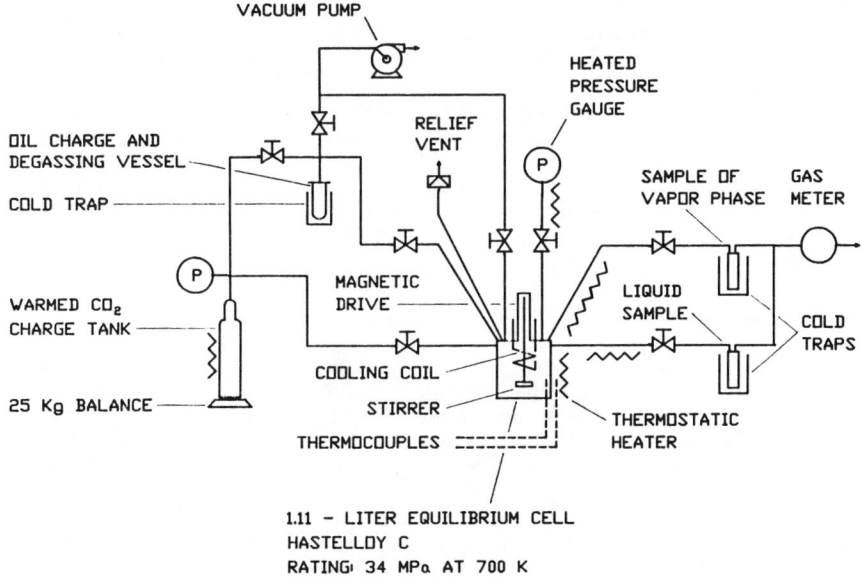

Figure 1. Key Constituents in Lemon Oil.

Figure 2. Vapor-Liquid Equilibrium Apparatus.

suitable for correlations and for generating coefficients for the Peng-Robinson equation. Detailed experimental data can be found in Coppella (10).

The results of an experiment at 308 K and 6.98 MPa are detailed here. The liquid phase contained 48 wt% (74 mole %) carbon dioxide and the vapor phase contained 99.5 wt% (99.8 mole %) carbon dioxide. A gas chromatogram for the lemon oil from the liquid phase sample is shown in Figure 3. Peak identification, retention, response value, and concentration are given in Table I. Composition of the lemon oil in the vapor phase is also listed. Average molecular weight of lemon oil is 137.9 in the liquid and 136.4 in the vapor.

The relative volatility, or selectivity factor, for each component in the presence of carbon dioxide is also given in Table I. Relative volatility is defined as the ratio of K-value for component i to K-value for limonene. The K-value for component i is defined as mole fraction i in the vapor (extract) phase to mole fraction i in the liquid (raffinate) phase.

By comparing the relative volatilities of various components, the ease of separation between these components can be determined. For example, a key terpene-oxy separation involves terpinolene with a relative volatility with respect to limonene of 0.7 and citronellal with relative volatility with respect to limonene of 0.4. The relative volatility of terpinolene to citronellal is then 0.7/0.4 or 2.

The second cut point in the separation is between geranylacetate and β-caryophyllene. This separation factor is 0.06/0.07 or 0.9. This is opposite that for geranial to β-caryophyllene: 0.2/0.07 or 3. Since there is some overlap in the volatilities of C_{12} oxygenated hydrocarbons and sesquiterpenes, some β-caryophyllene will be extracted into the desired C_8-C_{12} oxy product.

Solubility and Selectivity

Solubility diagrams were prepared for the phases that separated in the lemon oil extractions performed in this study with carbon dioxide. Such diagrams can serve only as guides, since solubility is composition-dependent and is a function of the amount extracted. The data are shown in Figures 4, 5, and 6 at 303, 308, and 313 K, respectively.

Extractions or extractive distillations with supercritical solvent need to be performed at as high as possible a solubility of oil in the extract or vapor phase in order to reduce the solvent or carrier gas requirement. From our lemon oil-carbon dioxide phase diagrams, it appears that the highest practical solubility level is 0.9 mole % (2.8 wt%) essential oil. This is obtainable at 313 K. At lower temperature, sensitivity of solubility to pressure requires that solubility be lower (e.g., 0.3 mole % at 308 K).

At 313 K and 8.4 MPa, the slope of extract phase solubility versus pressure is 0.06 weight fraction oil/MPa. For a 15-m tall extraction tower operated at a density of 0.5 g/cm³, the pressure at the bottom is higher than that at the top by 0.075 MPa. The solubility at the bottom will then be 0.15 wt% higher at the bottom

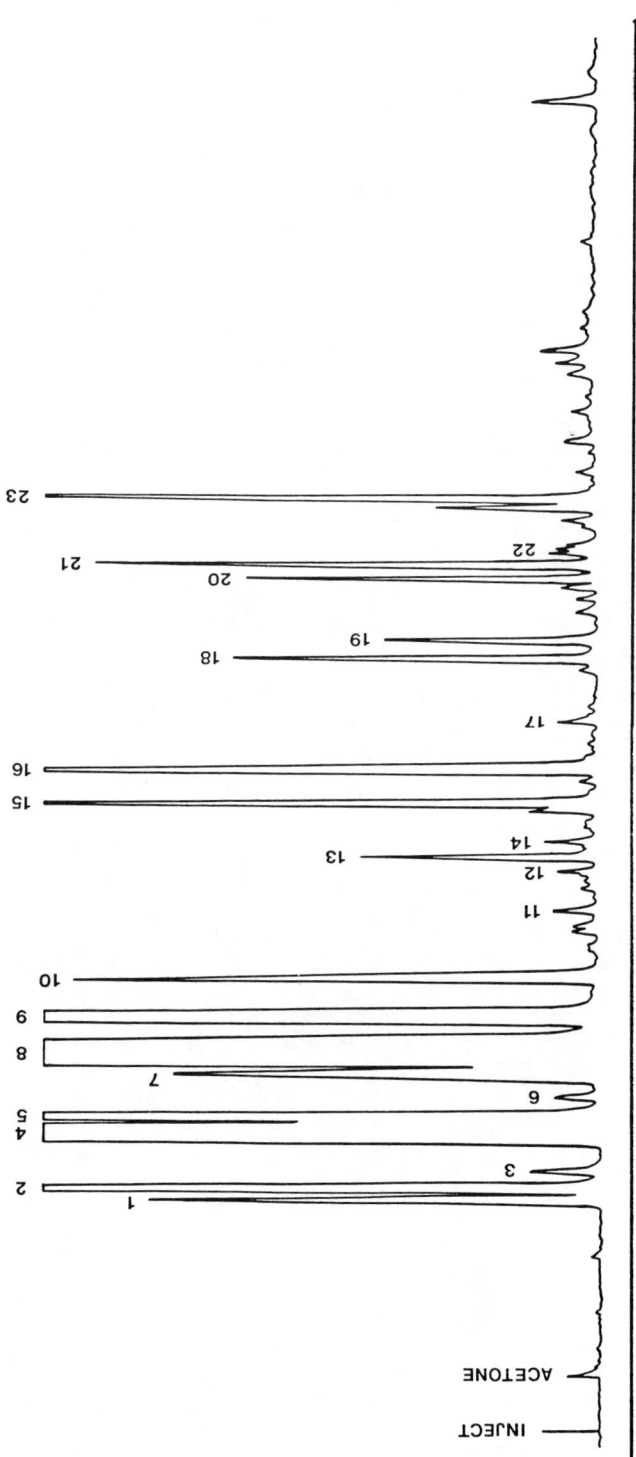

Figure 3. Gas Chromatogram of Lemon Oil in Liquid Phase Sample.

Table I. Lemon Oil Vapor-Liquid Analyses at Equilibrium in CO_2

Peak No.	Compound	Retention[a] Relative to Limonene	Weight[b] Relative Response	Mole % Vapor	Mole % Liquid	Volatility Relative to Limonene
ref	acetone	0.00	---	---	---	---
1	α-thujene	0.55	0.75	0.61	0.35	1.8
2	α-pinene	0.58	0.75	2.80	1.68	1.7
3	camphene	0.63	0.70	0.09	0.06	1.6
4	sabinene, β-pinene	0.76	0.74	17.27	13.06	1.34
5	myrcene	0.79	0.73	1.87	1.59	1.2
6	octanal, phellandrene	0.86	1.17	0.09	0.07	1.3
7	α-terpinene	0.92	1.19	1.21	1.10	1.1
8	limonene	1.00	0.75	67.14	68.22	1.0
9	γ-terpinene	1.09	0.78	7.59	8.45	0.91
10	terpinolene, linalool, nonanal	1.19	0.78	0.47	0.67	0.7
11	citronellal	1.40	1.06	0.02	0.04	0.4
12	terpinenen-4-ol	1.51	0.92	0.02	0.04	0.4
13	α-terpineol	1.55	0.92	0.06	0.20	0.3
14	decanal	1.60	0.97	0.01	0.08	0.1
15	neral	1.70	0.96	0.24	0.92	0.3
16	geranial	1.80	0.96	0.29	1.45	0.2
17	nonylacetate	1.95	1.24	0.00	0.03	---
18	nerylacetate	2.12	1.02	0.03	0.25	0.1
19	geranylacetate	2.18	1.02	0.009	0.15	0.06
20	β-caryophyllene	2.36	0.78	0.013	0.18	0.07
21	trans-α-bergamotene	2.40	0.78	0.016	0.26	0.06
22	α-humalene	2.53	0.71	0.00	0.05	---
23	β-bisabolene	2.60	0.77	0.013	0.38	0.03

[a] Retention time for limonene averaged 8.7 minutes
[b] Peak area multiplier

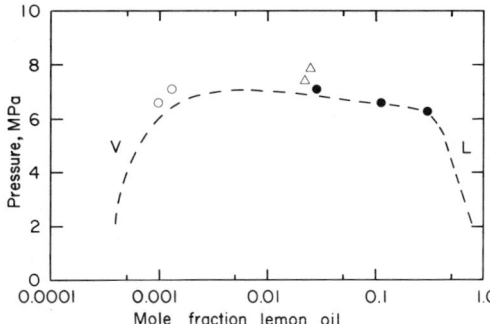

Figure 4. Phase Equilibrium for Carbon Dioxide:Lemon Oil at 303 K. Experimental Two-Phase: O - Vapor ● - Liquid. Δ - Experimental One-Phase. -- Peng-Robinson eq.

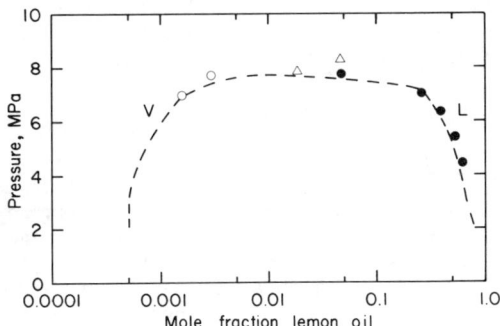

Figure 5. Phase Equilibrium for Carbon Dioxide:Lemon Oil at 308 K. Experimental Two-Phase: O - Vapor ● - Liquid. Δ - Experimental One-Phase. -- Peng-Robinson eq.

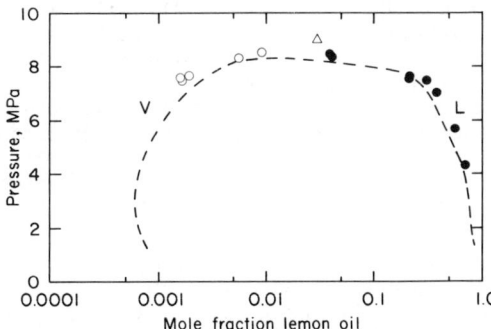

Figure 6. Phase Equilibrium for Carbon Dioxide:Lemon Oil at 313 K. Experimental Two-Phase: O - Vapor ● - Liquid. Δ - Experimental One-Phase. -- Peng-Robinson eq.

(ignoring composition effects). At 308 K and 7.69 MPa, an increase in pressure of 0.15 MPa causes the two-phase system to revert to a one-phase system. It is imperative that temperature and pressure profiles in the supercritical extraction tower be maintained accurately. Windows in the tower are recommended to confirm that operation remains in the two-phase domain.

The selectivities in the supercritical carbon dioxide extraction of terpenes from the oxy fraction, and the oxy fraction from sesquiterpenes, for lemon oil are shown in Figures 7 and 8, respectively. Figure 7 shows the relative volatility, or selectivity factor, of limonene to geranial as a function of oil solubility in the vapor phase. Operation of the extractor at 308 K and 1 wt% solubility (the highest practical level at this temperature) provides a relative volatility of 2. Operation at 313 K and 2.8 wt% solubility provides a relative volatility of 1.3. The selectivity factors are an order of magnitude lower than the vapor pressure ratios at 303 to 313 K.

Figure 8 shows the relative volatility of geranial to β-caryophyllene as a function of oil solubility in the vapor phase. At 313 K and 1 wt% solubility, relative volatility is 1.4. At 308 K and 1 wt% solubility, these constituents are inseparable by supercritical carbon dioxide extraction. The ratio of vapor pressures for this pair is in the vicinity of 2 (at 392 K, the lowest temperature with vapor pressure data).

Modeling of Equilibria

The carbon dioxide/lemon oil P-x behavior shown in Figures 4, 5, and 6 is typical of binary carbon dioxide:hydrocarbon systems, such as those containing heptane (Im and Kurata, 11), decane (Kulkarni et al., 12), or benzene (Gupta et al., 13). Our lemon oil samples contained in excess of 64 mole % limonene; so we modeled our data as a reduced binary of limonene and carbon dioxide. The Peng-Robinson (6) equation was used, with critical temperatures, critical pressures, and acentric factors obtained from Daubert and Danner (14), and Reid et al. (15). For carbon dioxide, $\omega = 0.225$; for limonene, $\omega = 0.327$, $T_c = 656.4$ K, $P_c = 2.75$ MPa. It was necessary to vary the interaction parameter with temperature in order to correlate the data satisfactorily. The values of d_{12} are 0.1135 at 303 K, 0.1129 at 308 K, and 0.1013 at 313 K. Comparisons of calculated and experimental results are given in Figures 4, 5, and 6.

Attempts to model the relative volatilities of the minor organic constituents to limonene using the Peng-Robinson equation proved unfruitful.

Proposed Process

Lemon oil can be concentrated by supercritical carbon dioxide extraction in the temperature range of 308 to 313 K. Pressure in the extractor will be in the range of 7.7 to 8.5 MPa. Solubilities of oil in the extract or vapor phase will range from 1 to 3 wt%. Selectivity factors near 1.4 will be obtained for the terpene-oxy split and the oxy-sesquiterpene split. The operation can be

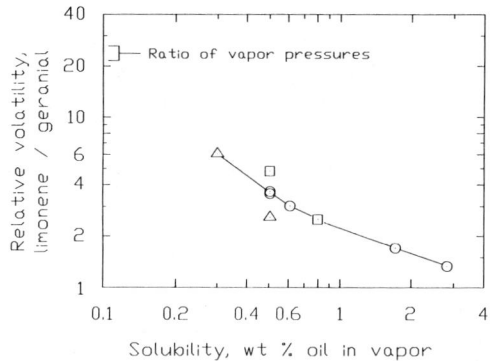

Figure 7. Lemon Oil:Carbon Dioxide Vapor-Liquid Equilibria.
Δ - 303 K, □ - 308 K, O - 313 K.

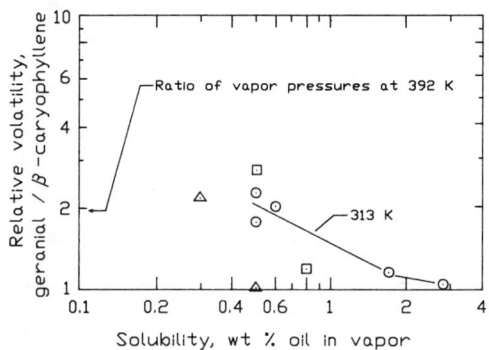

Figure 8. Lemon Oil:Carbon Dioxide Vapor-Liquid Equilibria.
Δ - 303 K, □ - 308 K, O - 313 K.

performed in a multistage extractor with reflux. Operation can be either in the batch or continuous mode. Solvent recovery is performed at conditions slightly below the critical point of carbon dioxide. Flash evaporation can be used; refluxing in an enriching column can be added to decrease the oil content in the recycled carbon dioxide. Residual carbon dioxide in the product oils is removed by flashing to atmospheric pressure.

Solubility limitations require that the solvent-to-oil feed ratio be high; an economic analysis is needed to establish process feasibility. The use of cosolvents or higher temperatures can be used to increase oil solubility and to decrease solvent-to-oil feed ratio. These process modifications, however, cause additional thermal degradation. With cosolvents, thermal degradation occurs during the distillation step needed to remove these unwanted components (and flavors) from the desired extract.

Literature Cited

1. Paulaitis, M. E.; Krukonis, V. J.; Kurnik, R. T.; Reid, R. C. Reviews in Chemical Engineering 1983, 1, 179.
2. Ely, J. F.; Baker, J. K. NBS Technical Note 1070 1983.
3. Gerard, D. Chem. Ing. Tech. 1984, 56, 794.
4. Stahl, E.; Quirin, K. W.; Glatz, A.; Gerard, D.; Rau, G., BBPCAX 1984, 88, 900.
5. Robey, R. J.; Sunder, S. "Application of Supercritical Processing to the Concentration of Citrus Oil Fractions", Annual Meeting of American Institute of Chemical Engineers, San Francisco, 1984.
6. Peng, P. Y.; Robinson, D. B. I.E.C. Fund. 1976, 15, 59.
7. Staroscik, J. A.; Wilson, A. A., J. Agric. Food Chem. 1982, 30, 507, 835.
8. Zeigler, E. "Production, Application and Analysis of Concentrates of Citrus Oils in the Food, Pharmacy, and Perfumery Industries"; VIII International Congress of Essential Oils, Cannes, 1980.
9. Staroscik, J. A., Sunkist Growers, Inc., Ontario, CA, personal communication, 1984.
10. Coppella, S. J. M.S. Thesis, The Pennsylvania State University, University Park, PA, 1985.
11. Im, U. K.; Kurata, F. J. Chem. Eng. Data 1971, 16, 412.
12. Kulkarni, A. A.; Zarah, B. Y.; Luks, K. D.; Kohn, J. P. J. Chem. Eng. Data 1972, 19, 92.
13. Gupta, M. K.; Li, Y. H.; Hulsey, B. J.; Robinson, R. L., Jr., J. Chem. Eng. Data 1982, 27, 55.
14. Daubert, T. E.; Danner, R. P., Eds. "Technical Data Book - Petroleum Refining", 4th ed.; American Petroleum Institute: Washington, DC, 1983.
15. Reid, R. C.; Prausnitz, J. M.; Sherwood, T. K. "The Properties of Gases and Liquids", 3rd ed.; McGraw-Hill Book Co.: New York, 1977.

RECEIVED August 27, 1986

Chapter 17

Near-Critical Separation of Butadiene–Butene Mixtures with Mixtures of Ammonia and Ethylene

D. S. Hacker

Amoco Chemicals Company, Naperville, IL 60566

The results of an experimental investigation are presented for the separation of mixtures of 1,3-butadiene and 1-butene at near critical conditions with mixed and single solvent gases. Ammonia was used as an entrainer to enhance the separation. Several non-polar solvents were used which included ethylene, ethane and carbon dioxide, as well as mixtures of each of these gases with ammonia in concentrations of 2, 5, 8 and 10% by volume. Each solvent and solvent mixture was studied with respect to its ability to remove 1-butene from an equimolar mixture of 1,3-butadiene/ 1-butene. Maximum selectivities of 1.4 to 1.8 were measured at a pressure of 600 psia and a temperature of 20 C in mixtures containing 5%-8% by volume of ammonia in ethylene. All other solvents showed little or no success in promoting separation of the mixture. The experimental results are reported for ethylene/ ammonia mixtures and are shown to be in fair agreement with VLE flash calculations predicted independently by a modified two parameter R-K type of equation of state.

A separation process is sought that can satisfy both our present economic and enviromental constraints. It would also provide an alternative to present practice that relies on expensive azeotropic or extractive distillation processes used in the recovery of products from low relative volatility streams. As an example, virtually all industrial butadiene recovery processes now rely on extractive distillation using acetonitrile or other equivalent agent to enhance the relative volatility of the C4 components. The use of supercritical or near critical separation of these streams may satisfy these requirements provided certain pressure, temperature and recompression criteria can be met. Such a process would also reduce the need for a complex train of distillation towers.

0097-6156/87/0329-0213$06.00/0
© 1987 American Chemical Society

Liquid ammonia has been suggested as a solvent for the C4 separation(1). A drawback to its use in the liquid state, however, is the need for costly refrigeration. Its use as a supercritical solvent would also be acceptable were it not for its high critical temperature (405.45 K). High temperature favors the polymerization of the butadiene; hence, its limitation in this role. In this study, a method was developed that seeks to circumvent this problem and yet achieve the desired separation of the C4's. Prausnitz(2) discusses the use of a mixture of supercritical solvents whose properties provide the optimal physical conditions for efficient extraction. It is equally possible to prepare mixtures of solvents that not only modify those critical properties of the individual solvent component, but also introduce the chemical features needed to maximize the separation of the feed mixture.

In view of the interest in this field, an experimental investigation was undertaken to determine the applicability of supercritical phenomena to the separation of butadiene from C4 mixtures. In particular, the separation of 1-butene from 1,3-butadiene is a key factor in the separation process. Results of these studies are considered in light of predictions obtained from a representative equation of state in the retrograde region of the SC solvent-solute VLE envelope.

Theoretical Discussion

The ability to separate specific solution components from the liquid phase with a supercritical or near-critical solvent component has been demonstrated for a few selected systems(3). Generally, the solvents are single component inorganic gases or light hydrocarbons. In common use have been carbon dioxide, ammonia, ethane, ethylene or propane. Weinstock and Elgin(4) used pressurized ethylene to promote the separation of a number of miscible aqueous-organic liquid mixtures. More recently, Paulaitis and others have used carbon dioxide to dehydrate ethanol(5,6). The separation of asphaltines with supercritical propane has been commercially demonstrated by Kerr-McGee(7). Starling, et al., have suggested that the retrograde regime may be exploited by the addition of sufficient solvent to separate a light liquid hydrocarbon mixture(8).

In addition, several equations of state have been developed to predict the VLE behavior of a subcritical liquid mixture with a supercritical component. These theoretical models are of current research interest. In addition, several approaches have been formulated to extend the analysis to multicomponent systems utilizing concepts of continuous thermodynamics(9, 10).

Reid and others(11, 12) have shown that supercritical solvents exhibit varying degrees of specificity towards a particular specie. Furthermore, the small number of SC solvents available limits the potential use SC extraction. The use of entrainers or mixtures of solvents, may remove the limitation imposed by the narrow choice of likely solvents. Moreover, it is possible that through the proper choice of entrainer and solvent the desired chemical activity can be adjusted to improve the selectivity of the solvent. For example, mixtures of solvent gases with entrainers can permit a modification of critical properties as well as chemical properties, so that P and T adjustment can be used to maximize some physical property of the system(2).

Brunner(13) studied the influence of entrainers on model mixtures by introducing acetone into a polyglyceride mixture, thus enhancing its separation with carbon dioxide. The entrainer served as a source of hydrogen bonding to augment the separation of the desired component. The principle of the entrainer has been widely known in liquid extraction processes in improving separation and was described by Treybal(14).

It is known that the addition of a strongly interactive third component to a miscible binary mixture can alter the phase behavior of this normally stable system. This occurs by a minimization of the Gibbs free energy. By modifying the pair potentials between the components comprising the system, changes in the pairwise activity coefficients will result(15). The addition of a fourth solvent component, if properly selected, can further alter the activity coefficients of the solute, preferentially increasing its concentration in a given phase(16, 17). An extension of these principles to multicomponent solid or liquid systems is still to be determined in the supercritical region.

We have applied some of these principles to the extraction of 1-butene from a binary mixture of 1,3-butadiene/1-butene. Various mixtures of sc solvents (e.g., ethane, carbon dioxide, ethylene) are used in combination with a strongly polar solvent gas like ammonia. The physical properties of these components are shown in Table I. The experimental results were then compared with VLE predictions using a newly developed equation of state (18). The key feature of this equation is a new set of mixing rules based on statistical mechanical arguments. We have been able to demonstrate its agreement with a number of binary and ternary systems described in the literature, containing various hydrocarbon compounds, a number of selected polar compounds and a supercritical component. It predicts quite well vapor-liquid-phase equilibria for a multi-component system in the retrograde region but cannot predict the formation of a second liquid phase.

The mixing rules used in the equation of state are defined as follows:

$$a_{lm} = (a_{11} \times a_{mm})^{1/2} (1-k_{lm})$$

and (1)

$$b_{lm} = [(b_{11})^{1/3} + (b_{mm})^{1/3}]^{2/3}$$

where a and b are the average Van der Waal configurational coefficients for mixtures, and l and m are the pairwise species which can be adjusted for the dipole contribution of ammonia-butene through the interaction parameters, k_{lm} and b_{lm}. All other k and b values of the interaction coefficients were set to 0 and 1, respectively. This model has some serious limitations in representing the effects of polar compounds, since quadrupole and strong interaction forces are ignored. Nevertheless, one can obtain a "ball park" estimate of the likely degree of separation.

TABLE I

Properties of C4 Components

Property	1-Butene	1,3-Butadiene
Molecular Wt.	56.11	54.09
Critical Temp, C	146.4	152.20
Critical Press.MPa	4.019	4.329
Critical Vol. cc/mole	4.276	4.083
Normal BP, C	-6.25	-4.411
Solubility Param.	4.7504xE04	4.8694xE04
Dipole Mom. Debye	0.34	0.0
Acentric Factor	0.1867	0.1932

Properties of Solvents

Property	Ethane	Ethylene	Carbon Dioxide	Ammonia
Molecular Wt.	30.07	28.05	44.01	17.03
Critical Temp,C	32.27	9.21	31.04	132.50
Critical Press, MPa	4.88	5.03	7.38	11.27
Critical Vol. cc	4.919	4.601	2.136	4.255
Normal BP, C	-88.60	-103.7	—	-33.43
Solubility Coeff.	3.9134xE04	3.932xE04	4.605xE04	9.239xE04
Dipole Mom.Debye	0.0	0.0	0.0	1.47
Acentric Factor	0.09896	0.085	0.2276	0.2520

Experimental Apparatus and Materials

All solvent gases were supplied as cylinder gases having a specified minimum purity of 99.0%. Butene, butadiene and ammonia were used as received without further purification.

Extraction experiments were carried out in a one-liter, stirred, stainless steel Autoclave vessel(A) rated at 5000 psig at 600 F. A complete schematic of the assembly is shown in Fig. 1. All connections with stainless steel 0.025 I.D. tubing were made with Autoclave Engineering Speedbite fittings. The presure vessel was maintained at constant temperature by means of external electrical heating and an internal cooling coil. Cooling was furnished by water chilled and circulated by a Blue-M freon refrigeration unit. An Autoclave magnetic stirring unit powered by an air motor (M) was used to ensure adequate mixing of the cell contents. The presence of phase interfaces were viewed through the windows of a 50 cc Jerguson gauge(J) (rated at 5000 psig at 72 F), in parallel with the main cell. The sight glass of the gauge which also serves as a level indicator, had an etched, direct reading scale, calibrated to measure the total volumes of the liquids in the cell.

Butene and butadiene were pumped into the the reactor through a common LP10 Whitey displacement pump(C1). A 30-pound nitrogen head was used to pressurize each hydrocarbon cylinder to maintain adequate pumping efficiency. Ammonia was pumped as a liquid through a second Whitey pump(C2), equipped with a refrigerated head to avoid vaporization during compression. The ammonia line was independently cooled to prevent vaporization during transfer to the reactor. A slight helium pressure was maintained in the vessel during charging to further ensure a minimum of vaporization at the liquid surface and ease the measurement of the liquid level.

Ethylene, carbon dioxide, or ethane were compressed to the desired pressure through a Haskel Model AG-62, gas compressor(C3), with a 25-1 compression ratio and a maximum rated outlet pressure of 9000 psig. The mass of the gases used in each run was determined by weight loss of the gas cylinder mounted on a Dayton digital scale. A redundant volume check was made by measuring the difference between the liquid levels before and after gas addition to the liquid charge and the total volume of the vessel. The pressure of the system and in each of the receiver vessels, E1 and E2, was measured with calibrated high pressure precision Bourdon tube gauges of appropriate range (G1,G2,G3).

The exterior of the Autoclave vessel wall was divided into three heating zones, each controlled by one Eurotherm 103 temperature controller(TC). The remaining sections were heat traced with self-limiting "Autotrace" heating tapes to prevent vapor condensation in the lines. Heating controls were adjusted manually and all temperatures were measured with copper-constantan thermocouples attached to the cell and receiver vessels. The thermocouple output voltages were indicated on a 10-point Acromag digital voltmeter. A calibrated platinum-10% rhodium thermocouple was used to measure the temperature of the contents in the autoclave reactor.

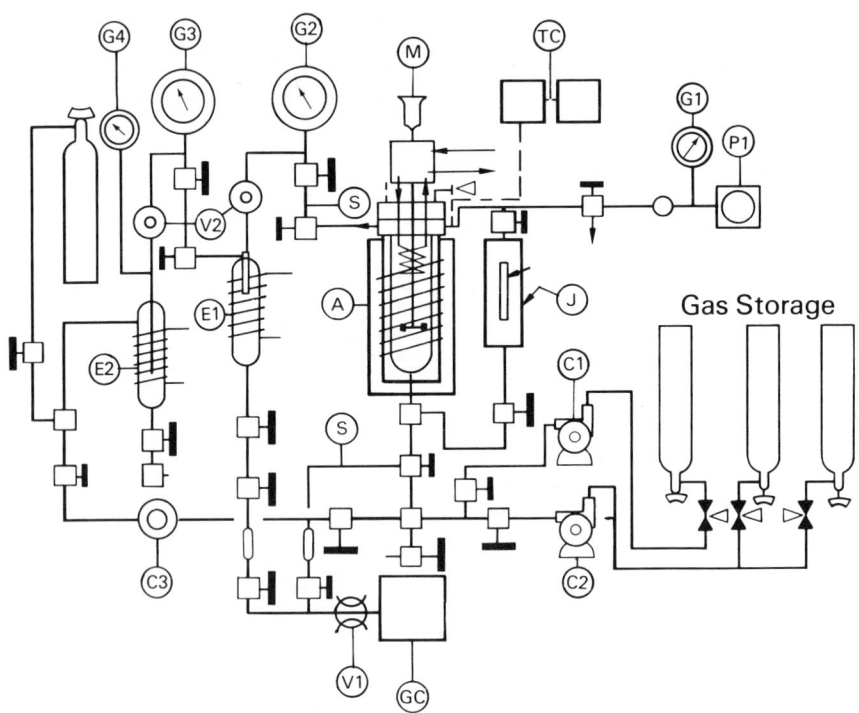

Figure 1. Experimental Apparatus

Procedure, Sampling and Analysis

Each experiment was conducted as follows: The system was evacuated to remove residual fluids. Nitrogen was then flushed through the system and the system once again evacuated. The hydrocarbon and ammonia were then pumped into the cell as liquids. The desired liquid was monitored from the level on the sight glass. The composition was determined by volume percent of the hydrocarbons and entrainer added. The solvent gas was then introduced. The temperature of the contents in the cell was adjusted to the desired value and stirred vigorously. The pressure of the solvent gas was slowly increased until the liquid interface disappeared and an opalescence regime appeared, indicating the attainment of the critical region. The gas pressure was then reduced by a slight depressurization in the chamber until the interface just reappeared. It was established that this procedure permitted the contents to be adjusted less than 2 to 3 percent of the critical pressure. Stirring was discontinued about two minutes before sampling, to allow the system to equilibrate the vapor and liquid phases.

An estimate of the volume of solvent added to the hydrocarbon mixture was obtained by calculating the difference in the liquid level before and after gas addition, which is approximately equal to the volume of gas dissolved in the liquid phase. This value is added to the volume above the liquid interface to obtain the total solvent volume added to the system. The ratio of ethylene/ammonia solvent mixture added to the volume of butadiene/butene solution was approximately 5:1.

The contents of the reactor were sampled before and after the introduction of the solvent gas. This was done by trapping approximately 0.5 cc of the mixture from the reactor volume in a precalibrated loop of 1/16 inch i.d. sample line located between two high pressure valves(S) adjacent to the vessel. The volume of the sample withdrawn was suffiently small to minimize any changes in the pressure (< 0.1 psi) of the main contents of the vessel. The high pressure sample was further expanded into a pre-evacuated 300 cc Hoke cylinder(E1) to about 5 atm. This volume of sample was again expanded to about 1 atm in a pre-evacuated, final 70-cc Hoke cylinder(E2). Portions of this volume were then injected by means of an automated Valco valve(V1) into the Varian 920 gas chromatograph(GC). All sample loops were also heat traced to prevent condensation in the lines. This procedure was also followed in sampling the liquid phase from the bottom of vessel as well. Syringe sampling through diaphram ports located at strategic positions in the system provided additional checks on the automated sampling procedures.

All samples were analyzed in an AllTech 1/8"I.D, 20 foot column, packed with VZ-7 resin, using helium as the carrier gas. This column proved more effective for the measurement of the C4 components than the Poropak- N column. Ammonia was not detected by this column and as a result some of the assumptions employed to treat the solvent phase are described in a later paragraph. The G.C. was run with the injector chamber set at 115 C, the column oven temperature at 60 C and the filament current set at 150 ma. Column analyses were performed by an Autolab Integrator. A Leeds and Northrup strip chart recorder was used to plot the output of the integrator and to provide a visual determination of the time trace and the magnitude peak heights for rapid comparison of the relative effectiveness of the separation. A sample strip chart record of the G.C. output is shown in Fig. 2.

Results

Tie lines of the system can be generated from the equilibrium compositions for each run and selectivities computed. The results of measurements obtained for the 5% by volume of ammonia/ethylene are represented in the binodal diagram in Fig. 3. Butene is represented as the distributed component between the solvent phase and the butadiene-rich phase. The ammonia-solvent gas mixture was considered to behave as a pseudo-solvent of fixed composition. The ratio of the integrated peaks for butene(i) and butadiene(j) was used to compute the selectivity, B (beta), defined on a solvent-free basis, as:

$$B = \frac{(y_i)\text{solvent phase} \times (x_j) \text{ heavy phase}}{(x_i)\text{heavy phase} \times (y_j) \text{ solvent phase}} \quad (2)$$

Representative results are shown in Table II and III for ethylene and ethylene/ammonia and ethane and ethane/ammonia solvent mixtures. Since our detection capability was limited to hydrocarbons gases only, ammonia could not be directly measured in either phase. However, the selectivity as defined in Eqn. 2 can be considered as a measure of the ratio of the distributed components only; hence, independent of the absolute concentration of the solvent components.

All extraction runs were made with equimolar mixtures of butene/butadiene using ethylene, ethane or carbon dioxide as the solvent with various concentrations of ammonia from 0 - 10 vol%. In the absence of ammonia, separation of the mixture of hydrocarbons was negligible. Table IV summarizes the data for pure solvents and shows a comparison with some selected data for ammonia/pure gas mixtures. Ammonia/ethylene mixtures appear to be more effective as solvent mixtures, with a marked separation of butene achieved. 2,3-Butadiene is concentrated in the heavy phase, in agreement with the findings reported for liquid ammonia (1).

A series of experiments was conducted to determine whether there exists an ammonia concentration for which maximum separation could be attained. At a pressure of 600 psia and 20 degrees C. and between 5 and 8 vol% ammonia, a maximum in the selectivity is obtained and is shown in Fig. 4. The selectivity, however, decreases with increasing temperature, with other variables held constant as shown in Fig. 5. The trend observed these experiments are in agreement with the findings of Brignole, et al.(19). An error of +/- 5% was estimated in the calculation of the experimental selectivities. The experimental results were compared with predictions obtained from the equation of state, assuming that the interaction parameter, $k_{1m} = 0.2$ for ammonia and 1-butene dominated the non-ideality of the system. A comparison of the calculated selectivities obtained for the ethylene/ammonia mixtures are given in Table II with the results in Table III for the ethane system.

17. HACKER *Separation of Butadiene-Butene Mixtures* 221

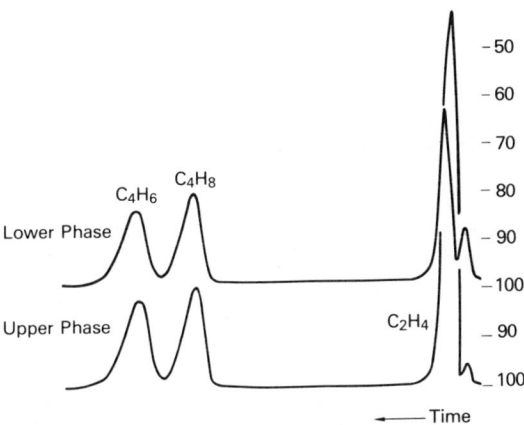

Figure 2. Typical Gas Chromatographs from a VZ-& Column with Column Temperature at 98 C. Ammonia Not Detected in this Column.

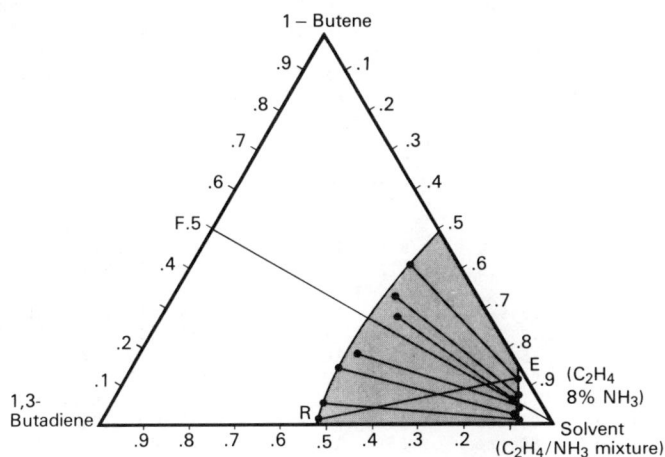

Figure 3. Ternary Phase Diagram for the System. 8% Ammonia Concentration in Ethylene. Pressure = 600 psia, Temp = 20 °C. Experimental Points Shown as Dots.

TABLE II. Butadiene-Butene-Ethylene-Ammonia Equilibrium Composition

TEMP C	PRES-SURE PSIA	NH_3 %	C_2H_4 y_i	SOLVENT BUTENE y_j	PHASE BUTA-DIENE y_k	C_2H_4 x_i	BUTADIENE BUTENE x_j	PHASE BUTA-DIENE x_k	BETA β
20	600	0	88.351	3.768	7.881	45.940	17.47	36.584	1.00
		0	88.425	3.763	7.839	38.338	19.765	41.901	0.99
		0	69.694	14.932	15.374	71.997	13.092	14.911	1.10
		2.3	82.199	8.828	8.973	58.461	17.483	20.879	1.17
		5.0	89.063	6.936	4.001	52.675	23.890	23.435	1.70
		5.0	88.879	6.903	4.218	40.109	30.149	29.742	1.61
		5.0	76.970	15.069	7.961	58.039	22.047	19.914	1.71
		5.0	75.469	15.505	8.448	55.886	23.201	20.913	1.61
		5.0	89.132	8.805	2.064	38.799	47.751	13.450	1.20
		5.0	90.259	8.066	1.675	34.527	50.693	14.780	1.40
		5.0	90.003	8.241	1.756	41.657	45.281	13.062	1.35
		8.0	91.066	2.629	6.305	46.473	12.884	39.303	1.27
		8.0	85.908	3.288	8.151	44.928	13.157	40.645	1.24
		8.0	86.668	11.710	1.622	47.847	41.382	10.771	1.88
		8.0	86.775	11.737	1.488	49.963	39.604	10.433	2.08
		8.0	69.579	19.803	10.618	49.100	27.951	22.949	1.53
		8.0	70.150	19.600	10.249	61.617	21.157	17.226	1.56
		8.0	92.786	2.932	4.238	48.052	17.872	34.075	1.30
		8.0	92.468	3.101	4.430	52.335	16.484	31.182	1.32
		8.0	90.478	7.865	1.657	37.229	48.002	14.769	1.46
		8.0	90.092	8.193	1.716	38.954	46.955	14.091	1.43
		8.0	71.455	8.932	19.613	44.076	13.800	42.124	1.39
		8.0	70.326	8.981	20.693	42.661	14.033	43.306	1.34
		8.0	71.296	9.006	19.698	40.781	14.424	33.795	1.42
		10.0	59.583	27.028	13.389	39.981	36.671	23.348	1.29
		10.0	60.217	26.028	13.389	41.815	35.692	22.493	1.32
		10.0	74.642	19.423	5.984	66.140	25.400	8.451	1.10
		10.0	74.960	19.222	5.818	65.523	25.784	8.694	1.10
20	800	2.3	70.166	13.873	15.960	34.716	28.791	36.493	1.10
		2.3	69.498	14.112	16.389	43.984	24.539	31.477	1.10
		5.0	76.947	12.740	10.312	74.065	14.520	11.415	1.05
		5.0	77.599	12.566	9.835	74.093	14.494	11.413	1.00
		5.0	85.538	3.497	10.965	68.631	7.017	24.352	1.11
		5.0	85.908	3.354	10.738	68.631	7.017	24.392	1.06
		10.0	81.026	7.044	11.930	67.741	11.236	21.021	1.10
		10.0	80.266	3.354	12.183	67.009	11.767	21.224	1.12
20	1100	0	69.694	14.932	15.374	71.997	13.092	14.911	0.90
		0	81.759	8.583	8.713	73.607	12.124	14.270	0.86
		5.0	84.760	8.380	6.860	77.316	11.457	11.227	0.84
		5.0	84.074	8.457	7.451	74.816	12.849	12.335	0.92
40	600	5.0	85.128	8.416	6.456	34.556	34.501	30.943	1.17
		5.0	86.617	7.812	5.571	29.325	37.302	33.372	1.25
60	600	5.0	75.099	11.814	8.756	19.215	42.621	38.163	1.21
		5.0	74.163	12.412	9.471	20.693	41.856	37.451	1.17

17. HACKER   *Separation of Butadiene-Butene Mixtures*   223

TABLE III. Butadiene-Butene-Ethane-Ammonia Composition

| | | | Solvent Phase | | | Butadiene Phase | | | |
| | | | | | | | | | |
Temp C	Pressure psia	NH3 %	C2H6 wi	Butene wj	Butadiene wk	C2H6 xi	Butene xj	Butadiene xk	Beta
20	900	4.18	76.258	10.925	12.817	74.367	11.568	14.065	1.0
		4.18	76.505	10.834	12.661	74.993	12.079	12.929	1.09
		4.18	74.654	11.232	14.113	75.137	12.667	12.169	1.31
	620	6.97	90.406	3.998	5.596	85.647	5.830	8.523	1.04
	640	6.97	90.107	4.123	5.770	83.324	6.804	9.871	1.03
17	725	6.97	87.224	5.220	7.556	88.497	4.817	6.686	1.00
	725	6.97	87.387	5.181	7.432	88.766	4.497	6.737	1.05
	1060	6.97	88.840	4.489	6.689	93.178	2.751	4.071	1.00
	1060	6.97	88.90	4.443	6.657	93.086	2.801	4.113	1.00
19	675	6.97	85.525	7.426	7.229	88.073	6.047	5.879	1.00
	675	6.97	86.137	7.105	6.758	87.789	6.295	5.914	1.00
18	600	6.97	88.024	6.135	5.841	85.503	7.379	7.117	1.01
	600	6.97	87.722	6.201	6.077	85.812	7.185	7.002	1.00
16	550	6.97	95.867	2.252	1.977	76.208	11.624	12.168	1.19
18	600	6.97	83.134	10.507	6.358	80.277	11.959	7.577	1.05
	600	6.97	82.572	10.834	6.594	80.176	12.228	6.044	1.02
20	700	6.97	87.532	7.907	4.561	84.122	9.834	6.044	1.06
	700	6.97	87.295	8.009	4.696	83.794	10.026	6.180	1.05
22	525	1.85	89.169	5.166	5.664	69.808	15.252	14.939	1.0
	525	1.85	93.286	3.828	2.886	67.462	16.489	16.049	1.28
	525	1.85	92.362	3.900	3.737	78.235	11.064	10.701	1.01
	1100	1.85	88.100	6.345	5.555	78.508	11.205	10.288	1.04
	1100	1.85	87.700	6.328	5.973	78.581	11.078	10.341	1.00
23	550	0	82.754	8.164	3.668	53.113	22.449	24.438	1.00
	550	0	92.243	9.091	4.089	55.308	21.121	23.570	1.00
	1075	0	77.928	10.663	11.438	72.545	13.088	14.367	1.00
	1075	0	77.279	11.078	11.693	75.175	12.095	12.730	1.01

TABLE IV. Separation of Equimolar Mixtures of Butene-Butadiene with Various Solvents

Experimental Values

TEMP C	PRESSURE psia	SOLVENT	SELECTIVITY
6.0	750	C2H4	1.23
40.0	1100	CO2	1.0
22.0	1000	*	1.0
22.0	700	*	1.0
20.0	500	*	1.0
23.0	1100	C2H6/5% NH3	1.0
18.0	700	C2H6/7% NH3	1.06
20.0	900	C2H6/4% NH3	1.09
23.0	550	C2H6	1.0

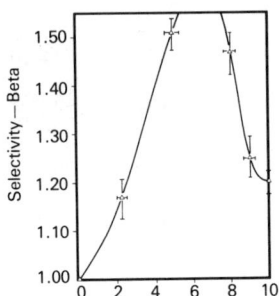

Figure 4. Selectivity as a Function of Ammonia Concentration at Temp = 20 C, Pressure = 600 psia.

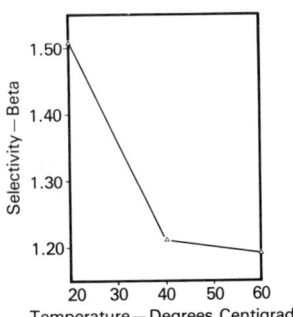

Figure 5. The Influence of Extraction Temperature on the Selectivity of 5% Ammonia - Ethylene Solvent on Mixtures of Butene and Butadiene.

Solvent to feed ratios as well as the effect of ammonia concentration in the solvent were independently varied to match the experimental data. The effect of increasing ammonia concentration at constant pressure and temperature in both ethylene/ammonia and ethane/ammonia solvent mixtures are shown in Table V.

Discussion

This study is by no means comprehensive and covers only a narrow range of variables. However, it does demonstrate the influence of entrainer in the improvement of separation over a single supercritical solvent. The increase in selectivity (1.4 to 1.8) for butene/butadiene mixtures is compared with the value of 1.63 obtained with liquid ammonia for the same binary system([1]). Moreover, it has been demonstrated that a mixture of a pure solvent and an entrainer permits an improvement in the separation at temperatures and pressures lower than would have been otherwise predicted with a single gas solvent([20]). For mixtures containing a highly polar component, such as ammonia, molecular size alone cannot account for the large selectivities observed in these experiments. At present, all theories are inadequate in explaining the chemical interactions between the entrainer and the mixture. The state of the art is comparable to liquid phase solvent extraction.

The distribution of the mixed solvents between the phases might provide some additional understanding of the reason for the effectiveness of the ammonia as an enhancing agent. The solubility of liquid ammonia in the liquid hydrocarbon phase has been shown to be high([21]) because of its strong basicity as compared with the non-polar hydrocarbon gases such as ethane and ethylene at normal conditions([22]). At higher pressures, however, solubilities of all compounds increase dramatically and, accordingly, the differences between the polar and non-polar solubilities are somewhat reduced([22]). Nevertheless, even for a relatively dilute polar mixture in a non-polar gas, the solubility of the polar component exceeds the non-polar component. At these pressures, it is likely to be completely absorbed preferentially into the liquid phase. As a result, it is estimated that the composition of the supercritical phase will contain little ammonia and predominantly solvent gas with extracted solute. For this mixture the solvent phase will consist of ethylene or ethane and butene with a small concentration of ammonia. This is confirmed in the VLE flash calculations for a 10:1 solvent extraction with a equimolar mixture of ethylene and ammonia. The distribution of ammonia to ethylene in the liquid phase is 2.5:1, whereas in the vapor it is 1:2.

The separation of the liquid components in the presence of a supercritical solvent occurs much as it does in liquid extraction with the entrainer, ammonia, concentrating in the liquid to increase the relative volatility of the butene to butadiene. The butadiene migrates to the ammonia-rich phase while the solvent gas phase or "vapor" will contain the butene.

The explanation for the higher selectivity of the ethylene/ammonia mixture over the ethane/ammonia system is somewhat surprising. The efficiency of sc extraction of a given solvent toward a particular solute is related to the supercritical

Table V

Separation of 1-Butene 2,3-Butadiene Mixtures
Computational Results by the Ely-Mansoori
Equation of State

$k_{ij} = 0.2$, all other $k_{mn} = 0$.
where i = ammonia , j = 1-butene

Mixture Ratio	Solvent-Feed Ratio	Temp. C	Pres. psia	Vapor yi	Liquid xi	Selectivity B
NH3/Ethylene						
0.25	20	27	735	0.011	0.405	1.54
		37	735	0.0208	0.460	1.87
	10	29	735	0.0175	0.385	1.29
		32	735	0.024	0.393	1.37
		37	735	0.0317	0.407	1.40
0.50	10	47	735	0.0332	0.641	2.86
		57	735	0.426	0.786	3.10
		37	588	0.339	0.635	3.94
		47	588	0.433	0.834	4.29
NH3/Ethane						
0.0	2	22	294	0.5501	0.228	1.04
0.50	10	47	588	0.0250	0.544	1.96
		57	735	0.252	0.606	1.65
		77	853	0.0348	0.931	1.10
	20	42	588	0.0129	0.579	2.26
		47	588	0.0173	0.679	2.48
1.0	10	117	1190.7	0.0331	0.900	0.91
		97	955	0.3636	0.512	1.13

solubility parameter, defined in terms of the reduced properties of the solute and solvent(23). The reduced temperature, defined in terms of the temperature at which extraction is carried out, should be close to unity to maximize separation(24).

Since both ethylene and ethane have reduced temperatures nearly equal to unity at the extraction conditions of 20 C, (T_r = .98) and ethylene (T_r = 1.04), their respective solvent capacities for butene should be about the same. This is the case as is reflected in the same values for the selectivity against butene for all pure solvent gases. One can conclude that the primary effect of the non-polar solvent is to increase the capacity of the "vapor" phase for the extracted solute near the critical. The influence of the second solvent provides only the option of modifying the physical parameters; namely, pressure and temperature, under which the optimal extraction is to be conducted. The evidence for this is the effect of the ammonia on the selectivity as calculated by the EOS in Table V. The higher values for the selectivities in the ethylene mixtures are pronounced. It can be concluded that the solvent mixture interaction parameters must dominate the solubility of butene in the vapor phase.

The simple model used in the EOS, however, does not completely indicate the presence of a maximum in the quantity of ammonia used as an entrainer. That this is in contrast to our experimental observations suggests that other factors such as chemical synergistic effects cannot be ignored and are likely to have a greater effect than anticipated. While this feature of the system was entirely unexpected, it does suggest that the H-bonding or other strong polar interaction that occurs with many selective solvents and solutes in liquid extraction, may be equally applicable in sc extraction where strong polar - polar interactions (as between ammonia and 1-butene) exist.

Conclusions

1. A maximum value in the selectivity of 1.4 - 1.8 can be achieved with a 5 - 8 vol% ammonia concentration in ethylene for the butadiene - butene separation. The higher selectivity found in the supercritical extraction is comparable to the selectivity reported for the same separation with liquid ammonia.
3. Ethylene/ammonia mixtures appear to be a more effective as solvents for the separation of a butadiene-butene mixture than are ethane/ammonia mixtures with the same concentration of ammonia. This is verified independently by the predictions of the Ely-Mansoori equation of state. The combination of enhanced solubility of the ethylene/ammonia mixture, according to Joshi and Prausnitz, as a solvent and the synergistic chemical effect between the highly polar ammonia and ethylene may explain the observed overall increased effectiveness of the solvent mixture.

Acknowledgments

I would like to express my appreciation to the Amoco Chemicals Co. for their support for this work. In particular, I wish to express my thanks for the interest shown in this work by D. E. Hannemann and P. G. Thornley and to Franke Brooks, my technician, for his industry and care in conducting these experiments.

Literature Cited

1. Poffenberger, N.; Horsley, L. H.; Nutting, H. S.; Trans. AIChE, 42:815, (1946)
2. Joshi, D. K.; Prausnitz, J. M.; AIChE Jour. 30:522, (1984)
3. Todd, D. B.; Elgin, J. C.; AIChE Jour. 1:20, (1955)
4. Weinstock, J. J.; Elgin, J. C.; Jour Chem Engr. Data, 4:3, (1959)
5. Paulaitis, M. E.; Kander, R. G.; DiAndreth, J. R.; Berichte der Bunsen-Gesellschaft f. Phy. Chem. 9:869 (1984)
6. Fogel, W.; Arlie, J. P.; Revue Francais Petrole, 39:617, (1984)
7. Brule, M. R.; Corbett, J. C.;, Hydrocarbon Process. 73, (1984) June
8. Starling, K. E.; Khan, M. A.; Watanasiri, S.; Fundamental Thermodynamics of Supercritical Extraction, Presented Annual Mtg. AIChE, San Francisco, CA, (1984)
9. Prausnitz, J. M.; Fluid Phase Equil., 14, 1, (1983)
10. Cotterman, R. L.; Dimitrelis, D.; Prausnitz, J. M.; Ber. Bunsen Gelsch. Phy. und Chem., 9, 796, (1984)
11. Reid, R. C.; Schmitt, N. J.; J. Chem Engr. Data, 31:204, (1986)
12. Vasilakos, N. P.; Dobbs, J. M.; Parisi, A. S.; IEC Process Design, 24, 121, (1985)
13. Brunner, G.; Fluid Phase Equil., Part II, 10, 289, (1983)
14. Treybal, R. E.;, "Liquid Extraction", 2nd Ed., McGraw-Hill, NY (1963)
15. Prigogine, I.; Bull Soc. Chem, Belg., 52, 115, (1943)
16. Eckert, C. A.; Greiger, R. A.; I&EC Process Design, 6, 250, (1967)
17. Rowlinson, J. S.; Liquids and Liquid Mixtures, 2nd Ed., Plenum Press London, (1969)
18. Mansoori, G. A.; Ely, J. F.; Density Expansion Mixing Rules, Unpublished, (1984)
19. Brignole, E. A.; Skjold-Jorgensen, S.; Fredenslund, J. M.; ibid. 9:801 (1984)
20. Brunner, G; Fluid Phase Equilibria, 10:289, (1983)
21. Patyi, L.; et al., Zhur. Priklad. Khimii., 51:1296 (1978)
22. Prausnitz, J. M.; "Molecular Thermodynamics of Fluid Phase Equilibria", Chap.10, Prentice Hall, NY (1969)
23. Gidding, J. C.; Myers, M. N.; et al., Science 162:67, (1968)
24. Paul, P. M. F.; Wise, W. S.; The Principles of Gas Extraction, Mills & Boon, Ltd., Monograph CE/5, (1971)

RECEIVED October 3, 1986

Chapter 18

Fractional Destraction of Coal-Derived Residuum

Robert P. Warzinski

U.S. Department of Energy, Pittsburgh Energy Technology Center, P.O. Box 10940, Pittsburgh, PA 15236

> An apparatus has been developed to fractionate coal-derived residuum by exploiting the solvent power of fluids near their critical points. Termed Fractional Destraction, the method fractionates residuum according to the solubility of its constituent components in a supercritical fluid. The novel aspect of the approach is the incorporation of a system to promote reflux of less-soluble components onto a packed bed. Fractionation of residuum will facilitate the determination of previously unattainable information concerning the composition and process-related behavior of this complex material. This paper describes operation of the unit to fractionate a residuum sample produced at the Wilsonville Advanced Coal Liquefaction Test Facility.

Describing the behavior of undefined mixtures, whether from natural or synthetic sources, often begins with the separation of these complex systems into effective pseudocomponents by distillation (1). Each pseudocomponent is then characterized as if it were a pure compound, and its characterization data are used in appropriate correlations. The presence of nonvolatile residuum poses a serious limitation to such methodology. For coal-derived liquids, heavy crude oils, tar sands, and shale oil, more than 50 percent of the fluid may not be distillable (1). Since this nonvolatile residue cannot be separated using conventional techniques, new methods of separation and characterization must be developed to provide the necessary information for design and operation of plants utilizing the fossil fuels mentioned above (2).

Apart from the need to fractionate residuum-containing fossil fuels for the measurement and prediction of thermophysical properties, other important problems could be resolved better through the study of residuum pseudocomponents. Two examples in the area of coal liquefaction are the role of

This chapter not subject to U.S. copyright.
Published 1987, American Chemical Society

residuum (a) in hydrogen utilization and hydrogen transfer and (b) in the manifestation of harmful biological effects.

Work at the Pittsburgh Energy Technology Center has been directed at the development of novel technology for the separation of fossil fuel residuum into effective pseudocomponents. In this respect, application of supercritical fluids in a manner similar to that reported by Zosel (3) is being developed. This approach is similar to conventional distillation in that an apparatus is used not only to extract the residuum but also to cause part of the residuum in the fluid phase to return as reflux onto a packed bed. This liquid reflux is caused by increasing the temperature of the supercritical fluid phase at constant pressure, thereby decreasing the density of the supercritical fluid and its carrying capacity for the residuum. Operation of a system in this region of retrograde condensation has recently been reported in the literature (4). Other recent investigations support the hypothesis that supercritical solubility is a density-driven phenomenon in the absence of strong associating forces in the solvent (5). The method is called either Supercritical Distillation or, as Zosel suggested, Fractional Destraction. This report describes the use of this technology to fractionate a coal-derived residuum from the Wilsonville Advanced Coal Liquefaction Test Facility.

Experimental

The experimental unit, called the Fractional Destraction Unit (FDU), has been designed to contact a 3-4 kg charge of residuum with a continuous flow of supercritical fluid at conditions up to 673 K and 27.6 MPa. The heart of the FDU is the Fractional Destraction Vessel (FDV) shown in Figure 1. The FDV consists of a modified 3.8-L 316 stainless steel pressure vessel onto which is attached a 78-cm column fabricated of 316 stainless steel (6.03-cm o.d., 1.11-cm wall). The column contains a packed bed and a condenser section. In the experiments reported here, the 30-cm bed section was packed with 0.41-cm stainless steel Pro-Pak protruded metal distillation packing from Scientific Development Co., State College, Pa. This is the same packing used in a conventional Podbelniak distillation column. The condenser consists of a removable 38-cm finger made of 316 stainless steel (2.67-cm o.d., 0.78-cm wall) that is heated by an internal cartridge heater to promote reflux. A triple zone furnace is used to control the temperature in the extraction zone. Temperature control on the column is accomplished with independently controlled band heaters on the packed bed and condenser sections.

The desired charge of residuum is placed in the extraction section of the FDV, and the entire unit is purged with nitrogen. The FDU is then brought up to the operating temperature before beginning solvent flow. The destraction fluid is then pumped as a liquid through a preheater to raise its temperature above the critical point before it is introduced into the FDV through the sparging device at the bottom. The pressure is controlled by a high-temperature Badger-Meter control valve located near the

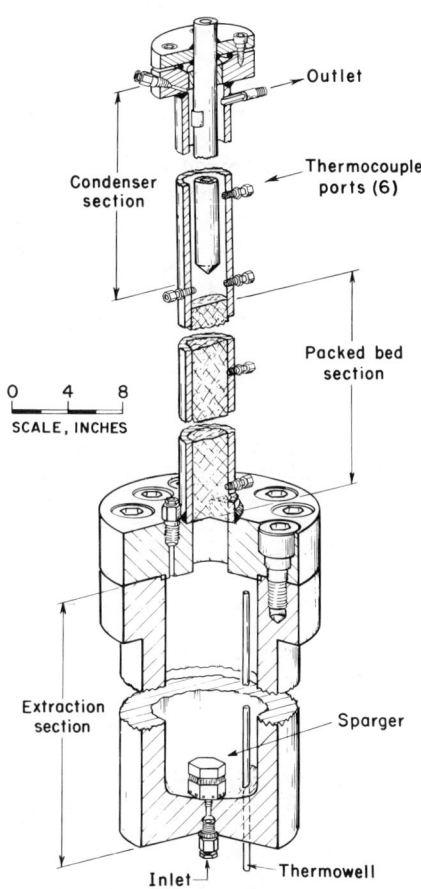

Figure 1. Sectional view of fractional destraction unit.

outlet of the FDV. After traveling up the column, the fluid stream, which now contains destracted residuum, exits at the top of the FDV and is partially depressurized through the heated control valve into a separator constructed of a 2.25 L (approximately 10.2-cm o.d., 42.5-cm long) 304 stainless steel cylinder. The separator is operated slightly above the critical temperature of the solvent (T_R=1.02) at a pressure of 0.8 MPa. Separation of the residuum from the supercritical fluid is accomplished by the reduction in pressure. The residuum is drained and collected from the bottom of the separator at periodic intervals, while the solvent is flashed to atmospheric pressure, condensed, and recovered as a liquid. The operation of the FDU will be reported in detail in a later publication.

Toluene (T_c = 591.7 K, P_c = 4.115 MPa) and cyclohexane (T_c = 553.4 K, P_c = 4.074 MPa) were used as the supercritical fluids in the work reported here. They were obtained in drum quantities at greater than 99 percent purity and used as received. Owing to inefficiency in the separator, some of the destracted residuum is collected with the solvent. This material is recovered in a rotary evaporator. The distilled solvent is then reused in the FDU. In the figures in this report that depict residuum overhead data, correction has been made for the residuum recovered in the spent solvent.

The residuum used in the work reported here was collected from the T102 vacuum distillation tower during Run 242 at the Wilsonville Advanced Coal Liquefaction Test Facility. This run was made using Illinois No. 6 coal from the Burning Star mine in what is termed a Short-Contact-Time Integrated Two-Stage Liquefaction (SCT-ITSL) mode (6). During the time this particular sample was collected, the T102 unit was operated at 594 K and 3.4 KPa. The residuum was crushed to minus 0.64 cm and mixed by riffling before use.

Discussion

Operation of the FDU with coal-derived residuum was preceded by tests on pure compounds and distillable coal liquids using n-pentane as the supercritical fluid (7). The results of tests with the coal-derived distillate showed that the FDU was basically performing as expected. Liquid reflux was generated by means of the hot finger when the device was operated at a temperature slightly above the critical temperature of the transport fluid. When reflux was established, fractionation based upon volatility was observed. Poorer separation was achieved in the absence of reflux.

The first work on the T102 bottoms involved operation in the non-reflux mode to obtain base-line data on the transport of this residuum in various hydrocarbon solvents. Using n-pentane, cyclohexane, and toluene at a T_R of 1.02 and at a P_R of 2, the residuum brought overhead was 23, 54, and 67 percent of that charged, respectively. Based on this information, further studies were performed using cyclohexane as the supercritical fluid, since a large portion of the residuum could be destracted at a temperature similar to or less than that in the T102 separator.

A two-step destraction procedure was developed to maximize the amount of T102 residuum brought overhead. In the first step, toluene is used in a manner similar to conventional supercritical extraction to produce a nearly ash-free material for subsequent fractionation. This first step is called the non-reflux mode because the column of the FDV is maintained at the same temperature as the extraction section. In the second step, called the reflux mode, the column and finger are heated to a higher temperature than the extraction section, which causes the density of the fluid to decrease as it travels up the column and thus promotes reflux of the less-soluble components. These two steps are described below.

In the non-reflux mode, the FDU was used to process five 800-gram charges of the T102 bottoms. Repetitive operation of the unit was performed to produce sufficient quantities of the final fractions for subsequent characterization and experimentation. Figure 2 summarizes the operation of the FDU during these five destractions. Shown is the amount of residuum brought overhead as a function of the time on stream at a P_R of 2. As previously mentioned, the residuum overhead data include residuum recovered from both the separator and the spent solvent. The FDV was maintained at a T_R of 1.02, with overall variance in temperature for the five tests being $\pm$ 5 K. Temperature variance during any one test was $\pm$ 2 K. The differences observed, especially in the final amounts destracted, appear to be due to this small temperature variance betweeen runs. The higher yields were consistently obtained at temperatures nearer the critical point. This phenomenon appears to be particularly sensitive to the temperature of the extraction zone when the fluid is first introduced. The fact that the initial dissolution of the residuum in the fluid influences the overall yield suggests that components in the residuum may be acting as cosolvents.

Table I contains the elemental analysis of the T102 bottoms, the material brought overhead with toluene, the residue remaining after the toluene destraction, and the starting coal used at Wilsonville during Run 242. The toluene overhead represented 66.8 percent of the material charged to the FDU, and the residue accounted for 27.9 percent. Other material collected from the FDU includes 4.2 percent in the spent solvent and 2.7 percent recovered during cleaning of the FDV and separator with tetrahydrofuran. The total material balance is 101.6 percent. This number also includes any residual toluene or tetrahydrofuran in the various samples. The overhead collected from the five destractions was ground and combined before use in the reflux mode experiments.

As previously mentioned, cyclohexane was chosen for the fractionation solvent for the second step, since it could transport sufficient quantities of residuum at reasonable temperatures. In the reflux mode, the column of the FDV is operated at a higher temperature than the extraction zone. As the carry-over of residuum decreases, the temperature of the column is reduced to cause the density in this region to increase and consequently more residuum is transported overhead.

TABLE I. Analysis of Feed Coal to Run 242, T102 Bottoms After Grinding and Mixing, and Materials Produced by Supercritical Extraction of T102 Bottoms With Toluene in the Non-Reflux Mode

	ILLINOIS[a] NO. 6 COAL	RUN 242[b] T102 BOTTOMS	TOLUENE OVERHEAD	TOLUENE RESIDUE
C	68.4	79.1	87.4	60.3
H	4.4	5.9	6.7	3.5
O	11.9[c]	4.0[c]	4.0[d]	7.1[d]
N	1.4	1.3	1.3	1.4
S	3.2	1.0	0.7	0.9
Cl	0.1	---	---	---
ASH	10.6	8.7	0.1	29.4
H/C	0.77	0.89	0.91	0.69
$\underline{M}_N$ (VPO, 353 K, PYRIDINE)	---	---	574	e

[a]Analysis from Wilsonville report, see reference (6).
[b]Approximately 8 percent unconverted coal content.
[c]Determination by difference. [d]Direct determination.
[e]Insufficient solubility in pyridine.

This is repeated until the column is at the same temperature as the extraction zone. In this work, four fractions were brought overhead by operating the column initially at 593 K and then decreasing the temperature to 578, 573, and 563 K as the residuum carry-over approached 1.0 gram per gram-mole of cyclohexane. This concentration value is calculated from the amount of residuum collected from the bottom of the separator after a 30-minute collection period and from the amount of cyclohexane pumped during that period. The column temperatures were selected both from density estimation and from actual experimentation. A more detailed discussion of the development of the operational parameters for the reflux mode will be presented in a future paper.

Figure 3 illustrates the difference between operation of the FDU in the non-reflux and the reflux modes with cyclohexane. This figure depicts the results in terms of the overhead concentration of residuum. The reflux mode data represent one of three replicate fractionations that were performed on the T102 toluene destraction overhead. Each point represents a 30-minute sample collection period. Owing to the limited quantity of toluene overhead produced, no non-reflux mode experiments were conducted using this material. From the earlier development work, however, several non-reflux mode experiments were performed on the T102 residuum sample from Run 242. The non-reflux mode data in Figure 3 were derived from one of these experiments and adjusted for comparability to the data from the reflux mode. The adjustment compensates for separator inefficiency and for residuum insoluble in supercritical

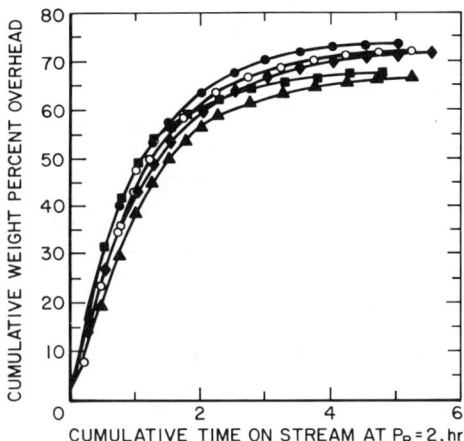

Figure 2. Comparison of five replicate destractions (non-reflux mode) of Wilsonville Run 242 T102 residuum using toluene.

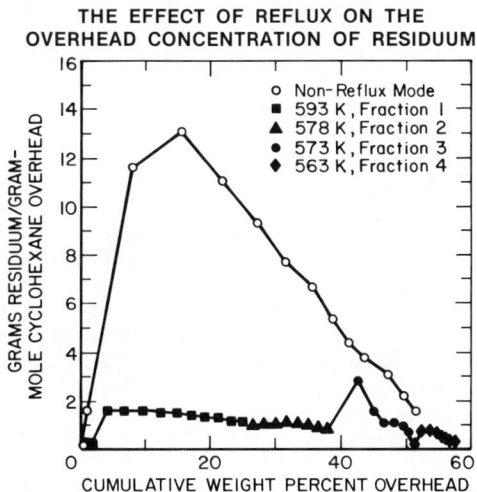

Figure 3. Comparison of non-reflux and reflux modes of operation using cyclohexane. Temperatures indicated are those of the column in the reflux mode. Pot temperature was 563 K in both modes.

toluene. All but the last three data points in the non-reflux mode data represent 15-minute sample periods. The only other difference between the two tests was the solvent delivery rate, which was 0.24 mole per minute for the non-reflux mode, and 0.43 mole per minute in the reflux mode.

As shown in Figure 3 the temperature of the column was initially 30 degrees higher than the pot, which produced the reduced carry-over rate by causing reflux to occur. An expanded view of the carry-over concentration is shown in Figure 4. This view shows more clearly how residuum carry-over is manipulated by changes in the column temperature. Figure 5 contains the same information for the three replicate cyclohexane fractionations performed on the toluene overhead. Conditions in all three experiments were similar; however, the small deviations due to inherent limitations of the process controllers produced some overlap between successive fractions. The amount of overlap is estimated at 2.6 percent between Fractions 1 and 2, 2.5 percent between Fractions 2 and 3, and 3.0 percent between Fractions 3 and 4. Also, 32 grams that should have been in the 573 K fraction were inadvertently collected in the 578 K fraction. The respective fractions and residues from the three fractional destractions were ground and combined before characterization.

Table II contains the analysis of the four overhead fractions and the residue. The overall material balance was 96.1 percent, with 57.2 percent of the charge being brought overhead (includes residuum recovered from spent solvent) and 31.7 percent remaining in the residue. An additional 7.2 percent was recovered upon cleaning the unit with tetrahydrofuran. The trends evident from the elemental analysis indicate that the components that have lower molecular weights, higher hydrogen-to-carbon ratios, and lower heteroatom contents are concentrated in the earlier fractions. Also note that the separator inefficiency is highest for the earlier fractions, indicating that the more volatile components are more readily carried through the separator with the solvent. The material carried through the separator with the solvent was not mixed with the respective fractions collected from the separator. Independent characterization of these samples will provide valuable insight into the operation of the separator.

The molecular weight data in Table II were determined by vapor pressure osmometry (VPO) and gel permeation chromatography (GPC). The molecular weight increases regularly from Fraction 1 to the residue. In comparison, the VPO molecular weights for six fractions collected in the non-reflux mode cyclohexane destraction shown in Figure 3 range from 441 to 471, with no consistent trend. The polydispersity values ($\bar{M}_W/\bar{M}_N$) shown for the four fractions in Table II are less than those for the non-reflux mode fractions, which ranged from 1.47 for the first fraction to 1.88 for the sixth fraction. This number is a measure of the breadth of the molecular weight distribution (9) and shows that operation in the reflux mode produces fractions with narrower molecular weight distributions than those prepared without reflux.

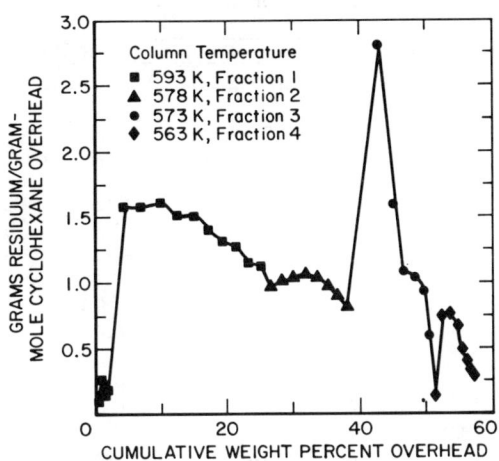

Figure 4. Fractional destraction of toluene overhead using cyclohexane.

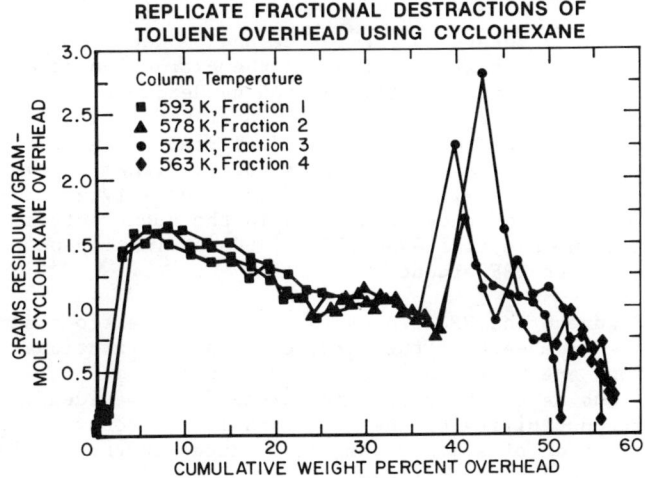

Figure 5. Comparison of three replicate fractional destractions (reflux mode) of toluene overhead using cyclohexane.

TABLE II. Characterization of Materials Produced From the Cyclohexane Fractional Destraction of the Toluene Overhead Sample in Table I.

	FRACTION DESIGNATION[a]				
	593 K	578 K	573 K	563 K	RESIDUE
C	87.8	87.3	87.4	87.0	87.0
H	7.6	7.5	7.2	6.5	5.9
O[b]	3.5	3.7	3.3	3.7	4.9
N	0.9	0.9	1.2	1.4	1.8
S	0.5	0.6	0.6	0.7	0.9
ASH	---	---	---	---	0.1
H/C	1.03	1.02	0.98	0.89	0.81
RESIDUAL CYCLOHEXANE, %[c]	0.5	0.5	0.8	0.6	0.5
RECOVERY, %[d]	23.9	13.0	13.8	6.4	31.7
SEPARATOR INEFFICIENCY, %[e]	27.5	17.8	8.4	11.1	---
MELTING RANGE, K	333-338	363-373	393-403	448-463	573
$\bar{M}_N$, VPO[f]	393	479	578	645	1226
$\bar{M}_N$, GPC[g]	407	449	527	564	715
$\bar{M}_W$, GPC	504	601	851	1010	1996
$\bar{M}_W/\bar{M}_N$	1.24	1.34	1.62	1.79	2.79

[a]The temperature refers to the reflux and packed bed zone. The residue is the material remaining in the extraction zone. [b]Direct determination. [c]Determination by Headspace Chromatography, see reference (8). [d]Total recovery of material based upon weight of toluene overhead charged to the unit. Except for the residue this includes both residuum recovered from the fraction receiver and from the spent solvent. [e]Percent of total material overhead recovered in the spent solvent. [f]Determination in pyridine at 353 K. [g]Determination on PLgel 100Å column with THF eluent.

In Figure 6 the GPC traces for the four fractions and for the residue from each of the three cyclohexane fractional destractions are overlaid. The trend to higher molecular weight distributions as the fractionation proceeded is evident, as well as the reproducibility of the fractionation process. Additional characterization of similar samples produced in the FDU is the subject of another paper (10).

Conclusions

The main conclusion to be drawn from the experimental data presented here is that fractionation of residuum through the use of a supercritical fluid system incorporating internal reflux produced by retrograde condensation results in sharper fractions than those obtained by ordinary supercritical extraction. The capability of the FDU to process coal-derived residuum in the

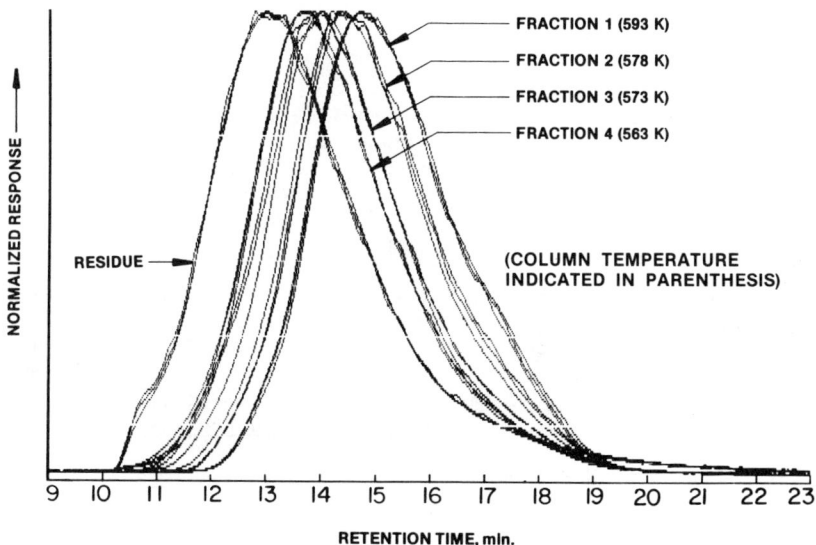

Figure 6. Comparison of the GPC results for the three replicate fractional destractions of toluene overhead using cyclohexane.

internal reflux mode has been demonstrated. The density-driven separation does appear to fractionate the residuum on the basis of volatility, with lower molecular weight species preceding larger ones.

Other methods of fractionation with supercritical fluids are conceivable. One such possiblility could involve manipulation of the pressure during the destraction or upon subsequent separation of the fluid and residuum. The relative merits of such possibilities remain to be explored. Successful development of such technology will result in the ability to fractionate and characterize material currently intractable by conventional methods.

Disclaimer

Reference in this report to any specific product, process, or service is to facilitate understanding and does not necessarily imply its endorsement or favoring by the United States Department of Energy.

Literature Cited

1. Brule, M.R.; Kumar, K.H.; Watansiri, S. Oil & Gas Journal Feb. 11, 1985, 87.
2. Brule, M.R.; Rhodes, D.E.; Starling, K.E. Fuel 1981, 60, 638.
3. Zosel, K. Angew. Chem., Int. Ed. 1978, 17(10), 702.
4. Brule, M.R.; Corbett, R.W. Hydrocarbon Processing June, 1984, 73.
5. Vasilakos, N.P.; Dobbs, J.M.; Parisi, A.S. Ind. Eng. Chem., Process Des. Dev. 1985, 24(1), 121.
6. Technical Progress Report by Catalytic, Inc., DOE Report No. DOE/PC/50041-19, July, 1983.
7. Warzinski, R.P.; Ruether, J.A. Fuel 1984, 63, 1619.
8. Loffe, B.V.; Vitenberg, A.G. "Head-Space Analysis and Related Methods in Gas Chromatography"; John Wiley & Sons, New York, 1984; p. 131..
9. Yau, W.W.; Kirkland, J.J.; Bly, D.D. "Modern Size-Exclusion Liquid Chromatography"; John Wiley & Sons, New York, 1979; p. 8.
10. Warzinski, R.P.; Strycker, A.S.; Lett, R.G. Prepr. Pap.- Am. Chem. Soc., Div. Fuel Chem. 1985, 30(4), 243.

RECEIVED June 25, 1986

FUEL APPLICATIONS

Chapter 19

Isotope Effects in Supercritical Water: Kinetic Studies of Coal Liquefaction

David S. Ross[1], Georgina P. Hum[1], Tiee-Chyau Miin[1], Thomas K. Green[2], and Riccardo Mansani[3,4]

[1]Fuel Chemistry Program, SRI International, Menlo Park, CA 94025
[2]Western Kentucky University, Bowling Green, KY 42102
[3]Eniricerche S.p.A., 20097 S. Donato Milanese (MI), Milan, Italy

> Coal liquefaction studies with Illinois No. 6 coal in
> supercritical water/CO systems have demonstrated that a
> suitable model for liquefaction includes a branch point.
> Thus coal is partitioned between reducing (i.e. liquefy-
> ing) steps, and steps where strictly thermal reactions
> consume convertible sites and yield unconvertible char.
> In supercritical H_2O it was found that the toluene-
> soluble products (TS) profile leveled off at about 50%.
> However in supercritical D_2O, liquefaction was consis-
> tently superior to that in the protio medium, and the
> ultimate convertibility was about 60%. This inverse
> isotope effect can be explained by a model in which the
> limits to conversion are solely based on the competition
> between the kinetics of the parallel routes to TS and
> char, respectively. Conversion is thus limited by the
> kinetics of the conversion system, this limitation being
> even more severe in conventional donor systems which are
> inherently less effective for liquefaction than is
> water/CO. In principle the conversion of coal virtually
> completely to TS in a single step is feasible in a system
> with sufficient reducing potential.

The chemistry of coal liquefaction is not very well understood, even after more than two decades of research into the kinetics and mechanism of the process. There have been a number of models for conversion proposed, most of them focused on the several liquefaction products, including preasphaltenes, asphaltenes, oils, and gases. A survey of some of the models has been presented (1), and a common feature among them is the multiplicity of paths connecting all of the components.

Kinetic model studies have invariably been carried out in organic donor media, and while the use of these media may be convenient in large scale conversion systems, they do not lend them-

[4]Current address: EniChem Polimeri, Milanofiori-Strada 2-Palazzo F7, Milan, Italy

0097-6156/87/0329-0242$06.00/0
© 1987 American Chemical Society

selves well to laboratory study of conversion. Not the least significant of the problems to be faced in such a study is the unavoidable participation of various interrelated and interlocked free radical chain reactions, most of which having no direct bearing on, and perhaps only a secondary relationship with, the liquefaction process. The complications presented by this network of incidental reactions, coupled with the multiple reaction paths to products, prompted the study of the kinetics of coal liquefaction using supercritical water as the medium.

BACKGROUND

The action of CO/water to reduce both low rank and bituminous coals to upgraded products has been known for more than half a century, having been introduced in the work of Fischer and Schrader (2). A review of the CO/water conversion process has appeared (3), and a summary of the characteristics of the medium is presented here. The critical temperature and pressure of water are respectively 374°C and 221.3 bar, at which point the medium has a density of 0.32 g/mL. Not surprisingly, at temperatures and pressures around that point, the medium can be a good solvent for organic solutes otherwise not soluble at ambient conditions. The high solubilities, however, are maintained at substantially lower temperatures; the solubilities of benzene and toluene for example are 12% and 7% respectively at 280°C (4). The critical curve for benzene/water passes through 294°C and 200 atm, at which point the components form a single phase, and the density of the medium is 0.74 g/mL, like that of a true liquid.

Similar behavior is reported for tetralin and naphthalene, where single phases with water are seen in the 300-340°C range. In these cases the critical temperatures for the two organics are greater than that of water. Thus the critical temperature of water is lowered by the addition of aromatics like those in coal and coal products.

Hydrothermal systems such as these can also support the dissolution of ionic salts. There is a tendency for the organic components in these systems to be salted out of solution; however conditions exist where a homogeneous system would contain both the dissociated ionic component and an aromatic component such as benzene (3).

At ambient temperatures CO is only slightly soluble in water. This behavior is in contrast to the case for carbon dioxide, which dissolves and is easily hydrated to carbonic acid and its conjugate bases. The simple solution of CO in water in accord with its Henry's law constant continues to about 250°C. At that temperature the rate of hydration becomes significant, and formic acid is formed. However the acid is thermally unstable, and subsequently decomposes to CO_2 and H_2.

EXPERIMENTAL

The coal used in this work was an Illinois No. 6 coal, PSOC 1098, supplied by Pennsylvania State University. The coal was ground and sieved under dry N_2 to pass -60 Tyler mesh and then dried under vacuum at 105°C overnight.

All reactions were performed in a 300-mL Magne-drive stirred Hastelloy C autoclave. The autoclave was loaded with about 5 g of dry coal and 30 g deionized water. Weighed quantities of KOH were added to the water to adjust the initial pH to the desired value. The system was sealed and purged with 500 psig N_2. It was then purged twice and charged with the appropriate pressure of desired gas.

The autoclave was heated with stirring to 400°C (± 5°C) for 20 min (± 1 min). Maximum pressures attained were in the range 4000-5000 psig. The full heat-up and cool-down times were both about 1 h.

The final, cold pressures were measured after reaction. Quantitative analyses of the product gases CO, CO_2, and H_2 were made with a Carle gas chromatograph using standard gases. The aqueous phase was pipetted from the autoclave and filtered. The nonvolatile products were quantitatively removed with tetrahydrofuran (THF) and transferred to a 500-mL round bottomed flask. THF was removed by rotary evaporation and 400-mL of toluene was added. This mixture was refluxed with stirring for 2 h in a Dean-Stark trap. Azeotropic distillation removed any water. After cooling, the mixture was filtered through a medium porosity filter (10-15 μm) to separate the toluene-insoluble (TI) from the toluene-soluble (TS) material. The toluene was removed by rotary evaporation, and both the TS and TI portions were dried overnight under vacuum at 105°C and weighed. The conversion to TS material in a dry, mineral-matter-free basis was calculated as

$$\%TS = (wt\ TS/wt\ DMMF\ coal) \times 100$$

RESULTS

Working Model. Our working model evolved from consideration of the profiles typically noted in the modeling literature. Products are seen to grow with time, and then level off at some level below quantitative conversion. At the same time the quantities of "unreacted coal" decline, and also level off. A picture of the coal organic matrix derived from this view of conversion would depict a collection of organic units, connected to each other through a series of links increasingly more difficult to break.

On this basis, conversion is limited by coal structure. And in terms of the conventional homolytic scission/H-capping view of conversion, increased yields of coal liquids are therefore obtainable only through increases in conversion temperature or residence time. Unfortunately, increases in the thermal severity of the process result in products reflecting the rise of dealkylation and aromatization reactions at higher temperatures. Thus increased product yields are brought about at a considerable cost to product quality (5).

The fact that the profiles of both the desired soluble product and the insoluble product level off at intermediate values suggest a simpler model for conversion. This scheme, presented in Figure 1, is our working model, and is tested in the study described here. In the scheme coal is partitioned in parallel, competitive routes between i) reaction with some reducing component in the system to yield TS, and ii) thermal loss of convertible sites to yield char.

In this work we have avoided consideration of specific mechanism for decrease in molecular weight with liquefaction, including in particular the perceived need for thermal scission of C-O and C-C bonds during conversion. Thus we view conversion simply as a process requiring some kind of nonspecific reduction chemistry.

This two-reaction scheme, if truly operative, suggests a view of the potential utility of liquefaction more optimistic than that derived from the conventionally accepted scheme. The yield of TS is a function of the rates of the two reactions. Therefore an increase in the rate of the TS route, or a decrease in the rate of char formation, would bring about increased TS yields. In turn, an increase in the rate of TS formation could be brought about with an increase in the reducing capacity of the system, rather than through an increase in reaction temperature. And thus in principle, increases in TS yield to product quantities representing all of the convertible portion of the starting coal could be obtained, and at no cost to product quality.

CO/Water Conversions. In contrast to the more-or-less fixed reducing capacity available in conventional donor systems, the CO/water system offered considerable latitude. We had earlier demonstrated that changes in the initial pH of the system brought about wide variation in the TS yields for Illinois No. 6 coal (6). In accord with the findings of several other groups including Appell, et al. (7), the conversions were found to be base promoted. The present study included a range of initial pH values, and focused on a comparison of the results in H_2O with those from a substitution in parallel experiments of D_2O.

The results for several runs in the two media are presented in Figure 2. The figure presents a plot of %-toluene soluble products vs quantity of CO consumed. There is a range of TS yields up to around 50% for the protio medium. These conversions were attained by ranging the initial pH from 7 to 13.

We have pointed out that the water gas shift reaction

$$CO + H_2O \longrightarrow CO_2 + H_2$$

parallels the conversion (6). Also presented in the figure are results for runs with H_2 in place of CO, and the results of these runs show clearly that H_2 is less effective in conversion. It is thus the CO, rather than the H_2 produced during the CO-conversions, that is responsible for the production of toluene-soluble product. Effects of mineral matter are also shown in the figure. Mineral matter is apparently necessary for the small degrees of conversion seen for the H_2/H_2O system, while the TS yields in the CO/water system are independent of the presence of mineral matter.

The major finding in this work is the isotope effect, as displayed in the figure. The deuterio system provides significantly greater TS yields than does the protio system. Moreover, the deuterio results level off at a substantially higher conversion level. This result, an inverse isotope effect, is not very common in isotope effect studies, and has recently been discussed by Keeffe and Jencks (8).

Products of Conversion. Conversion product analyses are presented in Figure 3. The procedure used here is that developed by

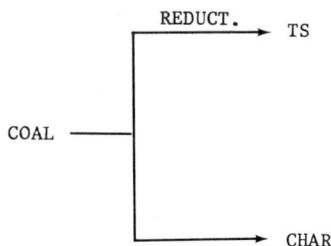

Figure 1. Working Model. TS = toluene soluble

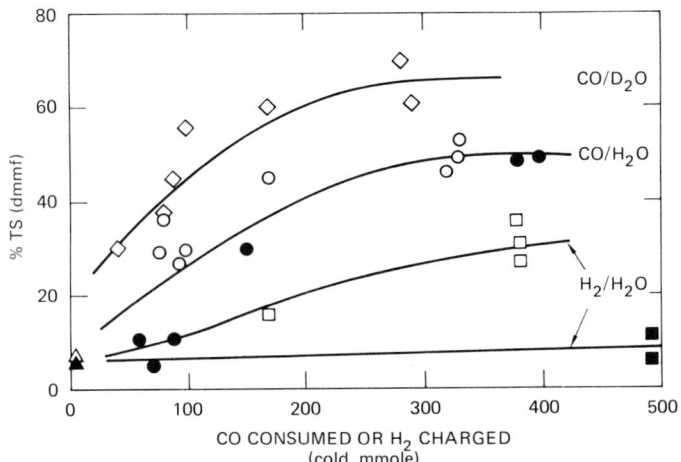

Figure 2. Conversions of PSOC-1098 and PSOC-26 at 400°C/20 minutes.
◇ CO/D_2O; ○ ● CO/H_2O; □ ■ H_2/H_2O; △ ▲ N_2/H_2O
(The filled symbols represent data for the demineralized coal.)

Farcasiu (9), with which we have separated the toluene-soluble fractions into subfractions of increasing polarity. The TS fractions for different coals yield different profiles, and yet we find here for the Illinios No. 6 coal that the TS fractions for conversion in water of 29% and 60% are virtually the same. Further, the TS fraction from the tetralin run is also essentially identical in its profile.

The H/C ratio of products from several CO/H_2O conversions are presented in Table 1. Within the scatter of the data, both the TI and TS fractions have ratios unchanging with over a range of conversions from 29 to 60%.

Table I. H/C Ratios for CO/H_2O Conversion at 400°C

%TS	H/C	
	TI	TS
29	0.58	0.95
30	0.56	0.95
30	0.53	0.92
44	0.63	0.95
45	0.64	0.92
59	0.61	0.94
60	0.63	0.93

DISCUSSION

Taken alone, the conversion profile from the protio work is consistent with the view that the conversion of coal is limited by its structure. Thus if the organic portion of coal contained a limited network of breakable links, the scission of which would liberate about 50% of the material to TS product, then runs with increasing conversion capacity would show increased conversion, leveling off at about 50% TS yield.

The inverse isotope effect, however, requires a different picture. Thus a simple change to the heavy medium brings about not only increased conversions, but a leveling off of conversion at a significantly higher level. Whatever the reduction mechanism, it is highly unlikely that isotopic switch from ^1H to ^2H would increase the inherent bond breaking capacity of the system. And so we conclude that the starting coal must contain many more breakable links than supposed above, but that some portion of the links are lost through other reactions. The proposed model in Figure 1 is consistent with this conclusion.

The full scheme for conversion is presented in Figure 4. In this scheme, formate is partitioned between reaction with coal, and a hydrogen ion transfer reaction with water to yield formic acid. The acid is unstable at the conversion temperatures (10), decomposing rapidly to carbon dioxide and hydrogen. Thus with the switch from protio to deuterio, the formic acid formation experiences a

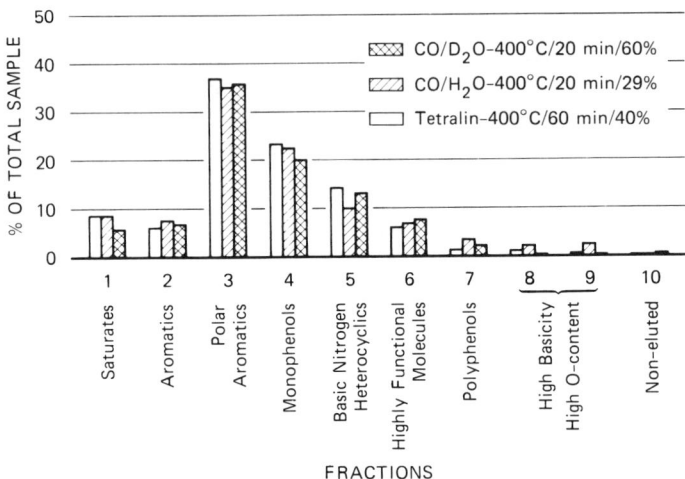

Figure 3. SESC Separations of TS Fractions.

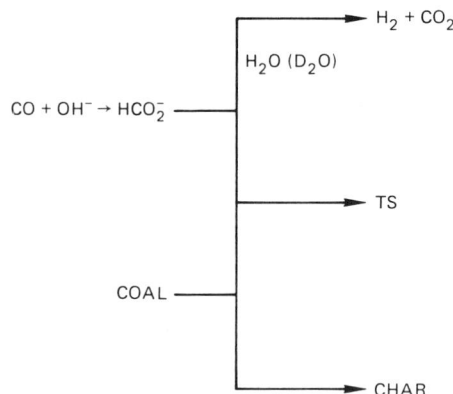

Figure 4. Overall Scheme for Conversion in Hydrothermal Systems

normal deuterium effect, i.e. protio > deuterio, and is slowed. The result is an increase in the steady-state concentration of formate, and an accordant increase in the TS yield.

These results, including most especially the product data, contrast decidedly with those discussed by Whitehurst et al., noted above (5). In the present work, we find that with increased reducing capacity and at constant temperature, the system gives increased yields of product, and at no cost to product quality.

In summary, we find that conversion is not limited by coal structure, but rather by the kinetics of the reducing step(s). Systems with even greater reducing capacity, and where the water gas shift reaction can be suppressed, should provide even higher conversions to toluene-soluble products. The products in turn should be no less rich in hydrogen than those from lower conversion runs.

It is still necessary to bring about an understanding of the specific reducing chemistry. From our present data we can conclude that the conventional thermal scission/H-capping sequence does not apply here. And since the TS product from tetralin conversion is no different from those from aqueous conversion, it would appear that the reduction in conventional donors breaks the same links broken by the hydrothermal system.

Thus the question of the nature of critical link scission in conventional conversions must be reconsidered. Brower has recently questioned the conventional scheme (11 a,b), and it is clear that the detailed mechanism of coal conversion is yet to be developed.

ACKNOWLEDGEMENT

We acknowledge the generous support of the U.S. Department of Energy. We also acknowledge the close cooperation of ASSORENI (Milan), who provided RM with support for a stay at SRI as an International Fellow.

REFERENCES
1. Mohan, G. and Silla, H., Ind. Eng. Chem. Process Des. Dev., 1981, 20, 349-358.
2. Fischer, F. and Schrader, H., Brennst.-Chem., 1921, 2 161-172; C.A. 15, 3193.
3. Ross, D. S. In Coal Science; M. Gorbaty, J. Larsen and I. Wender, eds., Academic Press, Inc., New York, 1984, pp. 301-338.
4. Schneider, G. M. and Jockers, R., Ber. Bunsenges. Phys Chem. 1978, 82, 576-582.
5. Farcasiu, M., Mitchell, T. O. and Whitehurst, D. D., Chem. Tech. 1977, 7, 680-686.
6. Ross, D. S., Blessing, J. E., Nguyen, Q. C. and Hum, G. P., Fuel 1984, 63, 1206-1210.
7. Appell, H. R., Miller, R. D. and Wender, I. presented before the Division of Fuel Chemistry of the American Chemical Society, 163rd National Meeting of the ACS, April, 1972.
8. Keeffe, J. R. and Jencks, W. P., J. Am. Chem. Soc., 1981, 103, 2457-2459.
9. Farcasiu, M., Fuel 56, 1977, 9-14.

10. Stenberg, V. I., Van Buren, R. L., Baltisberger, R. L. and Woolsey, N. F., J. Org Chem. 1982, 47, 4107-4110.
11. a) Brower, K. R., J. Org. Chem. 1982, 47, 1889-1893;
    b) Brower, K. R. and Pajak, J., J. Org. Chem. 1984, 49, 3970-3973.

RECEIVED July 17, 1986

Chapter 20

Effect of Solvent Density on Coal Liquefaction Kinetics

G. V. Deshpande[1], G. D. Holder[2], and Y. T. Shah

Department of Chemical and Petroleum Engineering, University of Pittsburgh, Pittsburgh, PA 15261

>Supercritical fluid extraction is an attractive process primarily because the density and solvent power of a fluid changes dramatically with pressure at temperatures near the critical. In complex supercritical extractions, such as the extraction of coal, the density of the supercritical fluid should also change the extractability of the coal. In this experiment a relatively inert supercritical fluid, toluene, was studied to determine the effect of density on the coal extraction/reaction process. Extractions were carried out for two to 60 minutes at reduced densities between 0.5 and 2.0 and at temperatures between 647 and 698K. The data obtained can be explained by the hypothesis that coal dissolution is required preceding liquefaction reactions and that the degree of dissolution depends upon solvent density and temperature.

Earlier efforts aimed at understanding supercritical extraction of coal used both flow and batch reactors. In the flow reactors ([1])-([5]), coal was packed into the reactor and the supercritical fluid was passed through the bed of coal until the condensed effluent was clear. The conversion was defined as the total weight loss of the coal due to extraction by the solvent. However, part of the coal will be soluble in the extracting fluid at supercritical conditions but will be insoluble in standard extracting solvents such as toluene or pyridine. Conversions measured in flow systems are over estimated because they include this normally insoluble fraction. In batch systems ([7])-([12]), the coal and solvent were placed in a reactor and heated together to the reaction temperature. The conversion of coal thus takes place during heat-up and during cool-down and the necessarily non-isothermal kinetic models which are

---

[1]Current address: Advanced Fuel Research, P.O. Box 18343, East Hartford, CT 06118
[2]Correspondence should be addressed to this author.

appropriate for such conditions cannot be developed with any degree of confidence. Thus both of these methods have limitations in evaluating reaction kinetics because the actual reaction time and the conversions are uncertain. The general trends observed in these studies do, however, provide some important insights; higher temperatures and higher densities result in higher conversions partially because more coal dissolves in the supercritical solvent as temperature and density are increased. A more thorough review of the literature on supercritical extraction of coal has been done by Deshpande (13).

More recent studies (14),(15) have employed a rapid injection autoclave, where coal is injected into a preheated supercritical solvent. After the reaction is over, the products are quenched by passing water through internal cooling coils. This method allows precise measurement of the reaction times corresponding to the conversions. It is used in this study.

Amestica and Wolf (12) in a study closely related to the one described herein, measured the conversion of Illinois No. 6 coal in toluene and ethanol. Their results clearly showed that conversions increased with temperature and solvent density but were not detailed enough to show the time dependence of the conversion. However, a result important to this study was that toluene converts coal to liquids without significantly reacting itself. After reaction, 98% of the toluene used was recovered versus only 73 - 85% of the ethanol in runs using it. Ethanol is a hydrogen donor and reacts extensively with the coal. While toluene probably reacts with coal to a small extent, its effect was primarily physical in nature. As such, it is a good candidate for studying the effects of a supercritical solvent on coal liquefaction kinetics since the enhancement effect of supercritical conditions is physical in nature.

Further evidence of the importance of the physical nature of the solvent is found in the work of Belssing and Ross (6) who correlated the coal conversion (pyridine solubles) with the Hildebrand solubility parameter, δ, which they defined as

$$\delta = 1.25 \ P_c^{1/2} \rho_r / \rho_\ell \qquad (1)$$

Here P_c is the critical pressure of the medium in atmospheres, ρ_r is its reduced density and ρ_ℓ is generalized reduced density of liquids, taken to be 2.66. They found the product pyridine solubility to be a linear function of δ. This implies that higher densities should produce higher yields and this concept is the motivation for the present study.

Experimental
Bruceton bituminous, a Pittsburgh Seam coal was used in the experiment. The chemical analyses of coal is given in Table I. The coal was dried in vacuum @ 343 K prior to use and stored in glass containers under nitrogen.

Table I. Analyses of Coals Used

	Bruceton Bituminous Coal	
	Old	New
Ash[a], %	4.90	3.88
Volatile Matter[a], %	–	37.01
Carbon[b], %	82.69	83.68
Hydrogen[b], %	5.56	5.40
Sulfur[b], %	1.46	1.11
Nitrogen[b], %	1.72	1.01
Oxygen[b,c], %	8.57	8.80

a – Moisture free basis
b – Moisture and ash free basis
c – By difference

Supercritical Coal Liquefaction Procedure. The experimental apparatus is shown in Figure 1 and consists of a 1-L stainless steel autoclave equipped with a Magnedrive stirrer and a coal injection system. The reactor is charged with a known quantity of toluene depending on the fluid density desired for the experiment and is heated at 3-4 K/min to the temperature desired. Once this temperature is reached, ambient coal is injected into the reactor from a coal reservoir using high-pressure Argon. The average weight of injected coal was 30g. Reaction times are measured from the time at which the coal is injected.

The liquid and solid contents were collected from the reactor and placed in an extraction thimble which was then placed in a soxhlet unit. The contents were then extracted with toluene until the extractant was clear after which the thimble was dried and weighed. The weight of the dried product is designated as toluene insolubles.

In this work, the conversion refers to the amount of toluene solubles present at the end of the reaction. The conversion products thus include gases, oils and asphaltenes (GOA). Some of the conversion products are due to simple dissolution or to heating of the coal, and some are due to conversion of toluene insolubles to toluene solubles. Only the latter portion of the conversion products should be affected by the toluene density. The former are inherently toluene solubles, having nothing to do with the density or presence of supercritical toluene. These are only combined with the products obtained by reaction for experimental convenience.

The conversion is experimentally determined as,

$$GOA\% = \frac{\text{Coal injected(g)} - \text{toluene insolubles(g)}}{\text{Coal injected(g)}} \times 100 \qquad (2)$$

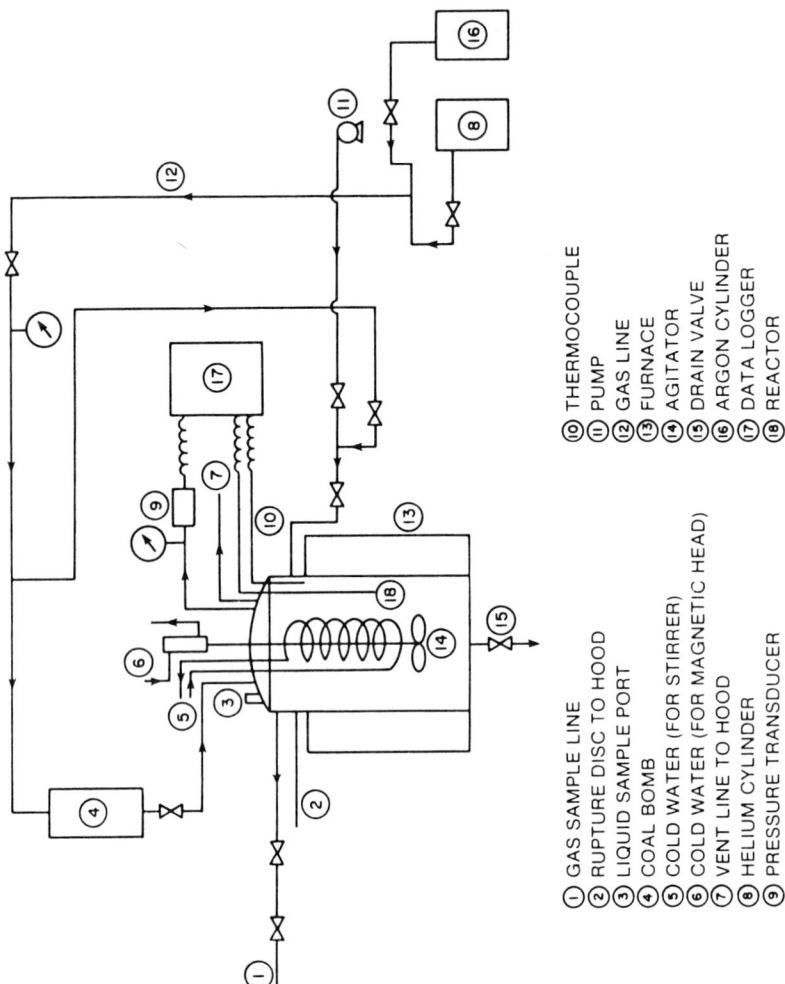

Figure 1: Experimental Set-up for Supercritical Extraction

where toluene insolubles refer to the total mass of reaction products which are insoluble in toluene at its boiling point. Toluene solubles, conversion products, and GOA are used interchangeably.

Toluene is a non-donor solvent and as such its effect on coal is one primarily of physical dissolution.

Discussion. Experiments were carried out with Bruceton coal and toluene at supercritical toluene densities in the range of 0.157-0.601 g/cc. The temperature range studied was 647-698 K and the reaction time was varied from two minutes to 60 minutes. The experimental results are explained by a kinetic model, the details of which can be found elsewhere (16).

The model is summarized briefly here. As coal is heated, part of it dissolves in the supercritical toluene and part does not. Note that a distinction must be made between the toluene solubles (GOA) in Equation 2 and the material soluble at reaction conditions, because the former refers to the material which is soluble at toluene's boiling point and the latter refers to the material soluble at the reaction temperature which is about 300 K higher. Thus, while 40% of the coal might be dissolved, for example, in the high temperature toluene just prior to quenching, the total conversion (GOA), as measured by Equation 2, may only be 15%. It is meaningful, therefore, to hypothesize that only the dissolved portion of the coal undergoes conversion from toluene insolubles to toluene solubles. As discussed below, this hypothesis offers a physical explanation of why the rate of conversion increases with toluene density, and it does not require that toluene be highly reactive. The model equations are

$$\frac{dC1}{dt} = -K_1 C1 \tag{3}$$

$$\frac{dA}{dt} = k_1 C1 - K_2 A^2 \tag{4}$$

$$\frac{d\,Char}{dt} = K_2 A^2 \tag{5}$$

$$C1/C = S \tag{6}$$

C	—	coal
C1	—	coal dissolved in supercritical solvent
C2	—	coal insoluble in supercritical solvent
A	—	gases, oils and asphaltenes
K_1, K_2	—	rate constants, min^{-1}, $min^{-1} gm^{-1} lit$
S	—	weight fraction which is soluble in toluene. It is correlated as a function of temperature and fluid density (16).

The results of experiments at 647K and reduced densities of 0.5 to 2.0 (toluene densities of 0.157 g/cc and 0.601 g/cc) are given in Figures 2 and 3. The experimental results show that toluene solubles (oils, asphaltenes, and gases) formed at low reaction times increases with both temperature and density. In general, solubility studies have shown that the amount of a solid

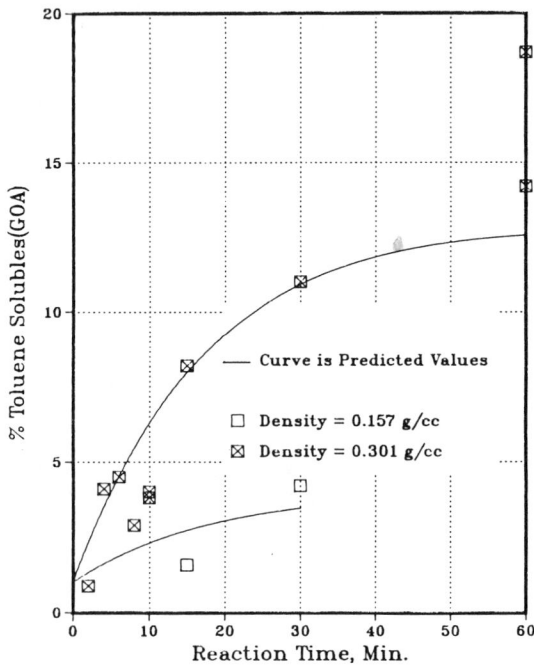

Figure 2: Effect of Reaction Time and Density on % Toluene Solubles (GOA) @ 647 K (Densities of 0.301 and 0.157 g/cc)

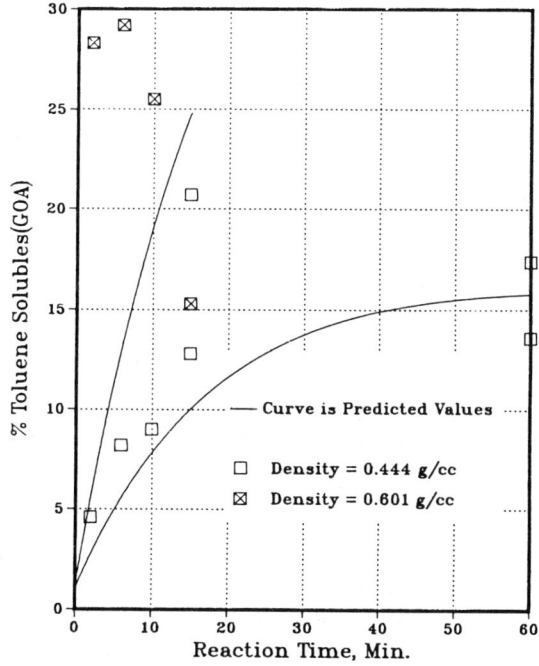

Figure 3: Effect of Reaction Time and Density on % Toluene Solubles (GOA) @ 647 K (Densities of 0.601 and 0.444 g/cc)

which can dissolve in a supercritical fluid increases with density and, generally, with temperature. Hence, we conclude that the conversion of coal to oils + asphaltenes + gases is in some sense limited by the dissolution of the coal in the solvent.

We hypothesize that the only part of the coal which undergoes reaction to gas, oils and asphaltenes is the dissolved fraction (including pyrolysis products), which increases with temperature and density. Increases in the dissolved fraction thus lead to higher conversions and faster reaction rates. These conversion products then participate in retrogressive reactions (Amestica and Wolf (12)) forming char, so that at longer times (15 minutes or more) the yield of gases + oils + asphaltenes decreases. At lower temperatures (647 K), the retrogressive reactions are insignificant and conversion does not decrease with time.

The above hypothesis is well explained by the results given in Figures 2 and 3. At a lower density of 0.157 g/cc the fraction of coal dissolved in the supercritical fluid is much lower than when the density of the supercritical fluid is 0.601 g/cc. Hence, the fraction of coal, which is converted to asphaltenes, oils and gases, is also lower.

The results of experiments at 673 K and reduced density of 1.0 and 1.5 (supercritical toluene density of 0.301 g/cc and 0.444 g/cc) are given in Figure 4. As observed at 647 K, the coal conversion to gases + oils + asphaltenes (toluene solubles) increase with reaction time and with the density of the supercritical fluid. The retrogressive reactions are more significant now and hence, the toluene solubles show a maxima in conversion with time.

The results of experiments at 698 K and reduced densities of 1.0 and 1.5 (supercritical toluene density of 0.301 g/cc and 0.444 g/cc) are given in Figure 5. As observed at 647 K and 673 K, the coal conversion to gases + oils + asphaltenes (toluene solubles) increases with reaction time and density of the supercritical fluid. The retrogressive reactions are more pronounced than that at 673 K but we still observe a maxima in the toluene solubles as a function of reaction time. The initial rate of formation of toluene solubles (as seen from the steepness of the curves of toluene solubles versus reaction time) is higher at higher density and higher temperature. Also note that the amount of toluene solubles at the maxima is higher when the temperature and density of the supercritical fluid is higher. This is consistent with the hypothesis that the soluble fraction of the coal increases with temperature and density of the supercritical fluid.

The results of the model simulation are given in Figures 6, 7 and 8. Figure 6 shows the effect of density on the kinetics of coal liquefaction at temperature of 698 K. It is very clear that conversion increases with an increase in density and it is consistent with the hypothesis made earlier on coal liquefaction with a supercritical fluid. Figure 7 shows the effect of temperature on the kinetics of coal liquefaction at density of 0.601 g/cc. It is evident that conversions are higher at higher temperatures and lower reaction times but at longer reaction times due to condensation reactions, a decrease in rate with temperature is observed.

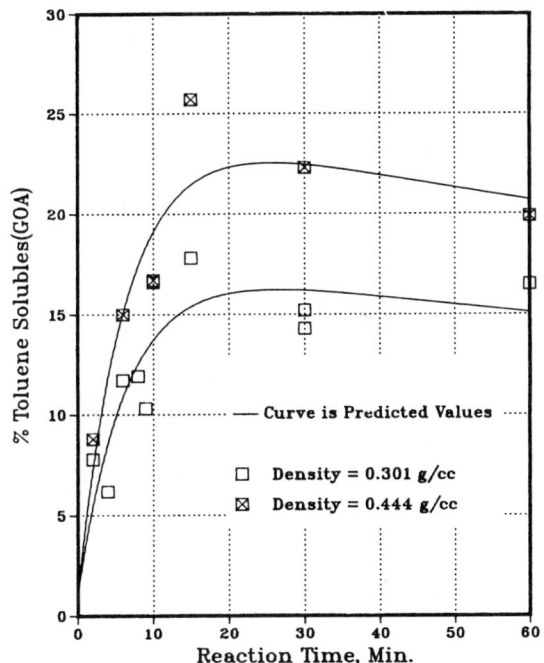

Figure 4: Effect of Reaction Time and Density on % Toluene Solubles (GOA) @ 673 K

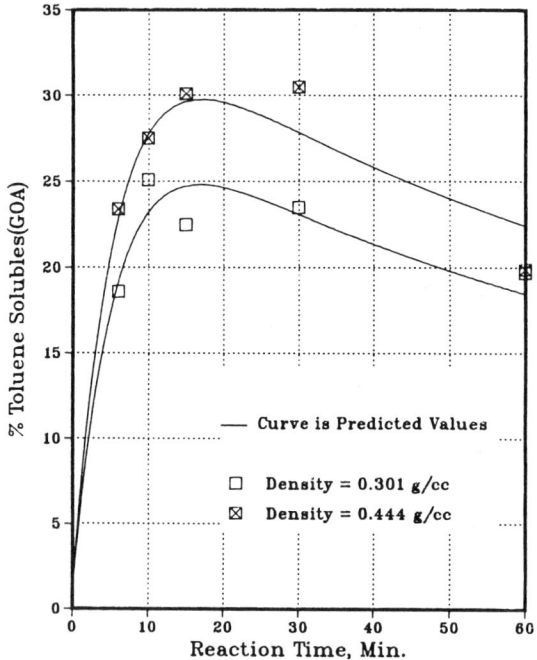

Figure 5: Effect of Reaction Time and Density on % Toluene Solubles (GOA) @ 698 K

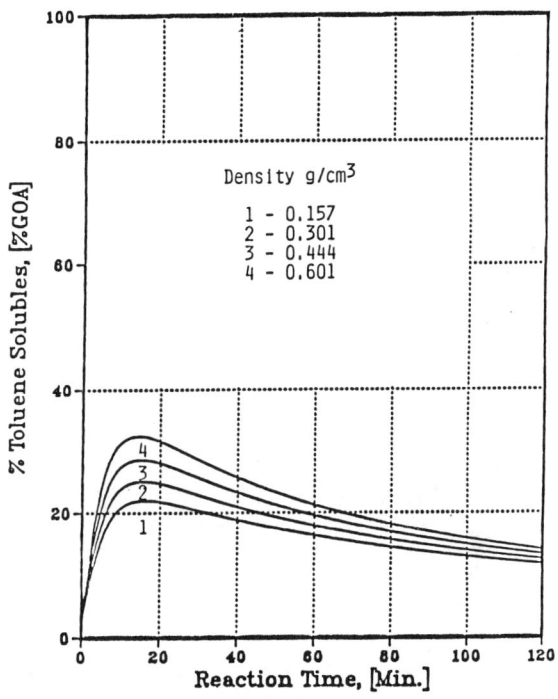

Figure 6: Effect of Density and Reaction Time on % Toluene Solubles at 698 K and Coal Concentration of 40 g/l

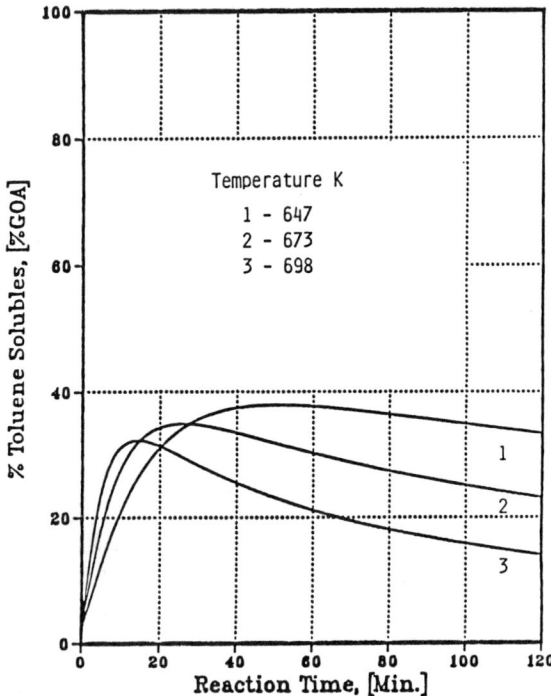

Figure 7: Effect of Temperature and Reaction Time on % Toluene Solubles at 0.601 g/cm^3 and Coal Concentration of 40 g/l

The coal/solvent ratio also affects the initial coal concentration and hence the concentration of intermediate products (toluene solubles) in the reactor. This is shown in Figure 8. These intermediate products undergo retrogressive/condensation reactions which follow second order kinetics. The retrogressive reactions become very significant at high temperatures and high densities. Hence, the conversion goes through a maximum and the conversions at longer reaction times are lower when coal/solvent ratio is higher. The initial part of the kinetic curve is not affected by the coal/solvent ratio. This is because retrogressive reactions are not predominant at low reaction times and the amount of coal which dissolves in the supercritical fluid is independent of the coal/solvent ratio.

Because coal is heterogeneous in nature, as the density is increased, heavier and heavier coal fractions (as opposed to more and more of the same fraction) are dissolved in the supercritical fluid. These would not dissolve if the density of the supercritical fluid were lower, and the supercritical fluid is, in general, unlikely to be saturated with any given fraction that is substantially dissolved (i.e. either zero or 100% of a fraction is dissolved). For example, if coal is considered to be composed of 100 fractions characterized by increasing molecular weight, more and more of these fractions dissolve as density is increased, but the solubility of a given fraction goes from (approximately) zero to 100% over a small change in density. If a given fraction, number 46 for example, is dissolved, then more of that molecular weight group would dissolve at that density if it were present, but none of the undissolved fractions would dissolve regardless of the amount of coal present. Hence, the fraction of the coal and not the amount of coal which dissolves in the supercritical fluid is a strong function of density and temperature of the supercritical fluid. In other words, if the amount of coal injected into the supercritical fluid at given temperatures and density was reduced to its half value, the absolute amount dissolved will fall by 50%.

If the amount of coal injected was increased indefinitely, then a point might be reached where the supercritical fluid is saturated with dissolved coal. If the amount of coal injected is increased beyond this value, the fraction of dissolved coal and the fractional coal conversion will start decreasing. The <u>amount</u> of dissolved coal in the supercritical fluid will be independent of the amount of coal injected in such a case. This phenomena is very well illustrated in Figure 9.

<u>Conclusions.</u> When coal is contacted with a non-donor supercritical fluid a part of the coal instantaneously dissolves in the supercritical fluid. The dissolved coal undergoes liquefaction reactions which are thermal in nature resulting in toluene soluble products being formed from coal. These products can subsequently undergo retrogressive reactions yielding insoluble material. Hence the toluene solubles show a maxima in conversion with time.

The fraction of coal which dissolves instantaneously in the supercritical fluid increases with an increase in the density and temperature of the supercritical fluid. This effect is similar to that generally observed for the solubility of a solid in a supercritical fluid. With an increase in density and temperature

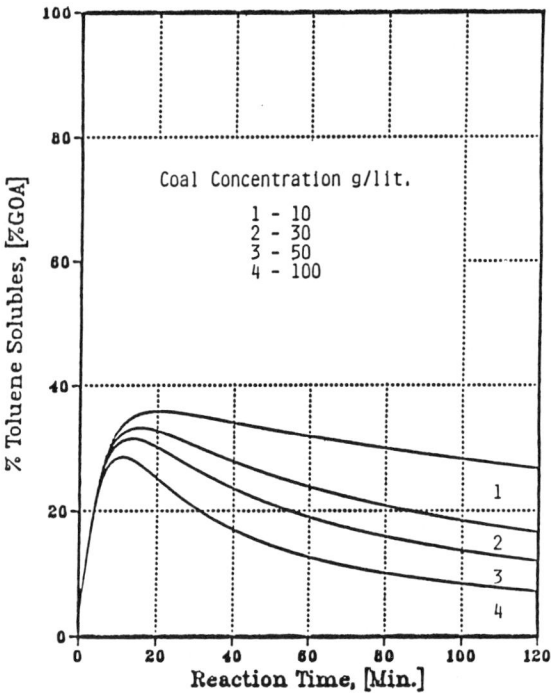

Figure 8: Effect of Coal Concentration and Reaction Time on % Toluene Solubles at 698 K and 0.601 g/cm^3

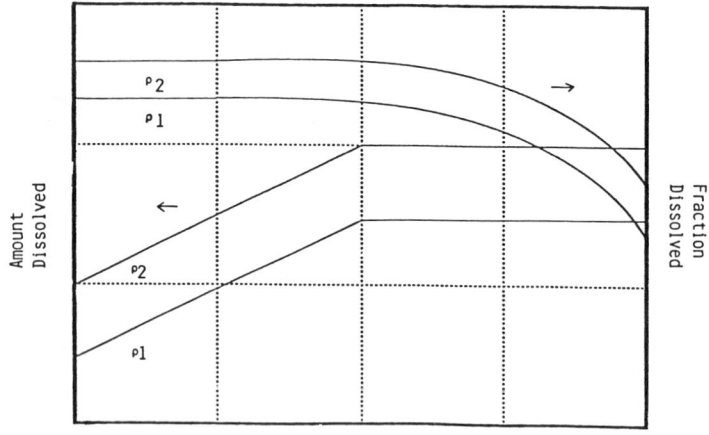

Figure 9: Effect of Coal/Solvent Ratio on Amount and Fraction of Coal Dissolved in a Supercritical Solvent

higher molecular compounds present in coal go into solution, resulting in an undersaturated solvent with the fraction, not the amount, of the coal which is being dissolved.

Acknowledgments

The authors wish to tank the U.S. Department of Energy under contract DE-FG22-PC71257 for support of this work.

Literature Cited

1. Slomka, B.; Rutkowski, A. Fuel Processing Technol. 1982, 5, 247.
2. Jezko, J.; Gray, David; Kershaw, J.R. Fuel Processing Technol. 1982, 15, 229.
3. Kershaw, J.R. Fuel Processing Technol. 1982, 5, 241.
4. Whitehead, J.C; Williams, D.F. J. Inst. Fuel (London) 1975, 48, 182.
5. Penninger, J.M.L. "Oil Extraction from Subbituminous Coal with Compressed Aqueous Gas Extractants-Stoichiometry and Mechanism of Extraction," presented at 1984 Annual AIChE Meeting, San Francisco, California, November (1984).
6. Blessing, J.E.; Ross, D.S. "Supercritical Solvents and the Dissolution of Coal and Lignite," ACS Symposium Series 71, American Chemical Society: Washington, D.C., 1978.
7. Bartle, K.D.; Calimli, A.; Jones, D.W., Matthews; R.S.; Olcay, A. Fuel 1979, 58, 423.
8. Tugrul, T.; Olcay, A. Fuel 1978, 57, 415.
9. Bartle, K.D.; Martin, T.G.; Williams, D.F. Fuel 1975, 54, 226.
10. Kershaw, J.R. S. Afr. J. Chem. 1977, 30, 205.
11. Barton, Paul Ind. Eng. Chem. Process Des. Dev. 1983, 22, 589.
12. Amestica, L.A.; Wolf, E.E. Fuel 1984, 63, 227.
13. Deshpande, G.V. Ph.D. Dissertation, University of Pittsburgh, Pittsburgh, PA, 1985.
14. Deshpande, G.V.; Holder, G.D.; Bishop, A.A.; Gopal, J.; Wender, I. Fuel 1984, 63, 958.
15. Towne, S.E. M.S. Thesis, University of Pittsburgh, Pittsburgh, PA 1983.
16. Deshpande, G.V.; Holder, G.D.; Shah, Y.T., Ind. Eng. Chem. Process Des. Dev., 1986, 25 (3).

RECEIVED June 25, 1986

Chapter 21

Extraction of Australian Coals with Supercritical Aqueous Solvents

John R. Kershaw and Laurence J. Bagnell

Commonwealth Scientific and Industrial Research Organization, Division of Applied Organic Chemistry, G.P.O. Box 4331, Melbourne, Victoria 3001, Australia

> Conversions between 42-68% were obtained for supercritical water extraction of Victorian brown coals at 380°C and 22MPa, considerably higher than using toluene under the same conditions. The conversions obtained with a bituminous and a sub-bituminous coal were much lower. Pressure had a marked effect on both the conversion and the extract composition, whereas temperature had only a slight effect. Considerably higher conversions were achieved using dilute sodium hydroxide rather than water. The composition of the products is discussed.

The extraction of coals with supercritical fluids is a promising route for the production of liquid fuels from coal. Generally, hydrocarbon solvents, notably toluene, have been used as the supercritical fluid. Supercritical water extraction has not received the same attention and only recently the first detailed study was reported. In that work, Holder et al. (1) obtained high conversions for extraction of a German brown coal (70-75%) and a Bruceton bituminous coal (ca. 58%) with supercritical water at ca. 375°C and 23 MPa. These high yields, however, contrast with the much lower conversions briefly reported (2-5) for other coals and there appears to be considerable variation in the extractive power of supercritical water with different coals, even allowing for the differences in the extraction procedure used. None of the above reports discussed in any detail the chemical nature of the products, nor how the products compare with those obtained from more conventional solvents.

The potential use of supercritical water appears especially attractive for the extraction of brown coals with their high water content, 50-70% for Victorian brown coals, thus removing the need for a coal-drying stage. The drying and extraction of these low rank coals would occur in a single process. The purpose of the present study was to investigate the feasibility of the extraction of Australian black and brown coals with supercritical water. The

0097-6156/87/0329-0266$06.00/0
© 1987 American Chemical Society

chemical nature of the products is discussed together with a comparison with toluene extraction of the same coals. The effect of addition of various co-solvents is also included.

Experimental

The analyses of the coals used are given in Table I.

Supercritical gas extractions Method A. Extractions were carried out for 1 h at temperature in a 1 l semi-continuous reactor (6). The reactor was charged with coal (50 g dry basis) and solvent (600 ml) and heated (7°C min^{-1}). When the temperature reached 300°C, solvent (1 l h^{-1}) was pumped via a dip tube, which acts as a preheater, into the bottom of the reactor and through the coal bed. A 15 micron filter was placed in the exit line. The pressure was controlled by adjusting throttling valves and the gaseous phase was condensed by a water-cooled condenser.

For the toluene extractions, the work-up procedure was as described previously (6). In the supercritical water experiments, most of the extract was insoluble in water, after cooling and lowering of the pressure, and precipitated out in the condenser and receiver from which it was collected by washing with acetone and then THF. The remainder of the extract was found in the aqueous suspension which was evaporated to dryness on a rotary evaporator and the residue extracted with acetone and THF. The solvents were removed under reduced pressure from the combined acetone and THF solutions to give the total extract. This was then extracted with hot toluene and the cooled solution filtered to give the pre-asphaltene fraction. After the toluene was removed under reduced pressure from the filtrate, the residue was re-dissolved in a small volume of toluene and a 20 fold excess of pentane added to precipitate the asphaltene which was filtered off. The pentane and toluene were then removed from the filtrate under reduced pressure to give the oil. For the NaOH extractions, the NaOH solutions were neutralised with HCl. The insoluble extract was washed with water and then extracted with THF. Removal of the THF gave the total extract.

The extraction residue was washed out of the reactor with acetone, filtered, washed with acetone and dried in vacuum oven. In calculating the conversion figures, any extract which was insoluble in THF was assumed to be unreacted coal, which had been carried over. The conversion and extract yield data are the average of duplicate determinations. Average variation between the duplicates was 1.8%.

Supercritical gas extractions Method B. Extractions were carried out in a 500 ml rocking autoclave fitted with a stainless steel liner. The internal volume of the autoclave with liner was 420 ml. The autoclave was charged with coal and solvent, heated to and maintained at temperature for 1 h. The residue was washed out of the cooled reactor with acetone, filtered, washed with pyridine and then acetone and dried under vacuum. In some cases, g.c. analyses of the gases in the cooled reactor were carried out.

TABLE I. Analyses of Coals Used

	Loy Yang (A)	Gelliondale (B)	Coolungoolun (C)	Morwell (D)	Yallourn (E)	Morwell Pale (F)	Yallourn Pale (G)	Millmerran (H)	Liddell (I)
Moisture	10.8[a]	7.3[a]	11.5[a]	52.6[a]	65.0[a]	56.5[a]	57.2[a]	4.9[a]	3.3[a]
Ash	0.6[b]	5.4[b]	2.2[b]	2.7[b]	1.3[b]	3.7[b]	1.6[b]	18.1[b]	29.0[b]
Minerals and inorganics		3.0[b]	2.0[b]	1.6[b]	1.0[b]	2.0[b]	1.3[b]		
Volatile matter	48.3[c]	51.4[d]	48.9[d]	50.2[d]	51.7[d]	56.0[d]	61.2[d]	50.7[c]	44.6[c]
C	71.0[c]	66.5[d]	70.7[d]	69.6[d]	67.3[d]	71.5[d]	71.2[d]	79.0[c]	80.6[c]
H	4.8[c]	4.7[d]	5.1[d]	5.0[d]	4.7[d]	5.8[d]	6.2[d]	6.3[c]	6.0[c]
N	0.6[c]	0.6[d]	0.6[d]	0.6[d]	0.6[d]	0.6[d]	0.5[d]	1.2[c]	2.1[c]
S	0.3[c]	0.8[d]	4.1[d]	0.3[d]	0.2[d]	0.7[d]	0.2[d]	0.7[c]	0.4[c]
O (by difference)	23.3[c]	27.4[d]	19.5[d]	24.5[d]	27.2[d]	21.4[d]	21.9[d]	12.8[c]	10.9[c]
H/C atom. ratio	0.81	0.85	0.87	0.86	0.84	0.97	1.04	0.96	0.89
Aromatic C (%C)			62				43		

[a] wt%; [b] wt% dry basis; [c] wt% dry ash free basis; [d] wt% dry mineral and inorganic free (dmif) basis.

Analytical procedures for the products were as described previously (7).

Supercritical Water Extractions

Good conversions were obtained for extraction of a number of Victorian brown coals in the semi-continuous reactor (method A) at conditions (380°C and 22 MPa) close to the critical temperature (374°C) and pressure (22 MPa) of water (see Figure 1). The conversion increased as the volatile matter content of these coals increased in a similar manner to the trends previously shown for toluene and 5% tetralin/toluene extraction (8) (see Figure 1). The highest conversions were obtained for the two pale lithotypes (F and G). The conversion of Yallourn coal was noticeably higher when extracted in a semi-continuous mini-reactor (see Figure 1) with a faster heating rate (ca 100°C min^{-1}) and only 10 minutes at temperature. This agrees with the work of Holder et al.(1) who obtained a conversion of about 58% for extraction of a German brown coal when the coal was present during the heating-up period, compared with 70-75% when the coal was injected into the hot reactor. The lower conversions obtained in the rocking autoclave (method B) (see Table II), when the volatile product is not removed from the autoclave, are also in agreement with these findings. Except for the low ash Loy Yang coal, data for the brown coals are given on a dry mineral and inorganic free (dmif) basis which is the preferred basis for recording data on these coals (9). The conversions were not as high for Millmerran (sub-bituminous) and Liddell (bituminous) coals; 34.7 and 25.5 wt% daf respectively. The asphaltene and pre-asphaltene contents of the extracts are high (see Figure 2), as is the case with supercritical gas extraction using hydrocarbon solvents.

Extractions of three of the coals were also carried out in a rocking autoclave (method B) mainly to allow analysis of the gases to be carried out. The results of these extractions are shown in Table II. The high carbon dioxide yields obtained were indicative

TABLE II. Water Extraction of Brown Coals in a Rocking Autoclave at 380°C

Coal	Loy Yang	Coolungoolun	Yallourn
Pressure (MPa)	24	23	26
Conversion	45.6[a]	39.8[b]	46.3[b]
Extract Yield	21.7[a]	21.3[b]	19.3[b]
CO_2	16.9[a]	9.7[b]	ca 19[b]
CO/air	0.3[a]	0.9[b]	N.D.[c]
$CH_4/C_2H_6/C_3H_8$	0.4[a]	0.7[b]	N.D.
CH_3OH	0.4[a]	0.05[b]	0.8[b]
CH_3COCH_3	0.6[a]	0.2[b]	0.8[b]
$CH_3CO\ C_2H_5$	0.1[a]	0.05[b]	0.1[b]
Phenols	0.3[a]	0.2[b]	0.5[b]

[a] wt% coal daf; [b] wt% coal dmif; [c] N.D.= not determined.

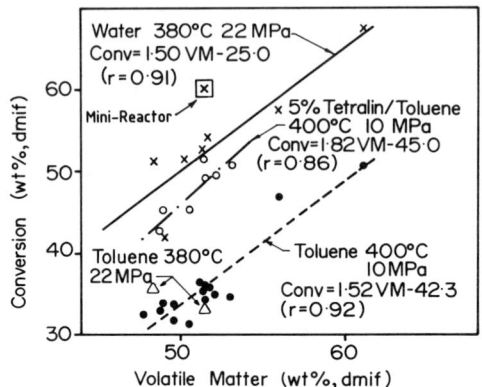

Figure 1. Variation in conversion with volatile matter content for Victorian brown coals.

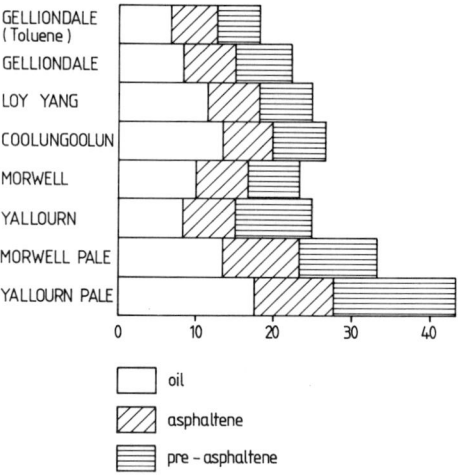

Figure 2. Extract composition for water (and toluene) extractions of brown coals at 380°C and 22MPa.

of the high carboxylic content of the brown coals. The hydrocarbon gas make was low as was the yield of carbon monoxide. Small amounts of methanol, acetone, ethyl methyl ketone and phenols were also formed. Studies of the pyrolysis of Victorian brown coals have shown that significant quantities of water are evolved together with the carbon dioxide (10). Presumably the difference between the conversion figures and the sum of the extract yield and other figures in Table II is mainly due to the loss of water (and also hydrogen sulphide in the case of Coolungoolun coal). The carbon dioxide yield was noticeably lower from Coolungoolun coal, which has the lowest oxygen content of the seven brown coals, than from the other two coals in Table II.

Product Composition

Analytical data for some of the oils, asphaltenes and pre-asphaltenes produced at 380°C and 22MPa from the brown coals are summarised in Table III. The supercritical water extracts were all similar except for the higher H/C atomic ratios and lower aromaticities of the oils and asphaltenes from coals F and G, which have higher H/C atomic ratios than the other coals. The extracts have a high oxygen and hydroxyl content, especially in the asphaltene and pre-asphaltene fractions, in keeping with the high oxygen content of these coals. The H/C atomic ratios of the oils, asphaltenes and pre-asphaltenes are higher than for most coal liquids. The relatively high H/C atomic ratios of the extracts indicates that the more aliphatic constituents of the coal are extracted under these conditions and supports the view that high conversions may adversely affect the liquid quality.

Analytical data on the residues from five extractions are given in Table IV. The high calorific value of the extraction residues, combined with their significant volatile matter content and relatively low ash content, indicates that they should be attractive materials for combustion.

The nature of the products obtained on supercritical water extraction of Victorian brown coals is further illustrated by the ^{13}C-NMR spectra of the various fractions obtained from Yallourn Pale (coal G). The solid state ^{13}C CP-MAS NMR spectrum of the starting coal shows an intense aliphatic signal centered at about 30 ppm (see Figure 3), indicative of long methylene chains in the coal. The spectrum of the extraction residue is predominantly aromatic while the pre-asphaltene is more aliphatic as indicated by the peak centered at approximately 30 ppm (see Figure 3). The aliphatic nature of both the oil and asphaltene is apparent from solution ^{13}C-NMR spectra of these fractions (Figure 4). The presence of long unsubstituted methylene chains ($>C_8$) is shown by the intense ε or inner methylene peak at 29.5 ppm, together with the α, β, γ and δ methylene peaks at 14, 23, 32 and 29 ppm respectively in the spectra of the oils and asphaltenes from all the extracts.

The oil from extraction of Yallourn Pale was further separated by elution chromatography on silica gel into three fractions, a pentane eluate, a toluene eluate and a chloroform/methanol eluate.

TABLE III. Analyses of the Extracts from Extraction of Brown Coals at 380°C and 22MPa

Coal Solvent	B[b] Toluene →	B	C Water	D	E	F	G →	D 0.5M NaOH	D 2M NaOH
Oil									
H/C	1.40[c]	1.41	1.35	1.34	1.37	1.41	1.48	1.36	1.34
O (wt%)[a]	9.2[c]	9.5	(9.3)	7.7	9.3	6.0	6.8	7.6	7.9
OH (wt%)	4.1[c]	5.1	3.7	4.4	4.7	2.2	2.7	3.3	3.5
Mol. Wt.	285[c]	271	314	291	287	326	353	373	361
Aromatic H (%)	16[c]	17	17	18	17	13	11	15	16
Aromatic C (%)	40[c]	44	47			36	31		
Asphaltene									
H/C	1.23	1.07	0.99	1.02	1.11	1.18	1.24	1.05	1.06
O (wt%)[a]	13.5	19.4	(18.3)	14.9	18.5	13.4	16.9	13.6	12.7
OH (wt%)	7.2	10.3	9.3	9.2	10.1	6.8	9.2	6.7	6.8
Mol. Wt.	470	374	415	369	323	433	417	432	429
Aromatic H (%)	28	36	34	32	33	24	24	30	29
Aromatic C (%)	53	69	66			51	50		
Pre-asphaltene									
H/C	0.88	1.03	0.96	0.93	0.93	0.96	0.94	0.86	0.85
O (wt%)[a]	14.8	19.6	(21.1)	17.1	19.7	19.1	21.5	16.0	15.6
OH (wt%)			8.3				10.8		
Mol. Wt.			469				724		
Aromatic H (%)			45	42	43	39	39	45	45
Aromatic C (%)			79				71		

[a] oxygen figures are by direct determination except for those in parenthesis which are by difference; [b] see Table I for coal type; [c] corrected for bibenzyl.

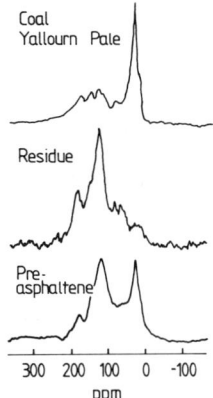

Figure 3. ^{13}C CP-MAS NMR spectra.

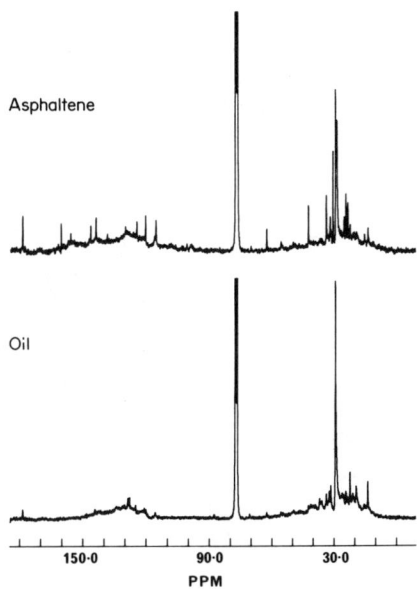

Figure 4.   62.9 MHz ^{13}C-NMR spectra of oil and asphaltene.

TABLE IV. Analytical Data for Residues from Water Extraction at 22 MPa

Coal	B	B	B	C	G
Extraction Temperature (°C)	380	420	460	380	380
Moisture (wt%)	10.8	10.6	9.8	6.0	N.D.
Ash (wt% dry basis)	8.9	9.2	9.4	3.6	5.0
Volatile Matter (wt% dry basis)	31.7	27.1	21.6	26.6	N.D
C (wt% dry basis)	71.8	73.9	76.2	76.3	67.7
H (wt% dry basis)	3.6	3.5	3.2	3.8	3.2
N (wt% dry basis)	0.7	0.7	0.7	0.7	1.4
S (wt% dry basis)	0.9	0.9	0.8	3.7	0.3
H/C	0.60	0.57	0.50	0.60	0.57
Specific energy (MJ/kg, dry basis)	27.5	28.1	28.9	30.4	N.D.
Aromatic carbon (% C)				84	85

These fractions correspond to predominantly aliphatic hydrocarbon, aromatic hydrocarbon and polar fractions. The principal components of the pentane eluate are n-alkanes with an average chain length of C_{17} by ^{13}C-NMR spectroscopy (Figure 5) and C_{20} by GC/MS. The latter is in good agreement with the molecular weight of 281. Branched chain alkanes, alkenes especially 1-alkenes and some cyclic compounds are also present in this fraction. The ^{13}C-NMR spectra of the aromatic hydrocarbon and polar fractions also show the presence of significant amounts of long alkyl chains (see Figure 5). The polar fraction of the oil has a high oxygen and hydroxyl content (see Table V).

TABLE V. Data for Oil Fractions from Yallourn Pale coal

	Pentane eluate (1)	Toluene eluate (2)	$CHCl_3/CH_3OH$ eluate (3)
Yield (wt% coal dmif)	2.9	4.7	9.3
(wt% oil)	17	28	55
H/C	1.84	1.33	1.46
O (wt%)	N.D.	3.3	10.6
OH (wt%)	N.D.	0.7	4.5
Mol. Wt.	281	344	365
Aromatic C(%C)	N.D.	45	37
Aromatic H(%H)	N.D.	11	10

The Effect of Pressure and Temperature

Pressure. The conversion and extract yield both increase considerably with pressure as do the asphaltene and pre-asphaltene content of the extracts (see Figure 6). The H/C atomic ratios of the oil and asphaltene fractions decrease, while the hydrogen and carbon (for the oil) aromaticities increase (see Table VI) with increasing pressure. This presumably indicates that the more aliphatic products are extracted preferentially at lower pressures.

TABLE VI. Effect of Pressure on Extracts from Loy Yang Coal. Water Extraction at 380°C

Pressure (MPa)		7	15	22
Oil	H/C	1.48	1.45	1.41
	Aromatic H (%H)	13	15	17
	Aromatic C (%C)	N.D.	41	44
Asphaltene	H/C	1.26	1.13	1.07
	Aromatic H(%H)	27	33	36

Temperature. Extractions of Gelliondale coal were carried out at 380°C, 420°C and 460°C. There were only small differences between the conversions, extract yields and nature of the products at these various temperatures (see Table VII). This is somewhat surprising as it was expected, especially in light of the recent work of Holder et al. (1), that the highest conversion would occur at the highest water density, namely at 380°C. However, this was not the case. Increased thermal fragmentation of the coal as the temperature is raised or significant extraction occuring before the final temperature (of 460°C) is reached, as indicated by the 60%

TABLE VII. Extraction of Gelliondale Coal with Water at Various Temperatures and 22MPa

Temperature (°C)	380	420	460
Conversion (wt% coal, dmif)	52.7	56.7	57.2
Extract Yield (wt% coal dmif)	22.6	23.1	20.5
Oil			
Yield (wt% coal dmif)	9.0	10.7	9.4
H/C	1.41	1.40	1.38
Aromatic H(%H)	17	18	18
Mol. Wt.	271	295	287
OH (wt%)	5.1	4.4	4.8
Asphaltene			
Yield (wt% coal dmif)	6.5	6.4	6.3
H/C	1.07	1.09	1.08
Aromatic H(%H)	36	35	36
Mol. Wt	374	349	350
OH (wt%)	10.3	10.0	11.6
Pre-asphaltene			
Yield (wt% coal dmif)	7.1	6.0	4.8
H/C	1.03	0.99	0.97
Residue			
H/C	0.60	0.57	0.50

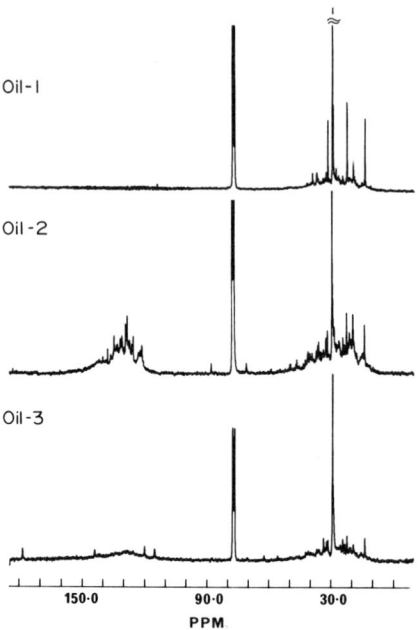

Figure 5. 62.9 MHz ^{13}C-NMR spectra of oil fractions.

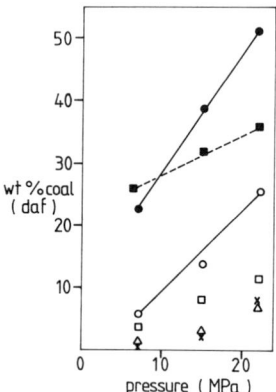

Figure 6. Effect of pressure on the extraction of Loy Yang coal at 380°C. ● Conversion; ○ extract yield; □ oil yield; Δ asphaltene yield; X pre-asphaltene yield for water extraction; ■ conversion for toluene extraction.

conversion of Yallourn coal in the mini-reactor after only 10 minutes, may explain our observation.

Generally, raising the temperature increases the aromaticity of coal liquefaction products. However, there were no significant changes in either the H/C atomic ratios or the hydrogen aromaticities (see Table VII) of the oils or asphaltenes with temperature. Though there was a decrease in the H/C atomic ratio of the extraction residue and possibly of the pre-asphaltene with temperature (see Table VII). The pre-asphaltene content also appears to decrease with increasing extraction temperature.

Comparison between toluene and water extractions

At 380°C and 22 MPa, the conversions for the brown coals using water were considerably higher (see Figure 1) but for black coals slightly lower (34.7 v 38.8 wt% daf for Millmerran and 25.5 v 30.4 wt% daf for Liddell) than with toluene. The difference between coals of various rank is further shown by extraction with water/toluene mixtures of a brown, a sub-bituminous and a bituminous coal at a constant gas density in a rocking autoclave. The presence of water was more advantageous for the extraction of the brown coal than for the higher rank coals (see Figure 7), though in all cases the presence of water increased conversion. The highest conversions were obtained for mixtures of the two solvents. Similar trends were noticed at 380°C and at 340°C. Extraction with water is more pressure dependent than with toluene (see Figure 6).

The major difference between the toluene and water extracts of Gelliondale coal at 380°C was the higher oxygen and hydroxyl content of the asphaltene and pre-asphaltene from the water extract (see Table III). The asphaltene from toluene extraction was less aromatic and had a higher molecular weight than the asphaltene from the water extraction. The oils, however, were similar. (The analyses of the oil from toluene extraction were corrected for 5% bibenzyl which is formed on thermolysis of toluene. Pyrolysis of toluene under similar conditions (11), leads to the formation of small amounts of bibenzyl. Though, the reaction of hydrogen and benzyl radicals from toluene with coal fragments is possible, it is unlikely that this would significantly affect the analysis of the products, given the small amount of bibenzyl formed at this temperature.)

Extraction with bases and tetralin/water mixtures

A considerable increase in conversion of Morwell brown coal occurs when dilute sodium hydroxide was used in place of water (see Table VIII and Figure 8). Dilute sodium carbonate and formate had a similar effect (see Table VIII) as did n-pentylamine (see Table IX), but no improvement was found with dilute hydrochloric acid, sodium chloride or ammonium hydroxide nor with 20% phenol/80% water and methanol/water mixtures. The conversion and extract yield increases with the molarity of the sodium hydroxide (Figure 8). The dissolution of low rank coals by sodium (or potassium)

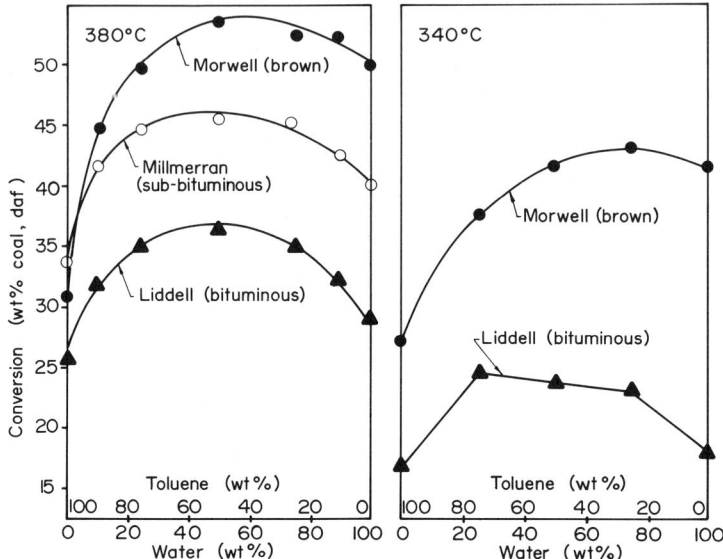

Figure 7. Extraction with toluene/water mixtures in a rocking autoclave at a gas density of 0.44 g/ml.

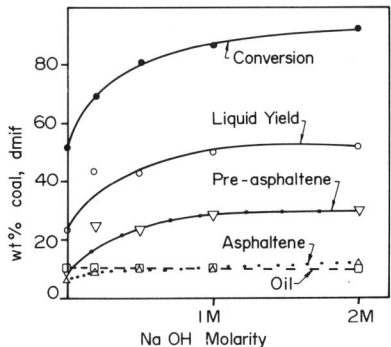

Figure 8. Sodium hydroxide extraction of Morwell coal in a semi-continuous reactor (method A).

hydroxide solutions (12 - 14) and primary aliphatic amines (14) has been previously documented and, therefore, an increase in conversion was expected. It is noticeable that the increase in extract yield with sodium hydroxide is mainly due to an increase in the pre-asphaltene fraction (Figure 8). The composition of the oils and asphaltenes from the sodium hydroxide extractions appear similar to those obtained from water extraction but the pre-asphaltenes from the alkali extractions have lower H/C atomic ratios and higher hydrogen aromaticities than from water extraction (see Table III).

Increase in conversion is also obtained with tetralin/water mixtures (see Table VIII) in an analogous manner to the increased conversion with the addition of small amounts of tetralin to toluene (see Figure 1). Similar findings have been reported for the extraction of Powhatan coal with tetralin/water and tetrahydroquinoline/water mixtures since this work was completed (15).

TABLE VIII. Conversions for Extraction of Morwell Coal (15 g dry) in a Rocking Autoclave (Method B) with Aqueous Solvents (150 g) at 380°C

Solvent	Conversion (wt% coal dmif)
Water	48.7
0.5 M NaOH	66.6
0.25 M Na_2CO_3	63.8
0.5 M HCOONa	67.0
10% Tetralin/Water	55.3
20% Tetralin/Water	59.8

TABLE IX. Conversions for Extraction of Yallourn Coal with n-Pentylamine (Method A)

Solvent	Conversion (wt% coal dmif) A[a]	B[b]
Water	54.4	56.7
2M n-Pentylamine	69.6	95.8

[a] residue washed with acetone; [b] residue washed with pyridine.

Conclusions

This study indicates that extraction with supercritical water could be an attractive route for liquefaction of Victorian brown coals (but probably not black Australian coals). The low cost and ready availability of the solvent (water), the relatively high H/C atomic ratios of the extracts, and also as no hydrogen or coal-drying are required, are positive factors. Higher yields can be obtained when a strong base or a hydrogen-donor is added to the water.

It has been suggested (14) that the breaking of ester bonds is important in the dissolution of low rank coals. Supercritical water may hydrolyse ester groups and thus be effective for extracting brown coals. The high conversions obtained with sodium hydroxide may be explained by the ability of this strong base not only to break ester bonds but also to assist in solubilising the resultant fragments. Similarly, the pentyl group on the resulting amide may help solubilisation when a pentylamine was used. However, it must be noted that the physical evidence for significant amounts of ester bonds in brown coals is poor.

Acknowledgments

The authors are grateful to Mrs. I. Salivin for OH and molecular weight measurements, to Mr.R.I. Willing for solution ^{13}C-NMR spectra, to Dr. D.J. Cookson (BHP Melbourne Research Laboratories) for the solid state ^{13}C-NMR spectra and to Dr. D.J. Brockway (SECV) for analyses of the extraction residues.

Literature Cited

1. Deshpande, G.V.; Holder, G.D.; Bishop, A.A.; Gopal, J.; Wender, I. Fuel, 1984, 63, 956.
2. Scarrah, W.P. In "Chemical Engineering at Supercritical Fluid Conditions"; Paulaitis, M.E.; Penninger, J.M.L.; Gray, R.D.; Davidson P. Eds.; Ann Arbor Science: Ann Arbor, 1983, pp. 395-407.
3. Modell, M.; Reid, R.C.; Amin, S.I. U.S. Patent 4,113,446, September 12, 1978.
4. Jezko, J.; Gray, D.; Kershaw, J.R. Fuel Processing Technology, 1982, 5, 229.
5. Vasilakos, N.P.; Dobbs, J.M.; Parasi, A.S. Amer. Chem, Soc., Div. Fuel Chem. Preprints, 1983, 28(4), 212.
6. Kershaw, J.R.; Overbeek, J.M. Fuel, 1984 63, 1174.
7. Kershaw, J.R. Fuel Processing Technology,, 1984, 9,235.
8. Kershaw, J.R.; Overbeek, J.M.; Bagnell, L.J. Fuel, 1985, 64, 1069.
9. Kiss, L.T.; King, T.N. Fuel, 1979 58, 547.
10. Schafer, H.N.S. Fuel, 1979, 58, 667 and 673.
11. Kershaw, J.R. S. Afr. J. Chem. 1978, 31, 15.
12. Lynch, B.M.; Durie, R.A. Aust. J. Chem., 1960, 13, 567.
13. Brooks, J.D.; Sternhell, S. Fuel, 1958, 37, 124.
14. van Bodegon, B.; van Veen, J.A.R.; van Kessel, G.M.M.; Sinnige - Nijssen, M.W.A.; Stuiver, H.C.M. Fuel, 1984, 63, 346.
15. Towne, S.E.; Shah, Y.T.; Holder, G.D.; Deshpande, G.V.; Cronauer, D.C. Fuel 1985, 64, 883.

RECEIVED June 25, 1986

Chapter 22

Hydrotreating in Supercritical Media

J. Y. Low

Phillips Petroleum Company, Phillips Research Center, Bartlesville, OK 74004

Our research involved the investigation of catalytic hydrotreating in supercritical media. The program consisted of essentially two parts-- feasibility studies and the effect of process parameters of hydrotreating in supercritical media. Hydrotreating of shale oil has revealed that this type of hydrotreating is very effective in heteroatom removal. For example, the nitrogen and sulfur contents (19000 and 7000 ppm, respectively) were reduced to 120 and 25 ppm in a single pass. It is also effective for residua conversion. The 650F+ fraction of a petroleum crude was processed to yield a product with less than 50% in the 650F+ fraction and less than 5% in the 850F+ fraction. Supercritical hydrotreating had been tested for processing heavy coal liquids, supercritical shale oil extract, and conventional retorted shale oil. We have investigated the effects of process parameters in supercritical hydrotreating. The parameters studied include temperature, pressure, liquid hourly space velocity, and crude concentration.

In recent years, supercritical fluid extraction has been a very popular technique for separations. Recently, the energy industries have extended this application to coal liquefaction and oil shale extraction (1-5). Phillips Petroleum Company has also found that supercritical extraction of oil shale improves the oil yield compared to in situ or above ground retorting, but produces a lower quality liquid product. Phillips is interested in developing technology to upgrade the supercritically extracted (SCE) shale oil to synfuel or clean motor fuels. One project was to investigate hydrotreating in the presence of the supercritical fluid which is used in the extraction step. If the hydrotreating step can be integrated with the supercritical extraction step, one can possibly take advantage of the pressure and heat available in the extraction step.

0097-6156/87/0329-0281$06.00/0
© 1987 American Chemical Society

This investigation consisted of essentially two parts. The first involved a feasibility study of catalytic hydrotreating in the presence of supercritical fluid. The second part of our investigation involved the parametric studies to see how reaction parameters affect supercritical hydrotreating.

Conclusions

From an extensive investigation of the potential of hydrotreating of shale oil using a solvent under supercritical conditions, the following conclusions are made:
- A high nitrogen-containing, heavy oil such as shale oil can be hydrotreated under supercritical conditions to yield very low nitrogen fuels and syncrude in one step, depending on the conditions used;
- Compared to no solvent used, the presence of a light solvent gives a better product and reduces coke formation on the catalyst surface;
- Compared to an aromatic solvent, a non-aromatic solvent such as heptane improves nitrogen removal and reduces hydrogen consumption (from 2600 scf/bbl of shale oil to 1200 scf/bbl);
- With Arabian topped crude, the solvent loses its enhancement effect for nitrogen removal if the solvent is less than 50 weight percent of the feed;
- For extensive nitrogen removal, a relatively long residence time (30 minutes or longer) is required at 850°F and 1400 psig;
- Sulfur is almost completely removed even with the mildest reaction conditions studied.

Experimental

Hydrotreating System. A bench-scale hydrotreating unit was used for these experiments, as shown in Figure 1. The reactor was a 316 stainless steel tube with an inner diameter of 1 inch and a length of 27.5 inches. The total volume is about 290 ml. The reactor was equipped with a thermocouple well (a 1/4" x 25" stainless steel tube) for temperature measurements. The reactor was first filled with about 90 ml of inert packing again to serve as the preheating zone for the oil and hydrogen. The temperatures were measured by thermocouples placed in the middle of each of the inert beds and the catalyst bed. The unit was run 24 hours per day, 7 days per week.

For the supercritical versus conventional hydrotreating experiments (the comparison experiments), a fresh catalyst was used for each of the above experiments. For the feedstock testings and reaction parameter studies, the same catalyst was used. Before data were collected, the fresh catalyst was lined out for one week (168 hours). When the feed or reaction parameter was changed, the catalyst was lined out for 24 hours before data were collected.

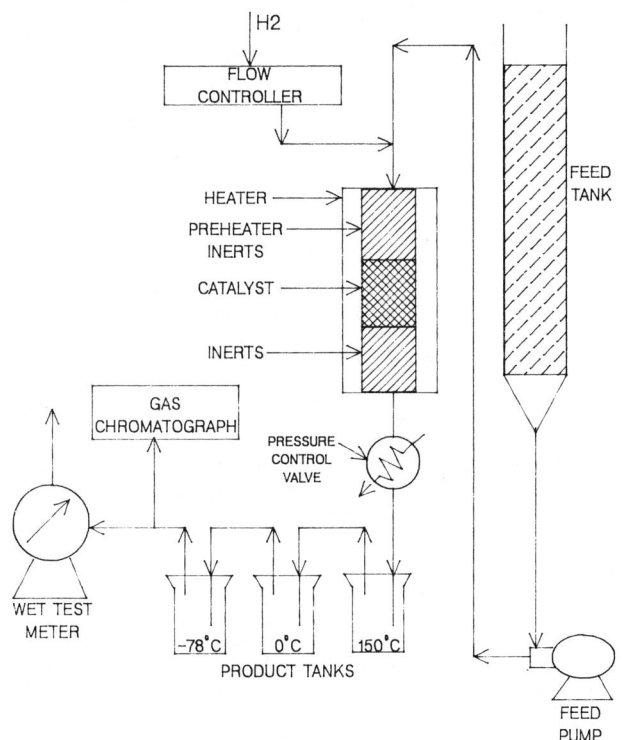

Figure 1. Supercritical hydrotreating system.

Feeds. During the course of the investigation, the following feeds were used: supercritical extracted lignite liquid, supercritical extracted shale oil, Paraho shale oil, and Arabian topped crude (650F+). Their properties are given in Table I. To make the feed mixture, the heavy oil was usually dissolved in a solvent such as toluene or n-heptane.

Catalysts. The Ni-Mo/Al$_2$O$_3$ catalyst (Nalco NM 502) was commercially available from Nalco.

Catalyst Presulfurization. The catalyst was generally heated to 300F with nitrogen purging, and then the nitrogen atmosphere was replaced with a flow of 10% H$_2$S in hydrogen. At the same time, the temperature was slowly increased to 600F and was kept at this temperature until the catalyst was completely sulfided. The reaction usually takes four hours at 600F and 100 liter of 10% H$_2$S in H$_2$.

Results and Discussion

In this report the results from the investigation of hydrotreatment of shale oil, Arabian topped crude, and lignite extract (under supercritical conditions) are discussed. These experiments were carried out to investigate the potential of hydrotreatment in the presence of a light solvent under supercritical conditions for shale oil upgrading and the effect of reaction parameters in supercritical hydrotreating. These experiments are exploratory in nature to find whether hydrotreatment under supercritical conditions has any advantage in the upgrading of high nitrogen heavy crudes, and if so, how do the major parameters affect the hydrotreatment under these conditions. The first few experiments were carried out with the shale oil obtained by supercritical extraction. This shale oil is a very waxy grease and almost fits the definition of a solid. As shown in Table I, it has a very high nitrogen content (2.3%) and a sulfur content of 1.0 wt%. The hydrogen content in the shale oil is relatively high with H/C atomic ratio of 1.48, equal to that of some petroleum crudes. Only a limited amount of this material was available, so only a few experiments were performed with this shale oil.

Table I. Feed Properties

Feed	>300F Wt%	\multicolumn{5}{c|}{Elemental Analyses}				
		C	H	N	S	H/C
SCE Shale Oil(1)	100	84.5	10.4	2.3	1.0	1.48
Paraho Shale Oil(2)	96.9	84.5	11.7	1.9	0.7	1.66
Arabian Topped Crude(3)	100	84.9	11.3	0.18	3.3	1.60
Lignite Extract(4)	100	82.0	9.3	0.91	0.47	1.36

(1) The SCE shale oil is a waxy black semi-solid.
(2) Paraho shale oil has about 70 vol% of 650F+ material.
(3) Totally 650F+ material.
(4) Hard solid at room temperature.

Supercritical Versus Conventional. A series of hydrotreating experiments were carried out under conventional conditions (without the use of a light solvent). The results are given in Table II, along with some results obtained from a supercritical hydrotreatment experiment. The experiments performed were not under identical conditions, but they are close enough that the results obtained are valid enough for comparison.

Table II. Comparison of Conventional and Supercritical Hydrotreating

Feed	(Conditions: 1400 psig, 850F, H_2 GSHV-300, Nalco Ni-Mo)	
	No Solvent Shale Oil (Neat)	With Solvent 20% Shale Oil In Toluene
LHSV	0.3	TOTAL - 1.6 SHALE OIL - 0.32
Gas Yield (Wt%)	25	10
Light (Wt%)	35	55
Heavy Oil (Wt%)	35	35
Coke (Wt%)	3.8	0.3

The results from the supercritical hydrotreatment experiments are superior in almost every respect to conventional hydrotreatment experiments (without the use of solvent). Under similar conditions, supercritical hydrotreating produced better products, for example: less gas yield (10% vs 25%), more of light oil fraction, <300F, (55 vs 35%) and less coke formed on the catalyst surface (0.3 vs 3.8% based on the feed). For conventional hydrotreating we had encountered reactor plugging problems when the unit was running more than 196 hours. This problem was not found with supercritical hydrotreating.

Supercritical Hydrotreatment of SCE Shale Oil. SCE shale oil was hydrotreated at high severity because of its high nitrogen content (Table I) and extremely high viscosity. The experimental results are shown in Table III. Based on the shale oil fed, the product distribution is the following: 12% gases, 52% boiling less than 300°F (calculated by difference) and 36% in the heavy oil fraction (>300°F). The elemental analyses of the heavy oil fraction indicate that better than 99% of the nitrogen was removed. The sulfur removal was equally high. The light boiling fraction (<300°F) usually has less than 10 ppm of sulfur and 10 ppm of nitrogen.

For Runs 2, 3, 4, and 5 the feed entered the hydrotreater directly from the supercritical extraction unit. The extract contained about 4% shale oil in toluene. Runs 2 and 3 were carried out at 842F, and Runs 4 and 5, at 750F. At the lower reaction temperature (750F), the yield of gases dropped to less than 2%, and the yield of heavy oil fraction increased by about 10%. The extent of nitrogen removal was reduced significantly at the lower temperature. However, the sulfur removal seemed to be unaffected by the lowering of reaction temperature from 842F

Table III. Supercritical Hydrotreatment of SCE Shale Oil*

Run No.(1)	Temp. °F	LHSV	Gases (C_1-C_4) Wt%	Heavy Oil (>300F) Wt%	N (ppm)	S(%)
1	842	0.5	12	36	52	0.03
2	842	1.6	--	46	150	0.03
3	842	1.6	--	44	150	0.02
4	750	1.6	<2	54	2300	0.08
5	750	1.6	<2	32(2)	2000	0.04

* A total operating pressure of 1400 psig and sulfided Ni-Mo Al_2O_3 catalyst were used in these experiments. The shale oil had the following elemental analyses: C, 84.5; H, 10.4; N, 2.3; S, 1.0.

(1) The feed for Run 1 consisted of 20 wt% shale oil in 80 wt% toluene, but for Runs 2-5, the feed was only 4% shale oil in toluene.

(2) In this Run the fraction was distilled up to 400F instead of 300F as usual.

to 750F. Thus, these experimental results suggest that the supercritically extracted shale oil can be processed to yield very low nitrogen and sulfur fuels or syncrudes.

Supercritical Hydrotreatment of Arabian Topped Crude. A series of experiments were performed with Arabian topped crude (650F+) to investigate the hydrotreatment of high sulfur crudes in the presence of a light solvent under supercritical conditions. The experimental results obtained are summarized in Table IV. The overall results are comparable to those obtained from the supercritical hydrotreatment of SCE shale oil (Run 1). The sulfur removal is very extensive (about 99% removal, reduced from 3.3% to 0.02%). The nitrogen content in the heavy oil fractions are relatively low, less than 60 ppm for runs with total liquid hourly space velocity (LHSV) of 0.5 (Runs 6 and 7) and about 770 ppm for Run 8 with an LHSV of 1.6. Thus, for extensive nitrogen removal, lower LHSV is needed.

The effect of crude oil concentration in the feed was investigated with Arabian 650F+ topped crude. Three concentrations (20, 50 and 80% of topped crude in toluene) were chosen, while the other experimental conditions were kept constant (see Figure 2 and Runs 9, 11, and 12 in Table IV). Both the nitrogen removal and heavy oil conversion are more extensive when the crude is hydrotreated at a more diluted level such as 20%. The heteroatom

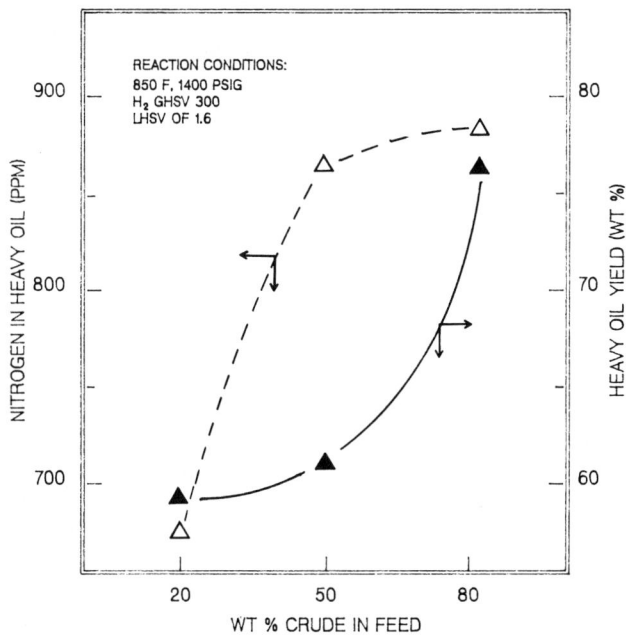

Figure 2. Effect of oil concentration in supercritical hydrotreatment of arabian topped crude.

removal is reduced at higher crude concentrations and levels off at 50% or higher. The yield of the heavy oil fraction increases with increasing concentration of crude in the toluene.

Table IV. Supercritical Hydrotreatment of Arabian Topped Crude (ATC) (850F, 1400 psig, H_2 GHSV-300, Nalco Ni-Mo Catalyst)

Run No.	Feed(1) Wt% ATC in Solvent	LHSV	Heavy Oil Wt%	N (%)	S (%)
6	20	0.5	34	0.006	0.02
7	20	0.5	34	0.004	0.02
8	20	1.6	59	0.068	0.04
9	20	0.5	33	2 ppm	0.02
10	20	0.5	36	3 ppm	0.03
11	50	1.6	64	0.094	0.17
12	80	1.6	72	0.094	0.13

(1) Solvent for Runs 9 and 10 was n-heptane; for others toluene was used.

Runs 9 and 10 were carried out to study the effect of an aliphatic solvent such as n-heptane in supercritical hydrotreating of topped crude. Comparing the results from Runs 6 and 7 in which toluene was used as the solvent, the nitrogen contents in heavy oil fractions are much lower for the heptane runs than the toluene runs. The sulfur contents are about the same. Thus, one can conclude that an aliphatic solvent is better solvent for hydrodenitrogenation than the aromatic solvent. The reason is that the aromatic solvent was competing for hydrogenation (about 4 wt% of the toluene was converted to methyl cyclohexane).

Supercritical Hydrotreatment of Lignite Extract. Lignite extract was hydrotreated in the presence of toluene under supercritical conditions (850F, 1400 psig, 20 wt% lignite in toluene, a hydrogen GHSV of 300, and a LHSV of 1 or 1.6). The results are tabulated in Table V. The elemental analyses of the heavy oil fraction have indicated the following changes (Runs 13 and 14): nitrogen, reduced from 0.91 to 0.14%, and sulfur reduced from 4700 ppm to about 100 ppm.

Table V. Supercritical Hydrotreatment of Lignite Extract

Run No.	LHSV	Heavy Oil Fraction (>300F)		
		Wt%(1)	N (ppm)	S (ppm)
13	1	30	1530	150
14	1	36	1400	90
15	1.6	43	1180	100
Lignite Extract Feed		100	9100	4700

Conditions: 850F, 1400 psig, 300 H_2 GHSV, 20 wt% lignite extract in toluene, Nalco Ni-Mo catalyst.
(1) Wt% is based on the feed.

Run 15 was carried out with a higher LHSV of 1.6, and the results are not too much different from Runs 13 and 14. The heteroatom removals are about the same. The heavy oil fraction, however, was increased to 43% from 30 and 36% (Runs 13 and 14). Thus, the conclusion is that the lignite extract, a solid, can also be upgraded to yield a syncrude by the supercritical hydrotreatment process.

Reaction Parameter Studies. Experiments were carried out with conventional shale oil (direct retorted Paraho shale oil) for the purpose of studying the effects of reaction parameters in hydrotreating under supercritical conditions. In one group of experiments, the space velocity was varied (1.6, 3.2, and 5) while the other reaction parameters were kept constant. For temperature studies, the following reaction temperatures were investigated: 700, 750, 800, and 850F. The pressure effect was examined at four levels -- 1000, 1400, 2000, and 2400 psig at 800F. We have also studied the effect of solvent to feed ratio and solvent types (saturate vs aromatics).

The liquid hourly space velocities (LHSV, including the solvent) of 1.6, 3.2, and 5 were investigated at 850F, 1400 psig, 20 wt% Paraho shale oil in toluene. The results are illustrated in Figure 3. With the LHSV of 1.6, the heavy oil fraction is only 34%, but when LHSV is increased to 3.2, the heavy oil fraction increased to 63% and did not change with further increase in LHSV. This seems to suggest that one-third of the shale oil can undergo a molecular weight reduction more rapidly (a residence time of about 12 minutes) than the second-third, which needs a residence time of up to an hour, while the last-third survives longer than one hour.

For heteroatom removal, the results reveal that for a LHSV of 1.6 the nitrogen content in the heavy oil fraction, which is about 34 wt% of the total products, is 360 ppm (1.9% in the feed). Thus, the overall nitrogen removal is greater than 98%. For LHSV of 3.2 the nitrogen content in the heavy oil is increased to about 5000 ppm. With a further increase in LHSV to 5, there is a greater increase in the nitrogen content. The sulfur removal is very rapid. Even with the LHSV of 5, there is only 53 ppm sulfur in the heavy oil.

Figure 4 illustrates the results from reaction temperature studies. The reaction temperaures used ranged from 700F to 850F, with an increment of 50F. The remaining reaction parameters were held constant, such as 1400 psig, LHSV of 1.6 and H_2 GHSV of 300. The sulfur seems to be removed rather easily even at 700F (from 7000 ppm in the feed to 200 ppm in the heavy oil fraction) or 97% removal. At higher temperatures the sulfur content in the heavy oil is much lower (100 ppm). On the other hand, the nitrogen removal requires a much higher temperature. This, of course, is not surprising. For the purpose of obtaining relatively clean syncrude, we probably have to operate at high temperature (800F) and low space velocity (<1.6).

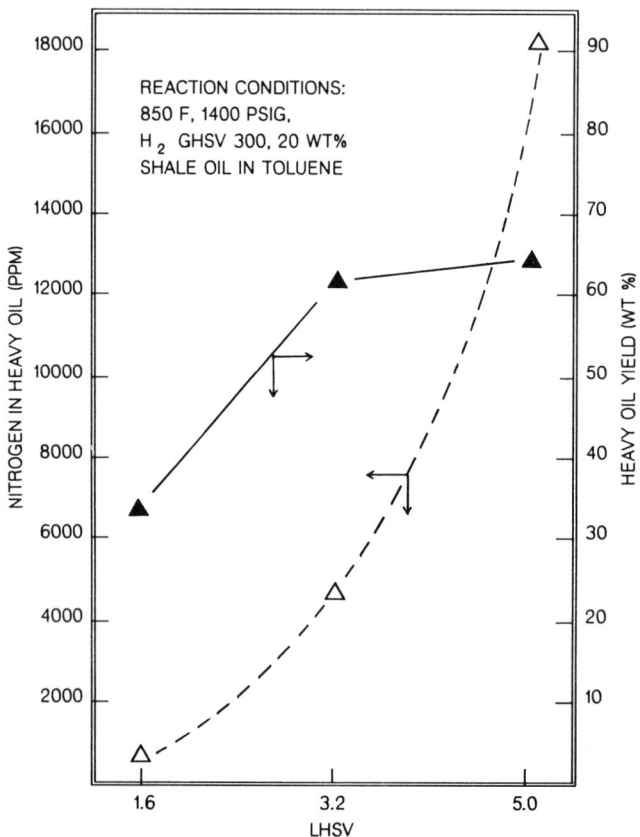

Figure 3. Space velocity effects on supercritical hydrotreatment of shale oil.

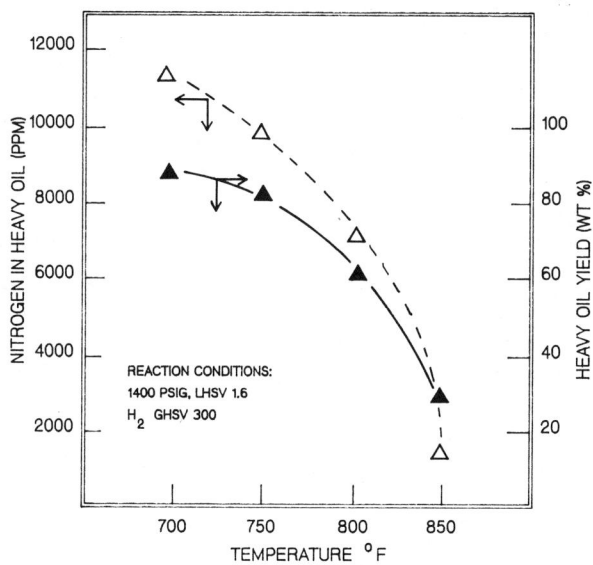

Figure 4. Effect of temperature on supercritical hydrotreatment of shale oil.

The yield of heavy oil increases with a decrease of reaction temperature (Figure 4). For example, with the reaction temperature of 850F, the heavy oil is only about 30% of the total products (96.9 wt% in the feed). While at the other extreme, that is, with the reaction temperature of only 700F, the heavy oil fraction represents 90% of the whole products. Thus, the higher temperature is necessary for the production of light products.

The effect of pressure on HDN and heavy oil yield in supercritical hydrotreating of shale oil is given in Figure 5. The experimental conditions used for these experiments are: 800F, hydrogen GHSV of 600, LHSV of 1.6, 20 wt% shale oil in toluene, and reaction pressure of 1000-2400 psig. It is observed that nitrogen removal increases with increasing reaction pressure. At the highest pressure studied (2400 psig), the nitrogen content in the heavy oil is only about 300 ppm. This is about 98% nitrogen removal. The nitrogen removal decreased with the decrease in reaction pressure. Pressure has some effect on the heavy oil yield. At the lowest pressure studied (1000 psig), the yield was 65 wt%. The heavy oil yield increases with the increase in the pressure. With the highest pressure studied (2400 psig), the yield was increased to 79 wt%. The increase in yield must be due to the suppression of hydrocracking by the pressure.

The use of a non-aromatic solvent as the supercritical hydrotreating solvent was studied first with Arabian topped crude and then with Paraho shale oil in hope of reducing hydrogen consumption. The results are tabulated in Tables IV and VI. Run 16 used toluene as the solvent, and Runs 17 and 18 used n-heptane. The experimental results have shown that nitrogen removal is slightly higher with n-heptane than with toluene (a more significant difference was found with Arabian topped crude as the feed, Table IV). The yield of heavy oil fraction is lower (37% for heptane and 57% for toluene) for the saturated solvent. This means hydrocracking was more extensive. The H/C atomic ratio of the heavy oil is also much higher for heptane than for toluene. Thus, hydrogenation is deeper with heptane as the solvent. Hydrogen consumption is also significantly reduced, 1200 scf/bbl of shale oil for heptane, versus 2600 scf/bbl for toluene. With everything considered, heptane is by far a better hydrotreating solvent than toluene.

Table VI. Supercritical Hydrotreatment of Shale Oil in Toluene or n-Heptane
(850F, 1400 psig, Nalco Ni-Mo Catalyst, 1.6 LHSV, 300 H_2 GHSV)

Run No.	Feed	Heavy Oil (>300F)				H_2 Consumption
		Wt%	N(ppm)	S(%)	H/C	Scf/Bbl
16	20% Shale Oil in Toluene	57	3149	<0.01	1.66	2600 ± 200
17	20% Shale Oil in Heptane	36	1595	<0.01	1.79	1200 ± 100
18	20% Shale Oil in Heptane	37	1890	<0.01	1.82	1200 ± 100

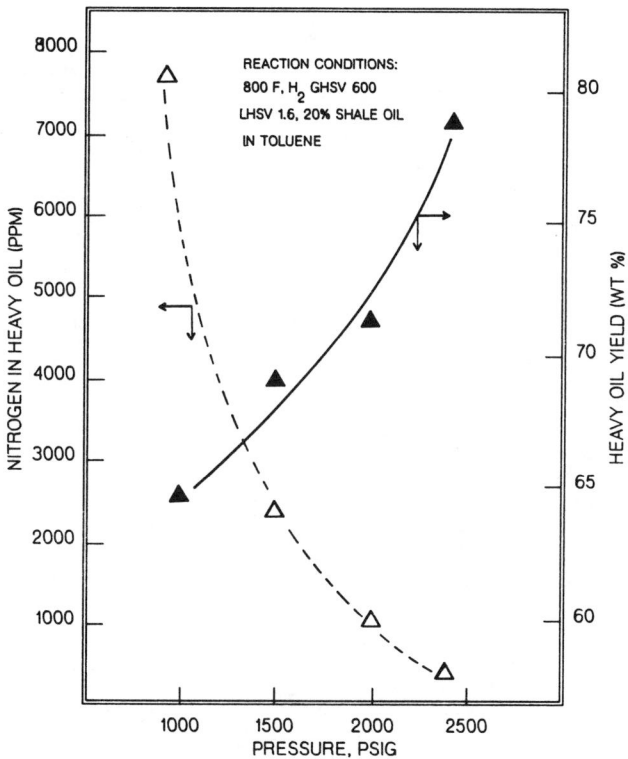

Figure 5. Effect of pressure on supercritical hydrotreatment of shale oil.

Summary

Extensive research has been done to study the potential of supercritical hydrotreatment for the upgrading of shale oil and the effects of major reaction parameters in supercritical hydrotreating. The experimental results have demonstrated that hydrotreatment in the presence of a light solvent and under supercritical conditions has some advantages over conventional hydrotreatment. Compared with a conventional process, the supercritical hydrotreating process yields a better product (less gas, more distillates) and produces less coke on the catalyst. Supercritical hydrotreating appears to be a very versatile process which can be optimized for a variety of different products. This process can be applied equally well for upgrading other heavy oils, such as coal liquids and high sulfur petroleum topped crudes, to yield low nitrogen and sulfur syncrudes and transportation fuels.

N-heptane was used in place of toluene as the process solvent. The saturated solvent seems to do better in terms of nitrogen removal and hydrogen consumption. With either solvent, sulfur removal was very extensive. From reaction parameter studies, a relatively high severity is required for extensive nitrogen removal.

Literature Cited

1. Blessing, J. E., and Ross, D. R., ACS Symposium Series 71, "Organic Chemistry of Coal", p. 15, 1978.
2. Whitehead, J. C., "Development of a Process for the Supercritical Gas Extraction of Coal", paper presented at the 88th AIChE National Meeting, June 1980.
3. Squires, T. G.; Aida, Tetsuo; Chen, Yu-Ying; Smith, Barbara F., ACS Fuel Chemistry Preprints, Vol. 28 No. 4, p. 228, 1983.
4. Compton, L. E., ACS Fuel Chemistry Preprints, Vol. 28 No. 4, p. 205, 1983.
5. Williams, D. F., U.S. Patent 4 108 760, 1978.

RECEIVED June 25, 1986

Author Index

Abraham, Martin A., 67
Aida, Tetsuo, 58
Antal, Michael Jerry, Jr., 77
Bae, Y. C., 2
Bagnell, Laurence J., 266
Barton, Paul, 202
Benmekki, E. H., 101
Bergstresser, T. R., 138
Brittain, Andrew, 77
Coppella, Steven J., 202
DeAlmeida, Carlos, 77
Deshpande, G. V., 251
Frye, S. L., 29, 172
Green, Thomas K., 242
Gulari, Es., 2
Hacker, D. S., 213
Holder, G. D., 251
Hum, Georgina P., 242
Johnston, K. P., 42
Jonas, J., 15
Jordan, J. W., 189
Kalkwarf, D. R., 29
Kershaw, John R., 266
Kim, Sunwook, 42
King, Jerry W., 150
Klein, Michael T., 67
Kumar, Sanat K., 88
Kwak, T. Y., 101
Lamb, D. M., 15
Low, J. Y., 281
Mansani, Riccardo, 242
Mansoori, G. A., 101
Miin, Tiee-Chyau, 242
Moradinia, Iraj, 130
Panagiotopoulos, A. Z., 115
Paulaitis, Michael E., 138
Ramayya, Sundaresh, 77
Reid, R. C., 88, 115
Ross, David S., 242
Roy, Jiben C., 77
Saad, H., 2
Shah, Y. T., 251
Skelton, R. J., 189
Smith, R. D., 29, 172
Squires, Thomas G., 58
Suter, U. W., 88
Taylor, L. T., 189
Teja, Amyn S., 130
Warzinski, Robert P., 229
Wright, R. W., 172
Yonker, C. R., 29, 172

Subject Index

A

n-Alkanes--See Solid n-alkanes

B

Batch reactors, use in supercritical extraction of coal, 251-252
Benzylphenylamine
  apparent rate constants in supercritical solvents, 72,73t
  effect of methanol density on conversion and product selectivity, 69,71f,72,74
  effect of water density on conversion and product selectivity, 69,70f,72,74

Benzylphenylamine--Continued
  experimental conditions for reaction, 68t
  major reaction products, 68t
  model solvolysis and pyrolysis through caged radical pair, 74,75f
  reaction in supercritical methanol, 69,71f,72
  reaction in supercritical water, 69,70f
  reaction pathways neat and with supercritical fluid solvents, 72,75f
  thermolysis, 69
Binary mixtures, theory, 91-93
Binary mixtures of toluene and meso-tetraphenylporphyrin, bubble and dew points, 145t,156f,147
Binary systems, phase-equilibrium behavior, 118-120

Biomass feedstocks, conversion
processes, 77-78
Biphenyl
solubility in CO_2 vs.
pressure, 180,181f
solute retention vs.
pressure, 180,182f
Butadiene recovery processes,
advantages of supercritical or
near-critical separation, 213

C

C_4 separation, liquid ammonia as
solvent, 214
Canonical partition function,
definition, 89
Carbon dioxide-hydrocarbon equilibria,
modeling, 210
β-Caryphyllene, structure, 203,205f
CO-water conversion
advantages, 245
effectiveness of CO, 245
inverse isotope effect, 247
isotope effect, 245
toluene-soluble products vs. CO
consumed, 245,246f
CO-water conversion process,
characteristics of the medium, 243
CO-water conversion product analyses
H/C ratios of products, 247t
separations of toluene-soluble
fractions, 245,247,248f
Coal conversion, vs. Hildebrand
solubility parameter, 252
Coal liquefaction
chemistry, 242
kinetic studies, 242-249
scheme for conversion in
hydrothermal
systems, 247,248f,249
Coal model compounds, reactant
decomposition kinetics, 67
Coals, chemical analyses, 252,253t
Compressed supercritical ethylene,
self-diffusion, 18-21,23f
Compressed supercritical toluene-d_8,
self-diffusion, 21-22,23f
Conversion, definition, 253
Critical densities, vs. corresponding
compositions, 5
Critical end points, location and
phase line, 22,26f,27
Cubic equations of state,
applications, 101-102

D

Data reduction for retention volume
measurements
computation of retention volume
data, 161-162

Data reduction--Continued
midpoint of frontal breakthrough
profile vs. elution peak
maxima, 161,163f
retention volume vs. adsorption
coefficient and surface area of
sorbent bed, 162
Decay rate of order-parameter
fluctuations
critical lines of three binary
systems, 5,6f
equation, 5
experimental procedures, 5
vs. calculated critical and
background contributions, 12,13f
vs. temperatures and
pressures, 5,7,9t
vs. transport coefficients, 3-4
Destraction procedure
comparison of modes, 234,235f,236
nonreflux mode, 233,235f
reflux mode, 233,235f
Diffusion coefficient in the
supercritical region, equation, 4
1,3-Dioxolane, hydration in
water, 82,84f
Dynamic renormalization group theory,
description, 3

E

Enhancement factor, calculation, 47
Enskog theory
analysis of density dependence, 20
coefficients, 20,23f
deviations, 20-21
Equation of state for 1-butene
extraction, mixing rules, 216
Equation of state modeling,
application to high-pressure phase
equilibrium behavior, 88
Ethanol dehydration in supercritical
water, conversion to
ethylene, 82,83t
Extraction of 1-butene
analysis procedure, 219-220
effect of ammonia
concentration, 225,226t
effect of different
solvents, 220,223t
effect of temperature on
selectivity, 220,224f
effectiveness of ammonia as
enhancing agent, 225
equation of state, 216
experimental apparatus, 217,218f
experimental procedure, 219
gas chromatographs, 220,221f
influence of entrainer, 225
influence of second solvent, 227

INDEX
297

Extraction of 1-butene--Continued
  sampling procedure, 219
  selectivity vs. ammonia
    concentration, 220,224f
  ternary-phase diagram, 220,221f
  ternary phase equilibrium
    composition, 222-223t

F

Flow reactors, use in supercritical
  extraction of coal, 251
Fractional destraction
  analysis of overhead fractions and
    the residue, 236,238t
  comparison of GPC results, 238,239f
  comparison of replicate
    destractions, 236,237f
  effect of temperature on carry-over
    concentration, 236,237f
  elemental analysis of feed
    coal, 233,234t
  experimental procedures, 230,232
  fractional destraction unit, 230
  fractional destraction
    vessel, 230,231f
  performance of fractional
    destraction unit, 232
  two-step destraction procedures, 233
  vs. conventional distillation, 230
Fundamentals of supercritical fluid
  adsorption
  adsorption of a high-pressure
    gas, 152
  breakthrough volumes, 156
  effect of pore structure on
    pressure, 154,155f, 156,157f
  effect of temperature on
    pressure, 154,155f
  plot of reduced-state
    variables, 156,158f
  pressure of adsorption maxima, 154
  typical adsorption isotherm for a
    supercritical gas, 152,153f

G

Gas-liquid equilibrium, measurement of
  dew and bubble points, 140
Gel permeation chromatography,
  determination of molecular
  weight, 236,238t
Geranial, structure, 203,205f
Gram-Schmidt reconstructions
  of hexane, 193,194f
  stack plot of C-H stretch
    region, 195,196f

H

Hard sphere diameters, calculation of
  theoretical Enskog
  coefficients, 20
Heterolytic reactions vs. homolytic
  reactions, 84
High-pressure gas chromatographic
  analysis of retention volume
  adsorbent/adsorbate
    classes, 160,161t
  apparatus, 156,159f,160
  column design and preparation, 160
  experimental assessment of the
    column void volume, 160
  pneumatic transport of the
    solute, 160
High-pressure phase equilibrium of
  aqueous solutions
  experimental procedures, 116
  schematic of equipment, 116,117f
Homolytic vs. heterolytic
  reactions, 84
Hydrotreating in supercritical media
  effect of nonaromatic
    solvent, 286t,293t
  effect of oil
    concentration, 286,287f,288
  effect of pressure, 289,292f
  effect of space velocities, 289,290f
  effect of temperature, 289,291f
  experimental procedures, 282,284
  feed properties, 284t
  hydrotreating system, 282,283f
  reaction parameter studies, 289-293
  supercritical vs. conventional, 285t
  treatment of Arabian topped crude
    oil, 286t,288
  treatment of lignite extract, 288t
  treatment of supercritical extracted
    shale oil, 285t,286

I

Interaction parameters between binary
  components, values, 127,128t
Intrinsic solvent strength,
  definition, 50
Isothermal compressibility of
  supercritical fluids,
  determination, 175-177

K

Kamlet and Taft scale, application
  and description, 30

Kinetic model for coal conversion
  description, 255
  effect of amount of dissolved
    fraction, 258
  effect of coal-to-solvent
    ratio, 263,264f
  effect of reaction time and density
    on toluene
    solubles, 255,256-257f,258
  effect of temperature, 258,259-260f
  model simulation, 258,261-262f,264f
Kinetic studies of coal liquefaction
  CO-water conversions, 245,246f
  conversion products, 245,247,248f
  experimental procedures, 243-244
  model, 244-245,246f

L

Lemon oil
  advantages of supercritical
    extraction, 202
  classifications, 203
  concentration, 203
  extractive distillation with a
    supercritical solvent, 203-204
  gas chromatogram in liquid phase
    sample, 206,207f
  proposed process for
    concentration, 204
Lemon oil-carbon dioxide equilibrium
  ease of separation, 206
  experimental conditions, 204,206
  relative volatility, 206,208t
  selectivities, 210,211f
  solubility diagrams, 206,209f
  solubility level, 206,210,212
Limonene, structure, 203,205f
Liquid ammonia, use as solvent for C_4
  separation, 214
Local solvent compression,
  determination, 51-52
Lorentz-Lorenz refraction equation,
  calculation of refractive
  index, 37

M

Mixed fluid solvent systems
  nature, 37
  spectroscopic vs. chromatographic
    measurements, 38
Mixing rules
  Peng-Robinson equation of state, 101
  Redlich-Kwong equation of state, 104
  van der Waals, 101

Mode-coupling theory
  description, 3
  dynamic renormalization group
    theory, 3-4
Model of coal liquefaction
  effect of kinetics of reducing
    step, 249
  effect of structure on
    conversion, 244
  schematic, 244,245f
  yield of toluene-soluble
    material, 244-245
Modeling of equilibria, CO_2-
  hydrocarbon systems, 210

N

Naphthalene
  solubility, 47,49f
  solubility vs.
    pressure, 178,179f,180
  solute retention vs.
    pressure, 180,181f
Naphthalene-CO_2 system
  critical end points, 22,24,26f,27
  pressure-temperature diagram, 22,25f
  solubility measurement, 24
Naphthalene solubility in
  supercritical CO_2, NMR
  measurements, 22-27
2-Nitroanisole
  absorbance maxima, 32,33t
  absorption spectra, 31
  effect of pressure on absorption
    maximum, 38,39f
  peak position vs. reduced
    density, 32,35,36f
  pressure dependence of wavelength of
    the absorption maximum, 32,34f
  values vs. reduced density, 31
NMR measurements of naphthalene
  solubility
  experimental solubilities, 24,25f
  isotherms, 24
NMR spectroscopy
  analysis of supercritical
    solubility, 17-18
  technique, 18

O

Onsager reaction field theory
  description, 35,37
  function vs. measured
    values, 35
Order-parameter fluctuations
  decay rate, 3-4

INDEX

Order-parameter fluctuations--Continued
  description, 3
  photon correlation spectroscopy, 3
  time-averaged intensity measurements, 3

P

Patel-Teja equation of state, data correlation for solid $\underline{n}$-alkanes in supercritical ethane, 132,134
Peng-Robinson equation of state, 102
  mixing rules, 104-105
  modification, 127
Pentane
  selection as supercritical fluid solvent, 139
  vapor pressures, 140t,141
Phase behavior of solids in supercritical fluids, applications, 138-139
Phase-equilibrium behavior for binary systems, carbon dioxide ethanol, 118,120f
Phase-equilibrium behavior for ternary systems
  compositions for water-acetone-CO_2, 118,121t
  concentration of acetone in the supercritical fluid phase, 122,126f
  distribution coefficient for water-acetone-CO_2, 122,124f
  salting-out effect, 118,122
  selectivity factor for acetone over water, 122,125f
Phenol blue
  transition energy in CO_2, 44,45f
  transition energy vs. CO_2 mixtures, 52,53f,54
  transition energy vs. density, 51,52f
Photoisomerizations, experimental procedures, 60
Porphyrins, constituent in crude oils, 139
Preparative-scale chemical class separation, procedure, 190-191
Pressure-filter liquids, definition, 190
Properties of supercritical fluids, effect of pressure and temperature, 59
Pure-component parameters
  determination, 90t
  estimation technique, 127,128t
  values, 90t
Pure components, theory, 89

R

Rapid injection autoclave, use in supercritical extraction of coal, 252
Rate constants
  correlation with transition energy, 48
  predicted value for Diels-Alder reaction, 48,49f
Redlich-Kwong equation of state, 102
Refractive index, calculation, 37
Regeneration of adsorbents, by supercritical fluids, 151,153f
Relative volatility, definition, 206
Residue class separation, results, 191,192t,193
Residue-containing fossil fuels, importance of fractionation, 229-230
Retention volume measurements
  data reduction, 161-162,163f
  high-pressure gas chromatographic apparatus, 156,159f,160,161t
  solute capacity factors vs. column pressure, 167,168f,169
  solute capacity factors vs. gas compression, 164-167,168f
  volume vs. CO_2 pressure, 162-166

S

Self-diffusion coefficients
  in deuterated toluene, 21-22
  in supercritical ethylene, 16
  in supercritical toluene-$\underline{d}_8$, 16
  predicted vs. measured values, 22,23f
Self-diffusion in compressed supercritical ethylene
  calculation of hard sphere diameter, 20
  choice of ethylene, 18
  coefficients vs. density and temperature, 18,19f,20
  Enskog coefficients, 20,23f
  Enskog theory, 20-21
  measurement conditions, 18
Self-diffusion in compressed supercritical toluene-$\underline{d}_8$
  coefficients, 21-22,23f
  polar gas model, 22
Separation of solution components from the liquid phase
  equations of state, 214
  influence of entrainers, 216
  solvents, 214

Solid n-alkanes
  heat of fusion vs. carbon number, 130,131f
  reasons for studying, 130
  solubilities, 132-136
Solid-liquid-gas equilibrium temperatures and pressures
  critical endpoint, 141
  measurement, 139
  pressure-temperature projection of vapor pressure curve, 141
  values for binary mixtures, 141t
Solid solubilities in supercritical pentane, measurements, 140
Solubilities for naphthalene, measurement, 17
Solubility of solid n-alkanes in supercritical ethane
  data correlation, 132,134
  enhancement factor vs. carbon number, 134,136f
  experimental and calculated solubilities, 134,135f
  experimental results at 308.15 K, 132t
  single-pass supercritical flow apparatus, 132,133f
  solid properties required in solubility calculations, 134t
Solute retention in supercritical fluid chromatography
  effect of pressure, 172-173
  effect of temperature, 173
  experimental apparatus and technique, 178
  vs. density, 180
  vs. pressure for biphenyl, 180,182f
  vs. pressure for naphthalene, 180,181f
  vs. solubility and pressure, 175
  vs. temperature, 178
Solvating power, scale, 30
Solvatochromic data, experimental procedures, 43-44
Solvatochromic probes
  absorbance maxima, 31-32
  experimental procedures, 31-32
Solvent effect on rate constants
  correlation with transition energy, 47,48
  prediction of activation volume, 48t
Solvent strength, description, 43
Solvent strength in the critical region, 44
Source and purity of n-alkanes, 132
Specific solvent strength, definition, 50
Spectroscopic solvatochromatic parameter, description, 43
Statistical-mechanics-based lattice-model equation of state
  applicability to mixture of different size molecules, 94,96f

Statistical-mechanics-based equation of state--Continued
  behavior, 90-91
  effect of solubility of chains, 99
  model behavior of solid-supercritical fluid binaries, 94,97f,98
  predicted vs. experimental data, 94,95f,96
  prediction for equilibrium fluid-phase composition, 98f
  sensitivity, 94
trans-Stilbene
  effect of temperature on CO_2 solubility, 60,63f
  in cyclohexane, 62t
trans-Stilbene photoisomerization
  concentration effects, 62,63f
  in CO_2, 62,65t
  pressure effects, 62,64f
  temperature effects, 62,64f
Supercritical CO_2 extraction of lemon oil, 203
Supercritical coal liquefaction procedure
  conversion products, 253,255
  experimental apparatus, 253,254f
Supercritical distillation--See Fractional destraction
Supercritical extraction
  description, 115
  factors influencing rate, 2
  recovery of polar organic compounds from aqueous solutions, 115-116
Supercritical extraction of coal
  use of a rapid injection autoclave, 252
  use of batch reactors, 251-252
  use of flow reactors, 251
Supercritical flow reactor
  schematic, 79,80f
  temperature, 79,81f
Supercritical fluid adsorption, fundamentals, 152-156
Supercritical fluid chromatography (SFC)
  advantages, 189-190
  analytical applications, 151
  application as a mobile phase, 189
  assumption of infinitely dilute solutions, 173
  hexane separation, 193,194f,195,196f
  measurement of physicochemical data, 151
  relationship among solute retention, solubility, and pressure, 175
  relationship between solute retention and temperature, 175-177
  retention factor, 173
  separation of polycyclic aromatic hydrocarbons, 195,197f
  solubility of a solute, 174

INDEX

SFC--Continued
  solute chemical potential, 173
  toluene separation, 195,198f,199
Supercritical fluid extraction,
  advantages over conventional
  extraction techniques, 15
Supercritical fluid extraction
  modeling, applications, 105-110
Supercritical fluid extraction of
  coal
  analyses of coals, 267,268t
  potential use, 266-267
  solvents, 266
  water vs. aqueous
    solvents, 277,278f,279t
  water vs. toluene, 277,278f
Supercritical fluid
  extraction-chromatography
  analytical procedures, 191
  disadvantages, 193
  Gram-Schmidt
    reconstruction, 193,194f
Supercritical fluid reactant addition
  system, diagram, 60,61f
Supercritical fluid reactor,
  schematic, 60,61f
Supercritical fluid solvents
  advantages, 42-43
  properties, 44,46t
Supercritical fluids
  applications, 58-59,150-151
  constraints for organic reaction
    investigations, 59
  critical properties, 32t
  desorption of adsorbates, 151,153f
  effect of additional components on
    solvating properties, 37-38,39f
  peak position vs. reduced
    density, 32,35,36f
  pressure dependence of
    wavelength of the absorption
    maximum, 32,34f
  solvent properties, 58
  solvent-solute interactions, 29-39
  use as solvents, 29
  values vs. reduced density, 35,36f
Supercritical gas extractions
  method A procedure, 267
  method B procedure, 267,269
Supercritical phase, definition, 2
Supercritical solubility measurements
  schematic of all, 17,19f
  use of NMR spectroscopy as an
    analytical technique, 17-18
Supercritical solvents, effects on the
  rates of homogeneous chemical
  reactions, 42-54
Supercritical solvents as a reaction
  medium, thermophysical
  properties, 78
Supercritical water
  ethanol dehydration, 82,83t

Supercritical water--Continued
  role of carbocation chemistry, 84-85
  thermophysical properties, 78-79
Supercritical water extractions of
  coal
  analysis of oil
    fractions, 271,274t,276f
  analytical data for
    residues, 271,274t
  ^{13}C-NMR spectra of
    fractions, 271,273f
  conversion vs. volatile matter
    content, 269,270f
  effect of pressure, 274,275t,276f
  effect of temperature, 275t,277
  extract analysis, 271,272t
  extract composition, 269,270f,271

T

Ternary systems, phase-equilibrium
  behavior, 118-126
meso-Tetraphenylporphyrin
  melting behavior, 147
  solid solubilities in various
    solvents, 147t,148
Theory of critical fluctuations
  effect of impurities, 10
  parameter values, 10,11t
  range of validity, 12
Thermodynamic model for vapor-liquid
  equilibrium calculations
  pressure-composition
    diagrams, 111,112f-113f
  theory, 111
Thermodynamic model of supercritical
  fluid extraction
  effect of mixing rules on solubility
    of solids, 105,106-110f
  theory, 105
Toluene
  effect on coal, 255
  selection as supercritical fluid
    solvent, 139
  vapor pressures, 140t,141
Transition energy
  influencing factors, 50-51
  of nonpolar solvents, 46t
  of phenol blue in CO_2 vs.
    density, 44,45f
  of phenol blue in CO_2 vs.
    pressure, 44,45f
  of phenol blue vs. CO_2
    mixture, 52,53f,54
  of phenol blue vs. density, 51,52f
  vs. reduced density, 46
Transport properties
  background, 4
  mass conductivity, 4

V

π^* values
  for supercritical CO_2, 35
  vs. reduced density, 35,36f
van der Waals equation of state
  general, 101
  mixing rules, 101-102
van der Waals mixing rules
  guidelines, 103
  theory, 102-103

Vapor-pressure osmometry,
    determination of molecular
    weight, 236,238t
Vapor pressures, for pentane and
    toluene, 140t,141
Vapor-liquid equilibrium apparatus
  diagram, 204,205f
  estimated relative errors, 204
Vapor-liquid equilibrium calculations,
    applications, 111,112-113f
Volume expansivity of a supercritical
    fluid, vs.
    temperature, 180,183,184f

Production by Paula M. Bérard
Indexing by Deborah H. Steiner
Jacket design by Pamela Lewis

Elements typeset by Hot Type Ltd., Washington, DC
Printed and bound by Maple Press Co., York, PA

K Mills

DATE DUE